DICTIONARY OF DEVELOPMENTAL BIOLOGY AND EMBRYOLOGY

DICTIONARY OF DEVELOPMENTAL BIOLOGY AND EMBRYOLOGY

Second Edition

FRANK J. DYE

Western Connecticut State University

WILEY-BLACKWELL

A John Wiley & Sons, Inc., Publication

Published by John Wiley & Sons, Inc., Hoboken, New Jersey
Published simultaneously in Canada

For general information on our other products and services or for technical support, please contact
our Customer Care Department within the United States at (800) 762-2974, outside the United States
at (317) 572-3993 or fax (317) 572-4002.

Wiley also publishes its books in a variety of electronic formats. Some content that appears in print
may not be available in electronic formats. For more information about Wiley products, visit our web
site at www.wiley.com.

Library of Congress Cataloging-in-Publication Data

Dye, Frank J. (Frank John), 1942-
  Dictionary of developmental biology and embryology / Frank J. Dye.—2nd ed.
     p. cm.
  ISBN 978-1-118-07651-4 (cloth)—ISBN 978-0-470-90595-1 (pbk.)
  1. Developmental biology–Dictionaries. 2. Embryology–Dictionaries. I. Title.
  QH491.D94 2012
  571.803–dc23
                                                                    2011030364

Printed in the United States of America

10 9 8 7 6 5 4 3 2 1

I dedicate this book to the memory of my high-school teachers, who unknowingly set the course of my life; my professors at Danbury State College (Western Connecticut State University), who raised my self-esteem; and especially to my professors at Fordham University, notably James Forbes and Alexander Wolsky, who fascinated me with embryology and developmental biology.

Preface

The motivation for the creation of the second edition of this work was a continued perceived need for such a book. Ten years have elapsed since the publication of the first edition, during which time developmental biology has rapidly progressed, in particular, developing the sub-discipline of evo-devo and giving birth to eco-devo. Unlike *The Dictionary of Cell Biology*, such a book does not exist for embryology and developmental biology; yet undergraduate students, graduate students, professors, and researchers would find such a book to be most convenient. Undoubtedly, most terms found in this book are accessible on the Internet; however, it is often not convenient to boot up the computer and initiate a search for the term in question. In addition to terminology, this book also presents historical vignettes of some major contributors, classical and modern, to embryology and developmental biology, as well as a number of concepts (as opposed to terminology per se).

This second edition will, perhaps, have omissions of some terms that you feel should be included. The author would greatly appreciate it if you would send suggested terms for possible inclusion in subsequent editions of this book to dyef@wcsu.edu.

I wish to acknowledge the contribution to the creation of this book of a sabbatical leave granted to me by Western Connecticut State University, during the Fall of 2006—it provided the momentum to get the ball rolling. Also, I wish to thank Dr. James Schmotter, president of Western Connecticut State University, for his support in obtaining the sabbatical leave. Additionally, I wish to thank Karen Chambers, Editor, Life Sciences, Wiley-Blackwell Publishing, for her belief in this project from the outset; Anna Ehler, Editorial Assistant, Life Sciences, Wiley-Blackwell Publishing, for her gentle nudging; Stephanie Sakson, Project Manager, Toppan Best-set Premedia Limited, for many helpful suggestions as we approached the completion of this project; and Sheree Van Vreede, who copyedited the manuscript. Finally, I thank the Connecticut State University System for providing an implantation site for the development of this developmental cell biologist.

Frank J. Dye
Western Connecticut State University

A

A23187   a calcium ionophore that increases the permeability of cell membranes to calcium ions; it has been used experimentally to trigger the cortical reaction of sea urchin eggs, which depends on an increase in the concentration of free calcium ions in the cytoplasm of the egg.

abaxial   the side of the leaf facing away from the meristem, and can be thought of as the ventral side. *See:* **adaxial**.

abdominal cavity   or peritoneal cavity, derived from the coelom and containing much of the viscera (internal organs).

ABO blood typing system   a system of typing human blood, for the purpose of blood transfusion.

aboral surface   the surface opposite the oral surface; in the sea urchin, this is the uppermost surface when the animal is in its normal, upright position. *See:* **oral surface**.

abortifacient   that which induces abortion, e.g., RU486.

abortion   the process of terminating a pregnancy with the demise of the conceptus.

abortus   that which results from an abortion: an aborted embryo or fetus.

abruptio placentae   premature separation of the placenta from the lining of the uterus.

abscisic acid (ABA)   a plant hormone; it has been implicated in the developmental arrest of the plant embryo during seed dormancy. The increase in ABA levels during early seed development, in many plant species, stimulates the production of seed storage proteins and prevents premature seed germination. The breaking of dormancy in many seeds is correlated with declining ABA levels in the seed. *See:* **viviparous mutants**.

abscission   the natural separation of flowers, fruit, or leaves from a plant.

accessory sex glands   in humans: in the male, seminal vesicles (2), prostate gland (1), Cowper's glands (2), and the glands of Littré (multiple); in the female, Bartholin's glands (2) and Skene's glands (2). In female insects, accessory glands open into the vagina and they commonly secrete an adhesive substance for attaching the eggs to an external object or for cementing them in a mass; in male insects, in many cases, the accessory glands produce substances that go into the formation of a spermatophore.

accutane   retinoic acid used to treat severe cystic acne. *See:* **retinoic acid**.

Acetabularia   a large, marine, single-celled, green algae used in classical experiments to gather evidence that the nucleus provides the information determining phenotype.

acetylated histones   histones enzymatically modified by the addition of acetyl groups; acetylated histones are relatively unstable and cause nucleosomes to disperse. *See:* **deacetylated histones**.

acetylation   a type of chemical modification of histones, meditaed by histone acetyltransferase enzymes, which destabilizes nucleosomes and allows for gene expression.

Achaete   an example of a transcription factor encoded by proneural genes in *Drosophila*. *See:* **proneural genes**, **Scute**.

achondroplasia   a typical, congenital dwarf, formed because of abnormal bone formation, specifically abnormal chondrification and ossification of the ends of the long bones; this birth defect is often inherited as a dominant trait.

acid blob   *See:* **trans-activating domain**.

acoelomate   animals lacking a coelom, e.g., Platyhelminthes (flatworms) and Nemertea (ribbon) worms.

acrania   the partial or complete absence of the cranium.

acrocentric chromosome   a chromosome in which the centromere is not equidistant from the two ends of the chromosome; therefore, the chromosome's two arms are of unequal length.

Dictionary of Developmental Biology and Embryology, Second Edition. Frank J. Dye.
© 2012 Wiley-Blackwell. Published 2012 by John Wiley & Sons, Inc.

acrocephaly also called oxycephaly is a condition in which the head is roughly conical in shape caused by premature suture closure.

acrogen a nonflowering plant that grows only at the apex, as in a fern.

acron the preoral, nonsegmented portion of an arthropod embryo; a marker for the anterior terminus of the *Drosophila* embryonic axis.

acrosin a protease exposed on the sperm head after the acrosome reaction; may be involved in sperm digestion through the zona pellucida.

acrosomal filament an elongated filament, extruded by the heads of the spermatozoa of some species (e.g., *Saccoglossus*, a hemichordate), as part of their acrosome reaction. The extension of the acrosomal filament is a consequence of the underlying polymerization of G-action to F-actin, resulting from increased intracellular pH (attributed to a release of hydrogen ions from the sperm head), which, in turn, is an early event of the acrosome reaction caused by altered spermatozoan plasma membrane permeability.

acrosomal vesicle *See:* **acrosome**.

acrosome an organelle, derived from the Golgi apparatus, found in the head of the spermatozoon, containing hydrolytic enzymes that play a role in the sperm reaching the plasma membrane of the egg; it is considered to be a modified lysosome. Also called the acrosomal vesicle.

acrosome reaction a change, during fertilization, undergone by the sperm of many types of animals that facilitates the breaching of the egg coats of the species by the spermatozoon; has been particularly studied in the fertilization of marine invertebrates, some of which produce dramatic acrosomal filaments as part of their acrosome reaction. In sea urchins, the acrosome reaction is initiated by a sulfated polysaccharide in the egg jelly coat; this allows calcium and sodium ions to enter the sperm head and to replace potassium and hydrogen ions; calcium ions mediate the fusion of the acrosomal membrane with the adjacent plasma membrane. In mice, the acrosome reaction occurs after the spermatozoon has bound to the zona pellucida; capacitation of mouse spermatozoa involves the unmasking of a specific enzyme on the sperm cell surface; the lock-and- key binding of zona *N*-acetylglucosamine by sperm glycosyltransferase links the spermatozoon and egg together; once this binding is complete, the acrosome reaction is initiated. During the acrosome reaction in the mouse, the sperm head loses the anterior part of the plasma membrane and the outer membrane of the acrosome, leaving the inner acrosomal membrane as the covering of the anterior part of the sperm head. From an evolutionary point of view, the primary role of the acrosome reaction seems to be the exposure or preparation of a specialized fusion surface on the sperm head.

acrotrophic ovariole an ovariole in which the trophocytes remain in the germarium but are connected with the developing oocytes by means of progressively lengthening protoplasmic strands that convey nourishment to them; found in Hemiptera and in some Coeloptera.

actin a ubiquitous protein in eukaryotic cells, occurring in especially abundant and organized form in skeletal muscle, where it is organized into thin filaments; in eukaryotic cells, in general, actin microfilaments make up one of the three major components of the cytoskeleton. Actin is enriched in the cortices of some eggs, e.g., amphibian eggs and sea urchin eggs.

actinomycin D an inhibitor of transcription; it prevents RNA synthesis by binding to DNA and blocking the movement of RNA polymerase.

activation of the egg *See:* **egg activation**.

activator a protein that positively regulates transcription of a gene.

active immunity immunity possessed by a host as the result of disease or unrecognized infection, or induced by immunization with microbes or their products or other antigens. *See:* **passive immunity**.

activin a member of the TGF-B family of growth factors, it plays a key role in the induction of dorsal mesoderm in *Xenopus* embryos; it is secreted by vegetal cells. *See:* **Brachyury**, **goosecoid**.

adaptive developmental plasticity when the organism alters its development in response to environmental cues such that the organism is more fit in that particular environment; e.g., plants grown in the shade often change their leaf structure and amount of branching. An inductive phenomenon. *See:* **niche construction**.

adaxial the side of the leaf that faces the meristem, and can be thought of as the dorsal side. *See:* **abaxial**.

adaxial cells   cells that flank the chordamesoderm and are the precursors of the somites in the zebrafish embryo.

adenosine triphosphate (ATP)   a, so-called, high-energy molecule, which is the almost ubiquitous source of immediate chemical energy for biochemical reactions.

adepithelial cells   small numbers of cells that migrate into imaginal discs, which are composed primarily of epidermal cells, and which, during the pupal stage, give rise to the muscles and nerves that serve the structures formed from the imaginal discs.

adherens junctions   a type of cell junction that binds one epithelial cell to another, and joins an actin bundle in one cell to a similar bundle in another.

adipocytes   also called fat cells; growth hormone triggers fat cell differentiation; fat cell precursors that have been stimulated by growth hormone become sensitive to IGF-1, which stimulates the proliferation of differentiating fat cells. A fibroblast-like precursor cell is converted into a mature fat cell by the accumulation and coalescence of cytoplasmic lipid droplets.

adrenal glands   a pair of endocrine glands found attached to the cephalic poles of the kidneys.

adrenal hyperplasia   abnormally excessive growth of the adrenal glands attributed to an inherited enzyme deficiency; this leads to overproduction of male hormones that causes masculinization in females and may cause early sexual development in males.

adrenaline   *See:* **epinephrine**.

adrenergic neurons   the sympathetic neurons derived from neural crest cells of the trunk and that produce the neurotransmitter norepinephrine. *See:* **cholinergic neurons**.

adult hemoglobin   the hemoglobin made up of two alpha polypeptide chains and two beta polypeptide chains; has a lower affinity for oxygen than fetal hemoglobin and releases oxygen in the placenta, which is taken up by fetal hemoglobin. *See:* **fetal hemoglobin**, **hemoglobin**.

adult stem cells   stem cells found in mature organs, usually involved in replacing and repairing tissues of that particular organ, and can form only a subset of cell types, e.g., spermatogonia, muscle satellite cells, and hemtopoietic stem cells.

adventitious   structures or organs developing in an unusual position, as roots originating on the stem.

AER   *See:* **apical ectodermal ridge**.

aequorin   a luminescent protein, derived from luminescent jellyfish (*Aequorea*), which emits light in the presence of free calcium ions; it has been used experimentally to visualize the release of free calcium ions into the cytoplasm of the sea urchin egg upon fertilization.

afterbirth   that which is born after the baby, namely the placenta, membranes, and part of the umbilical cord.

after-ripening   a process required by some dormant seeds, during which low-level metabolic activities continue to prepare the embryo for germination.

agamospermy   formation of seed without fertilization.

agenesis   this type of birth defect is the absence of an organ or other structure from early development; the precise kind of agenesis refers to the missing organ, e.g., renal agenesis.

aggregation   phase in the life cycle of *Dictyostelium discoideum*; the pseudoplasmodium of *D.d.* originates from the aggregation of many separate cells.

aging   the time-related deterioration of the physiological functions necessary for survival and fertility. *See:* **life expectancy**.

AGM region   aorta, gonads, and mesonephros region; the domain, near the aorta, where, in fishes, mammals, and frogs, the definitive hematopoietic cells are formed; mesodermal regions of the day 11 mouse embryo in which pluripotential hematopoietic stem cells and CFU-s cells can be found. These blood cell precursors will colonize the liver and constitute both the fetal and the adult circulatory system. Around the time of birth, stem cells from the liver populate the bone marrow, which then becomes the major site of blood formation throughout adult life.

Agnatha   the class of vertebrates consisting of the jawless fish; it includes two subclasses, Ostracodermi (extinct) and Cyclostomata (with two extant orders, lampreys and hagfish).

agnatha   without jaws. *See:* **gnathostomes**.

AIDS   acquired immune deficiency syndrome, a viral disease that may be transmitted from mother to offspring.

air space   *See:* **shell membrane**.

albinism   a hereditary absence of pigment from the skin, hair, and eyes.

albino   an organism lacking pigment.

albumen   the white of an egg, composed principally of albumin.

albumin   any of a group of plant and animal proteins that are soluble in water and dilute salt solutions.

albuminous seed   a seed with albumen, the nutritive tissue in a seed.

alcohol   a chemical teratogen.

alecithal eggs   eggs without yolk. In placental mammals, the yolk is virtually absent.

aleurone   protein granules stored in the outermost layer of the endosperm of the seeds of many grains.

aleurone layer   a border of protein-rich cells that surrounds the endosperm of many seeds.

algae   algae do not constitute a formal taxon (classification unit), but traditionally included the photosynthetic thallophytes, encompassing such plant divisions as chlorophyta, rhodophyta, chrysophyta, etc. The cyanobacteria (traditionally the "blue-green algae"), which are prokaryotes, were classified with the algae, which are eukaryotes; however, they are now classified with the bacteria, all of which are prokaryotes.

alkaline phosphatase   an enzyme involved in the deposition of calcium phosphate crystals during ossification.

allantoic diverticulum   an evagination or diverticulum of the hindgut, composed of endoderm and mesoderm. In human development, the endoderm of the allantoic diverticulum is rudimentary, but the mesoderm gives rise to important blood vessels of the umbilical cord.

allantoic stalk   *See:* **allantois**.

allantoic vesicle   *See:* **allantois**.

allantois   one of the four extraembryonic membranes; originates from the hindgut and is not conspicuous in human development. In the chick, the allantois originates within the body of the embryo and its proximal portion remains intraembryonic throughout development; the distal portion is not incorporated into the body. Late in the third day of incubation, it originates as a diverticulum from the ventral wall of the hindgut; its walls are composed of splanchnopleure. During the fourth day, the allantois pushes out of the embryonic body into the extraembryonic coelom; the proximal portion is parallel to and just caudal to the yolk stalk while the distal portion enlarges. The narrow proximal portion is the allantoic stalk while the distal portion is the allantoic vesicle. The allantoic vesicle extends into the sero-amniotic cavity, flattens, and encompasses the embryo and yolk sac. The mesodermal layer of the allantois (splanchnic mesoderm) fuses with the mesodermal layer of the serosa (somatic mesoderm); in the double layer of mesoderm, a rich vascular network develops, connected with the embryonic circulation by allantoic arteries and veins; this highly vascular fusion membrane is the chorio-allantoic membrane. In mammals, the allantois originates as a diverticulum from the hindgut almost as soon as the hindgut is established, from splanchnopleure; the proximal portion gives rise to the allantoic stalk while the distal portion becomes enormously dilated. The allantois early acquires an abundant blood supply by way of large branches from the caudal end of the aorta; as yolk sac circulation undergoes retrogressive changes the plexus of allantoic vessels becomes more highly developed and takes over entirely the function of metabolic interchange between the fetus and the mother. In mammals with a saccular allantois (e.g., the pig), the chorion is essentially a layer of allantoic splanchmopleure fused with a layer of serosal somatopleure. In primate embryos, where the lumen of the allantois is rudimentary, endoderm is not involved in the formation of the chorion; allantoic mesoderm and blood vessels, however, extend distally beyond the rudimentary lumen and spread over the inner surface of the serosa, establishing the same essential relationships as in less highly specialized forms. The fusion of the allantois with the inner surface of the serosa brings to this hitherto poorly vascularized layer (serosa) an abundant circulation. In mammals, the embryo is dependent on the chorion (serosa) for carrying on metabolic interchange with the uterus.

allele   most genes are present as two copies; each copy is called an allele; an allele is one of two (or more) forms of a particular gene.

allelic series   a set of alleles having different degrees of loss of function arranged in order by the severity of the abnormal phenotype.

Allman, George (1812–1898)   Anglo-Irish naturalist; coined the terms (in 1853) ectoderm and endoderm.

allogamy   cross-pollination.

allometric growth   *See:* **allometry**.

allometry   the quantitative relationship between a part of an organism and the whole organism or another part of it as the organism increases in size. Also known as heterogony. Allometry occurs when different parts of the organism grow at different rates and can play a role in forming variant body plans within a *Bauplan*. In whale development, enormous growth of the upper jaw forces the nose (blowhole) to the top of the skull.

allosteric enzymes   enzymes that have two functional sites: (1) the catalytic or active site that binds substrates as ligands and (2) the regulatory or allosteric site that binds allosteric effectors as ligands; the allosteric effectors may be positive or negative; i.e., they may activate or inhibit, repectively, the active site. *See:* **cooperativity**.

alopecia   congenital loss of hair.

alternation of generations   also called metagenesis; the phenomenon in which one generation of certain plants and animals reproduces asexually, followed by a sexually reproducing generation; a phenomenon generally exhibited by plants, wherein during their life cycle a haploid, multicellular, gamete-producing plant (gametophyte) alternates with a diploid, multicellular, spore-producing plant (sporophyte) (see Plate 17 in the color insert).

alternative RNA splicing   pre-mRNAs may contain numerous introns; by the selective recognition of these one can have alternative RNA splicing; this results in alternative proteins from the same gene. This is one way in which development can be regulated by RNA processing.

alveolar period   a period during lung development when the alveoli characteristic of the lung form.

alveoli   terminal air sacs of the lungs.

amelia   the congenital absence of limbs; a type of birth defect caused by the drug thalidomide.

ameloblasts   the epithelial cells in the enamel organ of the tooth germ that will secrete enamel.

Ametabola   those insects (Apterygota) in which the change of form from instar to instar is too slight and gradual to merit the term "metamorphosis."

ametabolous   insects, e.g., springtails and mayflies, that have no larval stage and undergo direct development.

AMF   formerly known as AMH. *See:* **anti-Müllerian factor**.

AMH   *See:* **anti-Müllerian duct hormone**.

amino acids   the building blocks of protein molecules.

ammocoete   the larval stage in the life cycle of the lamprey.

ammonia   in the life cycle of *Dictyostelium discoideum*, ammonia is produced by the migrating slug, which inhibits culmination; depletion of ammonia allows culmination to begin.

ammonotelic organisms   organisms, such as freshwater fishes and frog tadpoles, that excrete ammonia. *See:* **ureotelic organisms**.

amnio-cardiac vesicles   the pericardial regions of the coelom; in the chick embryo, these first appear as marked local enlargements of the coelomic chambers in the region of the anterior intestinal portal; later in development, they will form the pericardial cavity (see Fig. 19).

amniocentesis   a prenatal diagnostic procedure initiated by obtaining a sample of amniotic fluid from the amniotic cavity; this procedure provides fetal cells as well as fluid for analysis (see Fig. 1).

amniocyte   a cell found in the amniotic fluid.

amnion   one of the four extraembryonic membranes formed during the development of higher vertebrates, including humans; it makes up the wall of the amniotic cavity. The functional significance of the amnion is emphasized by the observation that it appears only in embryos of non–water-living forms (amniotes). In chick development, the amnion and serosa (chorion) are closely associated in origin so they are here considered together; they are both derived from extraembryonic somatopleure. At approximately 30 hours, the head of the chick embro sinks into the yolk and, at the same time, the extraembryonic somatopleure anterior to the head is thrown into a fold, the head fold of the amnion. Subsequently, two growth processes occur: (1) As the embryo increases in length, its head grows into the amniotic fold; and (2) growth of the somatopleure extends it over the head of the embryo; the head comes to lie in a double-walled pocket of extraembryonic somatopleure. The caudally directed limbs of the head fold of the amnion continue posteriorad as the lateral amniotic folds, which grow dorsomesiad and eventually meet in the midline, dorsal to the embryo. During the third day, the tail fold of the amnion appears. Continued growth of the

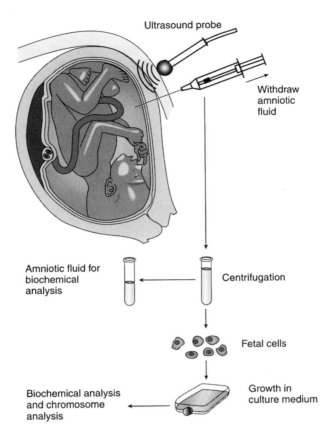

Ultrasound probe

Withdraw
amniotic
fluid

Amniotic fluid for
biochemical
analysis

Centrifugation

Fetal cells

Biochemical analysis
and chromosome
analysis

Growth in
culture medium

Figure 1. Diagram illustrating technique of amniocentesis. The amniotic fluid obtained by this technique contains both cells and liquid for prenatal diagnosis. Reprinted from Frank J. Dye, *Human Life Before Birth*, Harwood Academic Publishers, 2000, fig. 19-15, p. 173.

head, lateral, and tail folds results in their meeting and fusion above the embryo, providing the formation of a scar-like thickening, the sero-amniotic raphe, and the amniotic cavity, completely lined by ectoderm and continuous with the superficial ectoderm of the embryo. The outer layer of somatopleure becomes the serosa; the inner layer becomes the amnion. The cavity between the serosa and the amnion is the sero-amniotic cavity, a part of the extraembryonic coelom. The serosa eventually encompasses the embryo and all the other extraembryonic membranes. In all mammalian embryos, the amnion is formed at an exceedingly early stage of development; in some forms (humans), the amniotic cavity appears even before the body of the embryo has taken definite shape. In the pig, it is formed more leisurely (than in humans) and is easier to analyze; the first indication appearing shortly after the primitive streak stage; amniotic folds on all sides come together above the mid-dorsal region of the embryo; where the folds come together there persists for a time a cord-like mass of tissue between the amnion and the outer layer of the blastocyst; when this is obliterated, the embryo and its amnion lie free in the blastocyst. The amnion is attached to the body of the embryo where the body wall opens ventrally in the region of the yolk stalk; this ventral opening becomes progressively smaller giving rise to the umbilical ring; meanwhile the yolk stalk and the allantoic stalk are brought close together in the belly stalk (see Figs. 5, 6, 19, and 35).

-amnios-   this word part refers to the amnion and, in the context of human development, is a part of terms describing abnormal amounts of amniotic fluid. *See:* **oligohydramnios, polyhydramnios**.

amniota   with amnion. *See:* **amniotes, anamniota**.

amniote egg　*See:* **cleidoic egg**.

amniotes　vertebrates that develop with an amnion; reptiles, birds, and mammals. *See:* **anamniotes**.

amniotic cavity　a fluid-filled cavity characteristic of amniote development; begins to form on the eighth day of human development; during their development, both the embryo and the fetus float in this, their private aquarium (see Figs. 6 and 44). *See:* **amnion**.

amniotic fluid　the fluid that fills the amniotic cavity and in which the embryo and, later, the fetus float during their development; amniocentesis is a prenatal diagnostic procedure that begins with procuring a sample of this fluid (see Fig. 1).

amniotic folds　in insect development, folds that originate from the edges of and then enclose the germ band; when these folds grow and fuse, the germ band is enclosed in a space known as the amniotic cavity. Coverings membranes also result from this fusion: the outer layer or serosa, continuous with the extra-embryonic blastoderm, and the inner membrane or amnion, continuous with the margins of the germ band. The amniotic cavity, serosa, and amnion protect the growing embryo from injury. *See:* **amnion**.

amniotic sac　the sac that surrounds the embryo with amniotic fluid.

amoebocyte　an amoeba-like cell found in the body fluid of coelomates; it is frequently phagocytic.

amorphous　without any definite form; shapeless; lacking symmetry.

Amphibia　the class of vertebrates composed of amphibians; includes three extant orders, Anura, Urodela, and Apoda.

amphibians　members of the class of vertebrates (Amphibia) that includes frogs and salamanders.

amphiblastic cleavage　the unequal but complete cleavage of telolecithal eggs.

amphimixis　the union of paternal and maternal elements by syngamy.

Amphioxus　a cephalochordate traditionally studied in embryology/developmental biology courses as a representative of an evolutionarily primitive chordate; encompasses two genera, *Branchiostoma* and *Asymmetron*. Amphioxus lacks neural crest cells (see Plate 1 in the color insert).

amplexus　the physical interaction between male and female amphibians during mating, involving the embrace of the female by the male but not the intimacy of copulation.

ampulla　literally, "a flask": (1) the portion of the fallopian tube, closest to the ovary, in which fertilization normally occurs, and (2) the enlarged segment of the vas deferens, just before it receives the seminal vesicle (see Fig. 16).

anagenesis　*See:* **phyletic transformation**.

anal agenesis　literally, "the lack of the anal opening" (anus) but also used to refer to an abnormal opening as into the vulva in females or the urethra in males; the latter variety results from incomplete partition of the cloaca by the urorectal septum.

anal evagination　an outpocketing of the hindgut endoderm toward the proctodeal invagination of ectoderm.

anal membrane　the membrane formed of proctodeal ectoderm and hindgut endoderm; its normal breakdown creates the opening called the anus.

anal plate　*See:* **anal membrane**.

anal stenosis　a narrowing of the anal canal, probably resulting from urorectal septum deviation in partitioning of the cloaca.

analgesic　a drug that relieves pain.

analogous　structures that are similar in function but do not have the same origin, e.g., the wing of a bird and the wing of an insect.

analogous organs　body organs/parts of different species that have different evolutionary origins but have become adapted to serve the same purpose (e.g., wings of a bat and wings of an insect).

anamniota　no amnion. *See:* **anamniotes, amniota**.

anamniotes　vertebrates that develop without an amnion; fish and ampbibians. *See:* **amniotes**.

anaphase　the stage of mitosis or meiosis when the daughter chromosomes are moving toward opposite poles of the spindle (see Fig. 29).

anatomic position　standing erect with the arms at the sides and the palms forward (see Fig. 2).

-*andr*-　a man, e.g., as in polyandry.

androecium　collectively, all of the stamens in a flower.

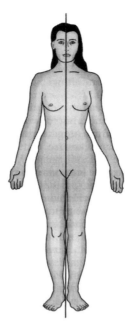

Figure 2. Anatomic position and bilateral symmetry. The body is erect, the arms are at the sides of the body, and the head, palms, and feet face forward. The line is in the median plane of the body, which is also the plane of bilateral symmetry. Reprinted from Frank J. Dye, *Human Life Before Birth*, Harwood Academic Publishers, 2000, fig. 1-1, p. 3.

androgen insensitivity syndrome   the result of a mutation in which XY individuals (chromosomally male) do not make a functional testosterone *receptor. See:* **testicular feminization**.

androgenesis   development of an embryo from a fertilized irradiated egg, involving only the male nucleus.

androgenic merogony   the fertilization of egg fragments that lack a nucleus.

androgens   male sex hormones, e.g., testosterone.

anencephaly   a type of birth defect in which most of the brain is missing; a term literally meaning "without a brain"; more specifically, it refers to the absence of the cerebrum, cerebellum, and flat bones of the skull. *See:* **anterior neuropore**.

aneuploid   the number of chromosomes in a cell that is not an exact multiple of the haploid number of chromosomes for the species. *See:* **euploid**.

angioblasts   mesodermal cells that organize themselves into capillary blood vessels.

angiogenesis   a type of blood vessel development that may occur in adults as well as in embryos, in which new capillaries sprout from existing vessels. In the adult, new vessels are produced only through angiogenesis; outside of female reproductive cycles, angiogenesis in the adult is largely controlled by pathological situations, such as wound healing and tumor growth. *See: **Bacteroides thetaiotaomicron**,* **vasculogenesis**.

angiogenetic clusters   also called blood islands. *See:* **blood islands**.

angiopoietin-1 (Ang1)   a protein involved in vasculogenesis; mediates the interaction between the endothelial cells and pericytes, smooth muscle-like cells recruited by the endothelial cells to cover them. *See:* **FGF2, vascular endothelial growth factor (VEGF)**.

angiopoietins   paracrine factors that mediate the interaction between endothelial cells and pericytes.

angiosperms   the flowering plants.

animal   has been defined (Slack et al.) as an organism that displays a particular spatial pattern of *Hox* gene expression.

animal cap   the upper portion of the animal hemisphere of the frog blastula, above the equatorial region. Isolated by itself, animal cap tissue will give rise to only ectoderm, but combined with vegetal cells of the blastula, it may be induced to form mesodermal tissue, such as muscle and notochord.

animal gradient one of two gradients of "principles" that are mutually antagonistic but that must interact with each other if normal sea urchin development is to occur. The animal gradient is centered at the animal pole. *See:* **animalized**.

animal hemisphere that half of the egg that has the animal pole at its center.

animal pole that point on the surface of an egg where the polar bodies are formed during the meiotic divisions of maturation (see Plate 3 in the color insert).

animalculist a preformationist who believed that the new individual is already preformed in the spermatozoon and that development involves essentially the growth of the pre-formed organism.

animalization the sea urchin embryo becoming animalized; may be caused by factors such as zinc, trypsin, and Evans blue dye.

animalized a term used in the context of experimental embryology of, primarily, sea urchins; when the sea urchin develops to excess those parts pertaining to the animal gradient, e.g., the tuft of cilia at the animal pole. If the vegetal gradient is weakened or suppressed, the animal gradient becomes preponderant and the embryo is animalized. *See:* **animal gradient**, **vegetalized**.

animal-vegetal axis of the embryo, an imaginary line joining the animal and vegetal poles.

animism the belief that inanimate objects possess a personal life or soul.

aniridia absence of the iris; caused by a mutation of the *PAX6* gene.

anisogamy sexual reproduction involving the fusion of unlike gametes.

ankyloglossia sometimes referred to as tongue-tie, refers to an abnormal attachment of the tongue to the floor of the oral cavity (mouth).

anlage the rudiment or primordium of a structure. *See:* **primordium**.

annual a plant that germinates from seed, flowers, sets seed, and dies in the same year.

anoikis rapid apoptosis that occurs when epithelial cells lose their attachment to the extra-cellualr matrix.

anovulation the lack of ovulation.

Antennapedia complex of *Drosophila*, comprises five homeobox genes, which control the development of the parasegments anterior to parasegment 5 in a manner similar to the bithorax complex. The Antennapedia complex controls parasegment identity in the head and first thoracic segment. The order of genes in the complex is the same as the spatial and temporal order in which they are expressed along the anteroposterior axis during development. *See:* **bithorax complex**.

Antennapedia **gene** a gene responsible for the specification of fly body parts; mutations in it lead to bizarre phenotypes; specifies the identity of the second thoracic segment in *Drosophila*. In the dominant mutation of this gene, legs rather than antennas grow out of the head sockets; in the recessive mutation of this gene, antennas sprout out of the leg positions.

anteriad toward the anterior.

anterior group *Hox* **genes** *Hox* genes expressed in the anterior region of bilaterans and located toward the 3′ end of *Hox* clusters; include the *Hox1* and *Hox2* genes in vertebrates.

anterior intestinal portal in the chick embryo, the opening from the midgut into the foregut (see Plate 2b in the color insert).

anterior neuropore the temporary opening at the cephalic end of the early neural tube. In the chick embryo, the anterior neuropore is almost closed at 33 hours. In human development, an improperly closed anterior neuropore gives rise to the birth defect known as anencephaly (see Fig. 30). *See:* **spina bifida**.

anterior organizing center classic embryological experiments demonstrated that there are at least two "organizing centers" in the insect egg, one is the anterior organizing center, and the other the posterior organizing center; in *Drosophila* embryos, the product of the wild-type *bicoid* gene is the morphogen that controls anterior development. *See:* **posterior organizing center**, **terminal gene group**.

anterior pole of the insect egg is that pole of the egg that lies in the ovariole directed toward the head of the parent insect.

anterior tip the anterior tip of the migrating pseudoplasmodium of *Dictyostelium discoi-deum* acts as an embryonic organizer in the sense that (1) a grafted tip will produce a new individual and (2) the tip is a source of an inhibitory gradient regarding the formation of tips.

anterior visceral endoderm   one of two signaling centers in the mammalian embryo; works together with the node ("the organizer") to form the forebrain; expresses several genes necessary for head formation, including genes for transcription factors, Hesx-1, Lim-1, and Otx-2, and for the paracrine factor Cerberus. *See:* **node**.

anteroposterior axis   the axis joining the head and tail; also called head-to-tail, craniocaudal, or rostrocaudal axis. *See:* **axes**.

anther   that portion of the stamen of a flower that produces pollen.

anther sac   one of the pollen chambers of the anther.

antheridiophore   a stalk that bears antheridia; found in some liverworts.

antheridium   (1) a male gametangium, a container of sperm, formed by, for example, ferns; (2) a type of reproductive structure in algae that produces the male gametes or spermatozoa, e.g., as in the filamentous green alga, *Oedogonium*; and (3) a unicellular or multicellular structure which produces spermatozoa; the male reproductive structure in moss and fern gametophytes (see Plate 16 in the color insert).

antheridogen   substance secreted by young fern gametophytes to stimulate the development of antheridia.

Anthropocene   successor of the Holocene epoch, an epoch characterized by a human-dominated environment.

anticlinal divisions   cell divisions that take place in planes perpendicular to the surface of the structure of which they are a part; these divisions increase the number of cells in a layer. *See:* **periclinal divisions**.

anticodon   the triplet of nucleotides in a transfer RNA molecule that hydrogen bonds to a codon of messenger RNA.

antifertilizin   a receptor molecule thought to exist on sperm heads of sea urchin sperm that would combine with fertilizin (sea urchin egg jelly coat) causing sperm agglutination experimentally; thought to function as the species-specific sperm receptor during fertilization.

antifungal symbiotic bacteria   bacteria found associated with embyos that protect the embryos from fungal infection, e.g., in the outer coats of embryos, such as those of American lobster embryos. *See:* **embryo defenses**.

anti-Müllerian duct hormone   one of two major hormones produced by the fetal testis (the other is testosterone); AMH destroys the Müllerian duct.

anti-Müllerian factor   TGF-ß family paracrine factor secreted by the embryonic testes that induces apoptosis of the epithelium and destruction of the basal lamina of the Müllerian duct, preventing formation of the uterus and oviducts. *See:* **anti-Müllerian duct hormone**.

antipodal cells   those three cells found opposite the end of the embryo sac that contains the egg and synergid cells; they have no known function.

antisense oligonucleotides   antisense oligonucleotides refer to short, synthetic oligonucleotides that are complementary in sequence to its cognate gene product and upon specific hybridization to its cognate gene product induces inhibition of gene expression. Oligonucleotides, as short as a 15 mer, have the required specificity to inhibit gene expression of a particular gene by annealing to the cellular mRNA. The mechanism of inhibition of gene expression is based on two properties; the first is the physical blocking of the translation process by the presence of the short double-stranded region, and the second is the presence of the RNA-DNA duplex that is susceptible to cellular RNase H activity. RNase H cleaves the RNA-DNA duplex region of the mRNA, thus preventing the faithful translation of the mRNA.

antisense RNA   an RNA complementary to a mRNA; the reverse of the original message; i.e., a transcript that is complementary to the natural one. When introduced into a cell, the antisense RNA binds to the normal message resulting in a double-stranded RNA, which is digested by cellular enzymes; this causes a functional depletion of the message, allowing developmental biologists to determine the function of genes during development. *See:* **redundancy**.

antral vacuoles   tiny fluid-filled spaces found among the follicle cells of a developing ovarian follicle.

antrum   the fluid-filled cavity of the mature graafian follicle (see Fig. 33).

Anura   tail-less amphibians; the frogs and toads.

anus   the caudal opening of the gut to the outside of the body.

aorta   the largest arterial blood vessel in the body, which carries blood away from the heart to various parts of the body. In mammals, the right fourth aortic arch degenerates and the left one gives rise to the arch of the adult aorta. In birds, the left fourth aortic arch degenerates and the right one gives rise to the arch of the adult aorta. Early in development the aorta gives rise to a segmentally arranged series of small vessels, the dorsal intersegmental arteries, which extend into the dorsal body wall. In the adult, three vessels originating from the dorsal aorta supply the abdominal viscera: the coeliac artery and the superior and inferior mesenteric arteries. The omphalomesenteric arteries develop as paired vessels; in the closure of the ventral body wall, they are brought together and fuse to form a single vessel. With the atropy of the yolk sac, the proximal portion of the omphalomesenteric artery persists as the superior mesenteric; the coeliac and inferior mesenteric arteries originate from the aorta independently.

aorta-gonad-mesonephros region   *See:* **AGM**.

aortic arches   a series of six arterial blood vessels that originate in early development in conjunction with the pharyngeal arches. The aortic arches lie embedded in tissue of the branchial arches of corresponding numbers. Although all six aortic arches do not persist, several give rise to important prenatal and adult arterial blood vessels. In the chick embryo, the endothelial lining of the heart is continued cephalad as the ventral aorta (beneath the foregut), which bifurcates to form the paired ventral aortic roots, which curve around the cephalic end of the foregut as the first pair of aortic arches and extend caudad as the paired dorsal aortae. In the chick embryo, at 33–38 hours, there is a single pair of aortic arches; at 50 hours, there is a second pair and capillary sprouts of the third pair; and at 60 hours, there is a third pair and capillary sprouts of the fourth pair. By the fourth day, two more pairs of aortic arches appear posterior to the four formed in the 55–60-hour chick embryo; The first, second, and fifth pairs eventually disappear. The members of the fourth pair of aoric arches have different fates on opposite sides of the body. In mammals, the left generally becomes greatly enlarged to form the arch of the adult aorta, whereas the right forms the root of the subclavian artery; the short section of the right ventral aortic root proximal to the fourth arch persists as the brachiocephalic artery from which both the right subclavian and right common carotid arteries develop. Regarding the fate of the members of the sixth pair of aortic arches: branches extend from both toward the lungs; after these pulmonary vessels are established, the right side of the sixth aortic arch loses communication with the dorsal aortic root and disappears. On the left side, the sixth aortic arch retains communication with the dorsal aortic root; the portion of it between the dorsal aorta and the point where it is connected to the pulmonary artery gives rise to the ductus arteriosus. *See:* **aorta**, **carotid arteries**.

Apaf-1   apoptotic protease activating factor-1; the mammalian homologue of CED-4 of *Caenorhabditis elegans*, which plays a role in mammalian apoptosis; i.e., it participates in the cytochrome *c*-dependent activation of caspace-3 and caspace-9.

apical dominance   regulation of the branching patterns of plant shoots by the shoot tip; the presence of a terminal (apical) bud inhibits the growth of lateral (axillary) buds beneath it; it is hypothesized that indole-3-acetic acid (IAA) coming from the shoot tip stimulates cells around lateral buds to make ethylene that, in turn, inhibits bud growth.

apical ectodermal cap   when a salamander limb is amputated, epidermal cells from the remaining stump migrate to cover the wound surface, forming the wound epidemis, which proliferates to form the apical ectodermal cap.

apical ectodermal ridge (AER)   a thickening of ectoderm found at the distal-most end of the developing vertebrate limb bud; interaction between the AER and the underlying mesenchymal mesoderm cells of the limb bud is necessary for limb development. The AER is induced by the underlying mesodermal cells.

apical meristem   a meristematic part of a plant, the activity of which leads to growth in length of a plant part, such as a root or shoot. *See:* **meristem**.

apical notch   the notch at the top of the cordate-shaped fern prothallus, at which there is meristematic activity (see Plate 15 in the color insert).

Apoda   legless amphibians; the caecilians.

apomixis (apogamy)   the formation of embryos from haploid eggs within embryo sacs and from cells that have not divided meiotically; seed production without fertilization.

apoplastic continuity   continuity that occurs outside the cell's cytoplasm, in the spaces between cell walls. *See:* **symplastic continuity**.

apoptosis   genetically programmed cell death; genetically programmed cell death is a normal part of development in many animals. Examples range from the resorption of the tail of the frog tadpole during metamorphosis to the sculpting of human fingers out of the original hand plate. Apoptosis may be initiated by a withdrawal of growth factors from the cell or by an active response to a signal.

apospory   development of gametophytes from somatic cells (of the nucellus), without the production of spores; the resulting gametophyte is diploid. *See:* **diplospory**.

appendage buds   in the three-day chick embryo, both the anterior and posterior appendage buds have appeared. The interior mass of each consists of closely packed cells derived from adjacent mesoderm and is covered by an ectodermal layer continuous with that of the general body surface.

appendicular skeleton   the skeleton of the arms and legs and the associated pectoral and pelvic girdles, respectively.

appendix   a small, blind, pouch attached to the cecum.

appositional growth   the growth of an object by the addition of material to its surface, as in the growth of the long bones.

Apterygota   one of two subclasses of insects; apterous insects, the wingless condition being primitive, with slight or no metamorphosis; includes the following insect orders: Diplura, Thysanura, Protura and Collembola. *See:* **Pterygota**.

aqueduct of Sylvius   the cerebrospinal fluid-filled cavity of the brain which connects the III and IV ventricles (see Fig. 7).

Arabidopsis   a small weed in the mustard family; *Arabidopsis* is used as a model system for angiosperm development because of its very small genome, its rapid life cycle, and its ability to obtain mutants of flowering in the laboratory.

archegonium   a female gametangium, a multicellular organ; an egg container formed by ferns and mosses; the female reproductive structure in moss and fern gametophytes (see Plate 16 in the color insert).

archencephalon   the forebrain, in the context of the regional specificity of the neural structures that are produced during development of the central nervous system.

archenteron   the cavity of the gastrula, the primitive gut. In the bird gastrula, no true archenteron is formed. *See:* **gastrocoel**.

archenteron roof   the dorsal wall of the primitive gut (archenteron); gives rise to the notochord that plays a role in induction of the embryonic axis.

archeocyte   a type of somatic cell in sponges that can differentiate into all the other cell types in the body of the sponge.

archeocytes   a type of amoebocyte in sponges that gives rise to germ cells and the totipotent cells of gemmules; all the tissues and cell types in the new sponge that eventually develop from the gemmule, including its sex cells, are derived from the archeocytes.

archetype   an original type or form from which others are derived.

architectural transcription factor   a protein that does not activate transcription by itself but helps DNA bend so other activators can stimulate transcription.

area opaca   when a chick blastoderm is removed from the surface of the yolk for observation, the periphery of the blastoderm has adherent yolk and appears darker (than the central region of the blastoderm) with transmitted light; this region is the area opaca. *See:* **zone of junction**.

area opaca vasculosa   that portion of the area opaca, proximal to the area pellucida, which has a darker (than the area opaca vitellina) and more mottled appearance; darker because it has been invaded by mesoderm and more mottled because of the formation of blood islands in that mesoderm. From mesoderm in this region, the yolk-sac blood vessels develop.

area opaca vitellina   that portion of the area opaca, distal to the area pellucida, which has a lighter (than the area opaca vasculosa) appearance because yolk alone underlies it; i.e., this region has not yet been invaded by mesoderm. The boundary between the area opaca vasculosa and the area opaca vitellina is established by the extent of the peripheral growth of the mesoderm.

area pellucida   when a chick blastoderm is removed from the surface of the yolk for observation, the center of the blastoderm has no adherent yolk (because the blastocoel was under it) and appears lighter (than the peripheral region of the blastoderm) with transmitted light; this region is the area pellucida.

area placentalis   the part of the trophoblast in immediate contact with the uterine mucosa in the embryos of early placental vertebrates.

Aristotle (384–322 B.C.) Greek philosopher-scientist, used the Greek word *embryon* (embryo) and wrote a general treatise on embryology. Some of the problems proposed by Aristotle remained unsolved for some 2000 years; e.g., he questioned whether the embryo was preformed and therefore only enlarged during development or whether it actually differentiated from a formless beginning; Aristotle believed the egg to be undifferentiated material that after fertilization began to organize and grow.

aromatase an enzyme important in temperature-dependent sex determination. This enzyme, temprature-regulated in several vertebrate species, converts testosterone into estrogen and controls the ratio of these hormones.

arrector pili muscle the smooth muscle fiber, derived from the dermis, associated with each hair of the skin.

"arrival of the fittest" as opposed to *survival* of the fittest, *arrival* of the fittest requires a theory of body construction and its possible changes,i.e., a theory of developmental change.

arteries blood vessels that carry blood away from the heart. Although this blood is generally rich in food and oxygen, the paired umbilical arteries carry depleted blood from the fetus out to the placenta.

arthrotome mesenchymal cells in the center of the somite that contribute to the sclerotome, becoming the vertebral joints, the intervertebral discs, and those portions of the ribs closest to the vertebrae.

artificial insemination introduction of sperm into the vagina by a means other than a penis.

Ascaris a roundworm parasite found in the intestines of certain mammals; a single female may possess more than 25 million eggs. The spermatozoa are, atypically, amoeboid and enter the primary oocyte before oogenesis is complete. *Ascaris* has been widely used in research and teaching to study maturation, fertilization, and early cleavage.

Ascaris megalocephala a nematode worm, in which the phenomenon of chromosome diminution was demonstrated by Theodor Boveri. Edouard Van Beneden (in 1883–1884) used *Ascaris* to demonstrate that equal numbers of chromosomes were contributed by the parents to the offspring; in *Ascaris megalocephala univalens*, it could be observed that each parent contributes a single chromosome to the pair possessed by the zygote, and in *Ascaris megalocephala bivalens*, the corresponding numbers are two chromosomes in each of the pronuclei and four in the zygote.

ascidians (sea squirts) organisms in the class Ascidiacea in the subphylum Urochordata; marine invertebrates belonging to the phylum Chordata, referred to as protochordates as they lack vertebrae.

asexual reproduction reproduction that involves only one individual and that does not involve meiosis (i.e., recombination of genes), e.g., vegetative propagation of plants or reproduction of amoebas. Three main types of asexual reproduction occur in metazoa: (1) fission, (2) budding, and (3) gemmule formation. Reproducing without sexual union.

assembly factor a transcription factor that binds to DNA early in the formation of a preinitiation complex and helps the other transcription factors assemble the complex.

astrocyte a type of glial cell found in the central nervous system; derived from the neuroepithelium (see Fig. 18).

astrotactin an adhesion protein, important for a neuron to maintain its adhesion to a glial cell.

asymmetric cell divisions cell divisions in which the daughter cells are different from each other because some cytoplasmic determinant(s) have been distributed unequally between them.

atavism the evolutionary reversion of a character to an ancestral state; appearance of a distant ancestral form of an organism or one of its parts as a result of reactivation of ancestral genes, e.g., hen's teeth.

ATP *See:* **adenosine triphosphate**.

atrazine a herbicide, believed to be responsible for gonadal malformations in many frog species; atrazine exposure causes developmental abnormalities of the limbs, gut, and head, as well as apoptosis of brain and kidney cells in *Xenopus*. Atrazine induces the enzyme aromatase, resulting in "chemical castration"; it also produces an immunodeficiency syndrome in amphibians, and evidence is accumulating that it causes immune deficiencies in mammals.

atresia imperforation; an abnormal closure of a normal opening or canal is an atresia or imperforation, e.g., esophageal atresia.

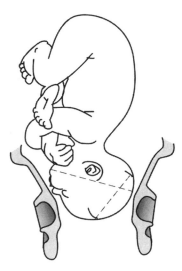

Figure 3. Alignment of the fetus in the uterus. Diagram shows examples of attitude (fetal position), lie (longitudinal), presentation (cephalic), and position (face right). Reprinted from Frank J. Dye, *Human Life Before Birth*, Harwood Academic Publishers, 2000, fig. 11-1, p. 91.

atretic follicles   during a given ovarian cycle, most of the follicles that begin to develop suddenly cease to grow and begin to degenerate; these degenerating follicles are called atretic follicles.

atrial septal defect   also referred to as ASD, a common congenital defect characterized by an abnormal opening in the septum between the right and left atria.

atrichoblasts   those cells of the plant epidermis that are nonhair cell precursors. *See:* **trichoblasts**.

atrioventricular canal   the passageway between the atrial and ventricular regions of the heart, which is subsequently divided by endocardial cushion tissue into right and left channels.

atrium   one of two kinds of chambers that make up the four-chambered heart. The atria pump blood through valves and into ventricles (see Fig. 20).

attitude   one of four descriptions of the fetus' alignment in the uterus; attitude is the posture that the fetus assumes in the uterus near the end of pregnancy (see Fig. 3). *See:* **lie**, **position**, **presentation**.

auditory nerves   the eighth pair of cranial nerves. These nerves connect the inner ear with the brain and provide the sensory input for hearing and balance.

auditory placodes   a bilateral pair of ectodermal thickenings on the head of the embryo, at the level of the hindbrain, which give rise to the auditory vesicles, each of which forms an inner ear labyrinth, whose neurons form the acoustic ganglion.

auditory vesicles   *See:* **otic vesicles**.

auricularia   echinoderm larva of the class Holothuroidea (sea cucumber).

autocrine growth factor   a growth factor that is made by the same cell (target cell) that responds to it, e.g., platelet-derived growth factor.

autocrine signaling   a mode of cell–cell communication in which signaling molecules (autocrine factors) attach to receptors on the same cell that produced them, e.g., the explosive proliferation of placental cytotrophoblast cells in response to platelet-derived growth factor (PDGF), which these cells themselves produce. *See:* **endocrine signaling**, **juxtacrine signaling**, **paracrine signaling**.

autogamous   self-fertilization.

autonomic ganglia   ganglia of the autonomic nervous system.

autonomic nerves   nerves of the autonomic nervous system.

autonomic nervous system   the portion of the nervous system involved with control of involuntary activity, such as that of the internal organs.

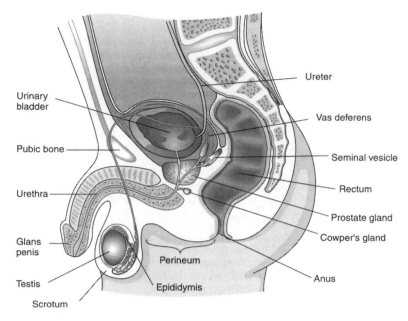

Figure 4. Sagittal section of the male pelvis. Reprinted from Frank J. Dye, *Human Life Before Birth*, Harwood Academic Publishers, 2000, fig. 5-1, p. 35.

autonomous specification   specification by differential acquisition of certain cytoplasmic molecules present in the egg; characteristic of most invertebrates. *See:* **conditional specification**.

autophilous   self-pollinated.

autopod   the distal bones (carpels-fingers/tarsals-toes) of the vertebrate limb, distal to the body wall. *See:* **stylopod**, **zeugopod**.

autopoietic   self-developing. *See:* **developmental symbiosis**.

autoregulation   the control of a gene by its own product.

autoschizis   self-cutting; a type of cell death exhibited by cancer cells treated with vitamin C and vitamin K_3, during which the cells generate enzymes that work like scissors, snipping the cells to bits.

autosomal mutations   mutations found on autosomes.

autosomes   all the chromosomes other than the sex chromosomes.

auxiliary sex glands   (see Fig. 4). *See:* **bulbourethral gland, Cowper's gland, epididymis, glands of Littré, prostate gland, scrotum, seminal vesicle, testis, ureter, urethra, vas deferens**.

auxin   a class of plant hormones; the principle one is indoleacetic acid (IAA). The first group of plant hormones to be identified; is responsible for events such as cell elongation, xylem regeneration in a wound, apical dominance, phototropism, adventitious root growth, gravitropism, and the stimulation, by seeds, of fruit development. Polar auxin transport is developmentally important because inhibitors of polar auxin transport stop embryo development between the globular and heart stages; it is also required for flower development and vascular differentiation. *See:* **indoleacetic acid**.

Aves   the class of vertebrates composed of birds; includes 27 extant orders.

avidin   a protein synthesized by the hen that prevents bacterial growth and is transported into egg albumin; avidn binds and inactivates biotin, a bacterial growth factor.

axes   during the development of bilaterally symmetrical organisms, three axes develop: (1) the anteroposterior axis; in *Drosophila*, this axis is laid down in the egg; (2) the dorsoventral axis; in *Drosophila*, this axis is laid down in the egg; and (3) the proximodistal axis. In *Drosophila*, the anteroposterior axis becomes divided into several unique regions on the basis of the overlapping and graded distributions of different transcription factors; the patterning of the dorsoventral axis involves two gradients, one of the dorsal protein and one of the decapentaplegic protein, with high points at the opposite ends. Patterning along the dorsoventral and anteroposterior axes divides the embryo into numerous discrete regions, each characterized by a unique pattern of zygotic gene activity.

axial patterning   in plants, establishes the apical-basal (shoot-root) axis, consisting of the shoot meristem, procambium, and root meristem; the axial pattern is established by the heart stage of angiosperm embryogenesis. *See:* **axiation**.

axial placentation   in plants, ovules attached to the central axis of an ovary with two or more locules.

axial protocadherin   one of two cell adhesion molecules that apparently direct the adhesive changes driving convergent extension. *See:* **paraxial protocadherin**.

axial skeleton   the portion of the skeleton consisting of the skull and jaws, vertebral column, ribs, and sternum.

axiation   the formation or development of axial structures, such as the neural tube; in anuran amphibians, the embryonic axis approximately corresponds to the egg (animal–vegetal) axis. Fate maps of several species of amphibian blastulas show that the presumptive ento-derm, mesoderm, and notochord map to different regions of the blastula surface (but note the different fate map in *Xenopus*), not coincident with the embryonic axis. During gastru-lation, entodermal, notochordal, and mesodermal cells become arranged anteroposteriorly in a linear fashion along the embryonic axis.

axillary   pertaining to the axilla, the depression between the arm and the thoracic (chest) wall; meristems found in leaf axils that produce axillary buds.

axoneme   the core of a flagellum or cilium, made up of a 9 + 2 array of microtubules.

axons   nerve cell processes, generally less numerous and longer than dendrites, which carry nerve impulses away from the cell body of the nerve cell.

azoospermia   the absence of sperm in semen.

AZT   azidothymidine; a drug used to slow the replication of the HIV virus in patients with AIDS.

B

B cells   *See:* **B lymphocytes**.

B lymphocytes   those lymphocytes (a subset of white blood cells) that produce antibodies.

backbone   a term in common usage referring to the bony vertebral column of vertebrates, including humans.

Bacteroides thetaiotaomicron   a bacterium vital, in mice, for the induction of host angiogenesis; germ-free mice, having no bacteria in their gut, have reduced gut capillary network formation compared with normal mice. *See:* **angiogenesis**.

Baer, Karl Ernst von (1792–1876)   Estonian naturalist and pioneer embryologist, born in Estonia, Russia, to a German family; did work on the embryology of the chick and on other animals; discovered the human ovum and notochord; showed that the various organs of vertebrates are derived from germ layers by differentiation. "Father of Modern Embryology" first described the mammalian egg in 1827; with his microscope, he could see that the mammalian egg was an undifferentiated mass containing no preformed creature; the idea that the mature egg or the early developing embryo possesses certain definite areas having different qualities, each of which contributes to the formation of a particular structure or several structures (as he outlined in 1828 and 1837). He became the first to trace the egg to the embryo. In 1828, he published the *Epistle*, in which he originally set forth his discovery of the ovum, and *Uber Entwickelungsgeschichte der Tiere* (Developmental History of Animals), which became a standard text of embryology; four major advances in embryology were included in this book: (1) the discovery of the mammalian ovum, (2) the germ layer theory, (3) the law of corresponding stages in the development of the embryo, and (4) the discovery of the notochord. "Father of Comparative Embryology"; he was the first to make embryology truly comparative. Together with Pander, formulated the germ layer concept as a structural fact for vertebrate embryology. Karl Ernst von Baer's celebrated work, *Development of Animals* (*Entwickelungsgeschichte der Tiere—Beobachtung und Reflexion*, the greatest classic in embryology, published in 1828, 1837, and 1888), added the final stroke against the preformation doctrine; he observed the egg and the actual developmental stages and showed differentiation to be a progressive process. von Baer enriched embryology in three directions: (1) he set a higher standard for all work in embryology and lifted the science to a higher level, (2) he established the germ layer theory, and (3) he made embryology comparative. He was able to show that in all animals except the very lowest, there originate in the course of development leaf-like layers, which become converted into the "fundamental organs" of the body; by extending his observations into all the principal groups of animals, he raised this concept to the rank of a general law of development. von Baer recognized four such layers; the outer and inner ones being formed first and subsequently budding off a middle layer composed of two sheets. For a long time after von Baer, the aim of embryologists was to trace the history of these germ layers. von Baer clearly saw that the fundamental anatomical features of the body are assumed by the leaf-like rudiments being rolled into tubes. He discovered the mammalian egg and that the notochord occurred in all vertebrate animals. According to Cleveland P. Hickman (1966), the very tiny ova of mammals escaped Regnier de Graaf's eyes, but von Baer brought mammalian reproduction into line with that of other animals by detecting them and their true relation to the follicles. One of the first achievements of the greatest embryologist of all time was this important discovery. According to Hickman, von Baer developed the biogenetic law, which states that the embryos of higher and lower forms tend to resemble each other more closely the further one goes back in their development. This is a conservative and sounder conception of the biogenetic law than that of Ernst Haeckel because von

Dictionary of Developmental Biology and Embryology, Second Edition. Frank J. Dye.
© 2012 Wiley-Blackwell. Published 2012 by John Wiley & Sons, Inc.

Baer did *not* maintain that the embryos of higher forms resemble the *adults* of lower organisms.

He illustrated the development of the embryo from leaf-like germ layers that may form tubes; (1) from the ectoderm, the formation of the neural tube and the outer tube; (2) from the endoderm, the digestive tube and associated glands; and (3) from the mesoderm, connective tissue, muscle, and other visceral parts.

In comparing corresponding stages in the development of different kinds of embryos, von Baer advanced four propositions that together became incorporated in the biogenetic law: (1) in development, general characters appear before special characters; (2) from the more general characters are developed the less general and finally the special; (3) in the course of development, an animal of one species diverges continuously from one of another species; and (4) a higher animal during development passes through stages that resemble stages *in the development of* lower animals. von Baer did not relate these observations to evolution; Ernst Haeckel made the connection and popularized the phrase "ontogeny recapitulates phylogeny."

von Baer made embryology a comparative science, and his fame is attributed almost wholly to his embryological work. Among the many honors that he received, he was elected a member of the Royal Society of London (Gardner, 1965). *See:* **Remak, Robert**.

Balbiani rings   large puffs along the lengths of polytene chromosomes of the larval salivary glands of *Chironomus*.

Baldwin effect   fixation of an environmentally induced phenotype by a new mutation. *See:* **genetic accommodation, genetic assimilation**.

Balfour, Francis (1851–1882)   British biologist; developed the best of Ernst Haeckel's work in his two-volume *Comparative Embryology* (1880–1881), which covers the development of vertebrates and invertebrates and makes critical comparisons. He transformed the work of the Haeckel school into an organized whole. In 1874, he published, with Michael Foster, *The Elements of Embryology*. Balfour saw developmental processes in the light of the hypothesis of organic evolution. The reading of ancestral history is a characteristic feature of the embryological work of Balfour's period. By using the biogenetic law and the germ layer doctrine, he produced a detailed analysis of the vertebrate urogenital system, answering many of the questions left by Rathke's study of the adder. Completed (in 1880–1881) the first comprehensive presentation of comparative embryology.

Balfour's law   the law that the speed with which any part of the ovum segments is roughly proportional to the protoplasm's concentration in that area.

balloon catheterization   a procedure used to open a blocked or partially blocked body tube by the expansion of a balloon-like device in the blocked portion of the tube; has been used to open blocked fallopian tubes, as well as coronary arteries.

banding techniques   chemical treatments of chromosomes that cause them to have a consistent pattern of bands (stripes); allows for the unequivocal identification of chromosomes and parts of chromosomes.

Barker hypothesis   certain anatomical and physiological parameters get programmed during embryonic and fetal development; changes in nutrition during this time can produce permanent changes in the pattern of metabolic activity, which can predispose the adult to particular diseases, e.g., undernutrition during first trimester = risk of hypertension and stroke as an adult, undernutrition during second trimester = risk of heart disease and diabetes as an adult, undernutrition during third trimester = prone to blood-clotting defects as adults. *See:* **fetal plasticity**.

Barr body   the chromatin (heterochromatin) of the inactive X chromosome in female mammalian cells.

Bartholin's glands   female auxiliary sex glands, the ducts of which open into the vestibule of the vulva. Their secretions provide lubrication at the time of sexual intercourse. Also called the greater vestibular glands.

basal   in phylogenetic terms, a group that branches at the base of a clade.

basal body temperature (BBT) method   a fertility awareness method, based on the knowledge that a woman's basal body temperature increases at the time of ovulation.

basal cell   one of the two cells resulting from the first division of the angiosperm zygote, which gives rise to the suspensor. *See:* **terminal cell**.

basal disc   (1) the basal disc of the slime mold *Dictyostelium discoideum*, which anchors the sorocarp to the substratum; (2) the "foot" of of *Hydra*, which enables it to adhere to the substratum; *Hydra* has been used in transplant studies of regeneration. *See:* **hypostome**.

basal lamina a tight extracellular layer formed by epithelial cells; together with the reticular lamina, it constitutes the basement membrane.

basal layer also called the stratum germinativum, the inner layer of the two-layered skin of the embryo at the end of the embryonic period; also the inner layer of the adult epidermis; a germinal epithelium that gives rise to all the cells of the epidermis. *See:* **periderm**.

basal level transcription a very low level of class II gene transcription (involving RNA polymerase II) achieved with general transcription factors and polymerase II alone.

basal placentation ovules positioned at the base of a single-loculed ovary.

base pair a pair of nitrogenous bases (each a part of a nucleotide in a nucleic acid molecule) hydrogen bonded to each other.

basement membrane a thin layer of extracellular matrix on which the epithelial cells of an epithelium rest; the basal lamina and the reticular lamina constitute the basement membrane.

base-pairing rules adenine hydrogen bonds to thymine in DNA, adenine hydrogen bonds to uracil in RNA, and cytosine hydrogen bonds to guanine in DNA and RNA.

basic helix-loop-helix (bHLH) a protein motiff observed in enhancer and promoter DNA-binding proteins (i.e., a distinct class of transcription factors); e.g., the muscle-specific transcription factors MyoD and myogenin contain this motiff. The helical domains of two interacting subunits are interrupted by nonhelical polypeptide loops. The bHLH proteins bind to DNA through a region of basic amino acids that precedes the first alpha-helix. One member of the dimer is found in all tissues of the organism, and the other member is tissue specific. Proteins that contain the HLH but not the basic part of the sequence form inactive dimers with other bHLH protins and inhibit their activity. Examples of bHLH proteins include E12 and E47, which are ubiquitous in vertebrates, the myogenic factor MyoD, and the *Drosophila* pair-rule protein hairy. An example of an inhibitor with no basic region is Id, which is an inhibitor of myogenesis.

basic helix-loop-helix (bHLH) family one of four major families of transcription factors; transcription factors in this family function in muscle and nerve specification, e.g., MyoD.

basic leucine zipper (bZip) transcription factors with a structure similar to that of basic helix-loop-helix proteins; are dimers, each of whose subunits contains a basic DNA-binding domain at the carboxyl end, followed closely by an alpha-helix containing several leucine residues. The two subunits of the protein interact through leucine-containing repeat sequences. CCAAT enhancer-binding protein (C/EBP) is a bZIP protein that can bind to the DNA sequence CCAAT; it plays a role in adipogenesis similar to that of bHLH proteins in myogenesis.

basic leucine zipper (bZip) family one of four major families of transcription factors; transcription factors in this family function in liver differentiation and fat cell specification, e.g., C/EBP, AP1, etc.

Bateson, William (1861–1926) British geneticist; studied the embryology of the worm-like marine creature *Balanoglossus*, discovering that, although its larval stage resembles that of the echinoderms, it also has gill slits, the beginnings of a notochord, and a dorsal nerve cord, proving it to be a primitive chordate; this was the first evidence that the chordates have affinities with the echinoderms. In 1894, used the word "homoeosis," rather than Loeb's term "heteromorphosis," to describe the spontaneous condition in *Palinurus penicillatus* (lobster). Although his sweet-pea crosses provided the first hint that genes are linked on chromosomes, Bateson never accepted Thomas Hunt Morgan's explanation of linkage or the chromosome theory of inheritance. In 1908, he became the first professor (Cambridge) of the subject he himself named—genetics. He produced the first English translation of Gregor Mendel's work on heredity.

battery a group of genes regulated by the same transcription factor.

Bauchstück term used by Hans Spemann, meaning the belly piece; refers to an unorganized mass of tissue, obtained from an isolated blastomere, of the two-cell frog embryo, that does not contain any of the gray crescent; an unorganized mass of ventral cells.

Bauplan (Baupläne) body plan.

B-catenin is the major candidate for the factor that forms the Nieuwkoop center in the dorsalmost vegetal cells of the amphibian blastula; it is a transcription factor that can associate with other transcription factors to give them new properties; the B-catenin/Tcf3 (a ubiquitous transcription factor) complex appears to bind to the promoters of several genes whose activity is critical for axis formation; a protein that can act as an anchor for cadherins or as a transcription factor (induced by the Wnt pathway); it is important in the specification of the germ layers throughout the animal phyla.

***Bcl-2* genes**   a family of genes, including genes that are homologues of the *ced-9* genes of *Caenorhabditis elegans* and play important roles in apoptosis in mammals.

Beadle, George Wells (1903–1989)   American biochemical geneticist, who showed that the eye color of *Drosophila* is a result of a series of chemical reactions under genetic control; he subjected the red bread mold *Neurospora crassa* to X-rays and studied the altered nutritional requirements of, and therefore enzymes formed by, the X-ray-induced mutants; developed, with Edward Lawrie Tatum, the idea that specific genes control the production of specific enzymes (one gene–one enzyme hypothesis), virtually creating the science of biochemical genetics. Beadle and Tatum shared the Nobel Prize for physiology or medicine in 1958 with Joshua Lederberg.

Beermann, Wolfgang (1921–2000)   made contributions to our understanding of the significance of polytene chromosomes. He showed (in 1952) that the banding patterns of polytene chromosomes were identical throughout the insect larva and that no loss or addition of any chromosomal region was observed in different cell types. Beermann found that polytene chromosomes of *Chironomus* larvae had regions that were "puffed out"; unlike the banding patterns, the puffs appeared in different places on the chromosomes in different tissues and their appearances changed as development progressed. He provided evidence (in 1961) that these puffs represent a local loosening of the polytene chromosomes and that the puffs are sites of active RNA synthesis.

Beneden, Edouard Van (1846–1910)   Belgian cytologist and embryologist; in 1877 proposed the creation of the phylum Mesozoa, to cater to the transition between single-celled and multicelled organisms, and in 1887 demonstrated the constancy of the number of chromosomes in the cells of an organism, decreasing during maturation and restored at fertilization; demonstrated that equal numbers of chromosomes were contributed to the offspring at the time of fertilization. In *Ascaris megalocephala univalens*, each parent contributes a single chromosome to the pair possessed by the zygote. In *bivalens*, the corresponding numbers are two chromosomes in each of the pronuclei and four in the zygote.

BFU-E   *See:* **burst-forming unit, erythroid (BFU-E)**.

bicoid   a key maternal gene in early *Drosophila* development, necessary for the establishment of anterior structures in the embryo; it establishes a gradient in some substance whose source and highest level are at the anterior end of the early embryo; this substance is the bicoid protein.

bicoid **mRNA**   in Drosophila, has been shown to be present in the anterior region of the unfertilized egg, where it is attached to the cytoskeleton; this mRNA is not translated until after fertilization. *See:* **nurse cells**.

bicoid protein   *bicoid* mRNA is localized in the anterior end of the unfertilized egg of *Drosophila*; after fertilization, it is translated (bicoid protein is absent from the unfertilized egg) and the bicoid protein diffuses from the anterior end, forming a concentration gradient along the anteroposterior axis. Historically, the bicoid protein gradient provided the first reliable evidence for the existence of the morphogen gradients that had been postulated to control pattern formation. The bicoid protein is a transcription factor that acts as a morphogen; it switches on certain zygotic genes at different threshold concentrations, thus initiating a new pattern of gene expression along the axis. The bicoid protein is a member of the homeodomain family of transcriptional activators that activates the *hunchback* gene by binding to regulatory sites within the promoter region.

bicornuate uterus   in mice, partial fusion of embryonic mullerian ducts gives rise to a uterus that is mostly double (consisting primarily of two horns); consequently, mice are said to have a bicornuate uterus.

biennial   a plant that lives 2 years, usually forming a basal rosette of leaves the first year and flowers and fruits the second year.

bifurcate   to split into two parts; branch.

bifurcation point   a developmental threshold, e.g., as when allometry generates evolutionary novelty by small, incremental changes that eventually cross the bifurcation point; when such a threshold is crossed, a change in quantity becomes a change in quality; e.g., such a mechanism has been postulated to have produced the external fur-lined pouches of kangaroo rats.

bilaminar embryonic disc   an amniote embryo prior to gastrulation; consists of epiblast and hypoblast layers (see Fig. 5).

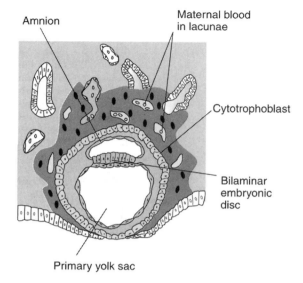

Figure 5. Bilaminar embryonic disc. Reprinted from Frank J. Dye, *Human Life Before Birth*, Harwood Academic Publishers, 2000, fig. 8-6, p. 65.

bilateral cleavage   the first cleavage plane divides the egg into bilateral halves; i.e., one blastomere of the two-cell stage will give rise to the left half of the body of the embryo, and the other blastomere will give rise to the right half of the body of the embryo. Characteristic of tunicates, such as *Styela*.

bilateral symmetry   a type of symmetry exhibited by some organisms in which only one plane, the midsagittal plane, divides the organism into mirror image halves. In tunicate eggs, the first cleavage furrow is in the plane of bilateral symmetry. In frogs, the plane that includes both the primary and secondary axes of the fertilized egg is the plane of bilateral symmetry. Frequently, but not invariably, in frogs, the first cleavage plane coincides with the prospective plane of bilateral symmetry; thus, each of the first two blastomeres receives half the gray crescent and represents one side of the future embryo (see Fig. 2).

Bilateria   most metazoans, which exhibit bilateral cleavage and bilateral symmetry; does not include sponges, cnidarians, or ctenophores.

Bildungstrieb   "developmental force"; a mechanical, goal-directed, force postulated to exist by Johann Friedrich Blumenbach; thought to be inherited through the germ line, development was postulated to proceed through this predetermined force inherent in the matter of the embryo.

bindin   an acrosomal protein exposed on the acrosomal process of sea urchin spermatozoa as a result of the acrosome reaction; for which there is a bindin receptor, a glycoprotein, on the vitelline membrane of the egg. Bindin is capable of binding to dejellied eggs; evidence indicates the existence of a species-specific bindin receptor on the vitelline envelope; a large glycoprotein complex has been isolated from the vitelline membrane of the sea urchin that binds bindin in a species-specific manner.

bindin receptor   a glycoprotein receptor on the vitelline membrane of the sea urchin egg, which binds the bindin protein exposed, by the acrosome reaction, on the surface of the acrosomal process of the spermatozoon.

binding site   that portion of the surface of an enzyme to which ligands bind; e.g., allosteric enzymes have two kinds of binding sites: the active (catalytic) site to which substrates bind and the allosteric site to which positive allosteric effectors (activators) and negative allosteric effectors (inhibitors) bind. Developmentally important transcription factors have DNA binding sites by which they bind to specific sites on DNA.

binocular vision   using both eyes synchronously so as to see one visual image, as opposed to double vision.

biochemical cytodifferentiation   cell differentiation involving an elaboration of biochemicals rather than an elaboration of structural features.

biofilms   mats of bacteria that regulate the life cycles of many marine invertebrates, determining where and when larvae can settle and undergo metamorphosis.

Biogenesis   the theory that life always originates from preexisting life. *See:* **spontaneous generation**.

Biogenetic Law   refers to Ernst Haeckel's aphorism (from 1875), "ontogeny recapitulates phylogeny." The concept that the successive *developmental* stages of any animal duplicate the *adult* stages of that animal's ancestors, in the same succession. This law is not generally accepted. *See:* **Baer, Karl Ernst von**; **paleogenesis**.

bioinformatics   computational biology; the application of the power of computational biology to solutions of complex biological data analysis; the building and manipulation of biological databases; the application of information technology, statistics, and mathematics to biological problems involving large volumes of data with complex interrelationships; it provides the foundation for much modern biomedicine and biotechnology.

biological mother   a woman carrying a pregnancy when the conceptus is the product of her egg and a spermatozoon.

bipinnaria larva   echinoderm larva of the class Asteroidea (sea star).

bipotential stage   *See:* **indifferent stage**.

birth   the process by which an infant normally leaves the uterus (womb) through the birth canal. With the beginning of muscular contractions, at the onset of labor, amniotic fluid is squeezed into the thin part of the chorionic sac (consisting of the decidua capsularis, chorion laevae, and adherent amnion, only), which is the preliminary dilator of the cervical canal. With more powerful contractions, the investing membranes rupture at this region and the embryo is freed from fetal envelopes; however, the placenta is still attached, and this retention of the placenta is vital. Continued contractions force the fetus into the cervical canal; once the fetus passes the cervical canal, it moves promptly through the vagina. With the tying off of the umbilical cord, the newborn infant is for the first time an independently living individual. Usually, 15–20 minutes after delivery of the fetus, uterine contractions loosen the placenta and decidua from its walls and finally expel them. The placenta plus the associated torn remnants of ruptured amnion, chorion laeve, and umbilical cord constitute the so-called afterbirth.

birth canal   the conduit, partially the uterine cervix and partially the vagina, through which the baby passes during birth.

birth defect   an abnormality in development that originates before birth and is usually, but not always, apparent at birth. *See:* **congenital anomalies**.

birthday   the time at which a neuron divides for the last time.

bisphenol A (BPA)   ubiquitous plasticizing compound and endocrine disruptor with numerous adverse developmental effects.

bithorax complex (BX-C)   of *Drosophila*, comprises three homeobox genes: *Ultrabithorax*, *abdominal-A*, and *Abdominal-B*. These genes are expressed in the parasegments in a combinatorial manner, which specifies the character of the parasegments. The bithorax complex controls the development of parasegments 5–14. The pattern of activity of the bithorax complex genes is determined by gap and pair-rule genes. Parasegment 4, of *Drosophila* embryos, is sort of a "default" state, which is modified in all the parasegments posterior to it by the proteins encoded by the bithorax complex. It is because the genes of the bithorax complex are able to superimpose a new identity on the default state that they are called selector genes. Differences between segments may reflect differences in the spatial and temporal pattern of HOM gene expression. The order of genes in the complex is the same as the spatial and temporal order in which they are expressed along the anteroposterior axis during development. *See:* **Antennapedia complex**.

bivalent   a pair of homologous chromosomes in synapsis during prophase I of meiosis (see Fig. 26).

bladder   *See:* **urinary bladder**.

-blast   a word part denoting a formative cell or a germ layer, e.g., myoblast, neuroblast, or epiblast.

blast cell   a primitive cell; usually, the least differentiated member of a line of blood-forming elements.

blastema   a mass of dedifferentiated cells; (1) the part of the parental organism giving rise to a new individual in asexual reproduction, as in budding of colonial tunicates; (2) a regeneration blastema is formed at the end of the stump of an amputated part of some animals, e.g., in salamanders, from which limbs have been removed, and in planaria, from which parts have been removed. When a part or limb is lost, the first stage of repair is that the epidermis from the edges of the wound spreads over the wound and covers the open surface. Subsequently, the epidermis covering the wound bulges outward as a mass of cells accumulates under the epidermis; the mass of cells, together with the epidermal covering, is the regeneration blastema. The regeneration blastema provides cells that proliferate and differentiate into various cells types as part of the regeneration process; the cells of the blastema have regained their embryonic plasticity.

blastocoel the cavity of the blastula or blastocyst; it has been suggested that the blastocoel represents an unfilled yolk cavity, which, in ancestral forms, was laden with stored food. The blastocoel seems to serve two functions: (1) it permits cell migration during gastrulation, and (2) it prevents cells beneath it from interacting prematurely with cells above it (see Plate 10 in the color insert).

blastocyst an early embryonic form of mammals, containing two types of cells (trophoblast cells and inner cell mass cells) and a fluid-filled cavity (blastocoel or blastocyst cavity). This embryonic form is the equivalent of the blastula in other animals. However, an important difference is that all the cells of a blastula contribute to the developing organism, but only the inner cell mass, not the trophoblast, contributes to the developing mammal. The trophoblast contributes to the development of extraembryonic membranes. In the inner cell mass stage, the outer layer of the blastocyst is the trophoblast; when the endoderm has been differentiated, it is called the trophectoderm; when the mesoderm splits and the somatic layer becomes associated with ectoderm, three different names are used: extraembryonic somatopleure, trophoderm, and serosa.

blastocyst cavity the blastocoel or cavity of the blastocyst (see Plate 7 in the color insert and Fig. 44).

blastocyte an embryonic cell that is undifferentiated.

blastoderm when the blastodisc has undergone a few cleavage divisions and has a cellular nature (although it may be partially syncytial), it is then, conventionally, referred to as the blastoderm. In hens' eggs, as cleavage continues, the blastoderm consists of central cells, which are relatively smaller and completely defined, and peripheral cells, which are relatively larger, flattened, and not walled off from the yolk. By the time the hen has laid the egg, the blastoderm contains some 60,000 cells. In *Drosophila* the cellular blastoderm consists of some 6000 cells and is formed within 4 hours of fertilization (see Plate 3 in the color insert).

blastodermic vesicle the blastocyst.

blastodisc the thin plate of relatively yolk-free cytoplasm found near the animal pole of macrolecithal, heavily telolecithal, eggs, such as those of fish and birds. In the embryos of these animals, cleavage is initially confined to the blastodisc and the cleavage pattern is described as being meroblastic discoidal. In hens' eggs, the blastodisc is a whitish, circular area, approximately 3 mm in diameter. *See:* **germinal disc**.

blastogenesis development that starts from a blastema.

blastokinesis in insects, the movements of the embryo of the lower insects (Exopterygota) from the ventral surface of the egg, toward the dorsal surface, and back to the ventral side of the egg; of questionable significance.

blastomeres the cells of an embryo undergoing cleavage. In amphibian development, the blastomeres, after the midblastula transition, are a much more complex population than they were in earlier cleavage (see Plate 10 in the color insert).

blastopore the opening from the gastrocoel (archenteron) to the outside of the embryo; in protostomes, the blastopore gives rise to the mouth, and in deuterostomes, the blastopore gives rise to the anus. In the chick, the amount of yolk is so large that it prevents the formation of any open blastopore; the pre-primitive streak is the symbolic homologue of a blastopore.

blastopore lips because of a small, eccentrically located blastocoel, during gastrulation in frogs, the blastopore is formed piecemeal; first the dorsal lip, then the two lateral lips, and, finally, the ventral lip; i.e., the lips of the blastopore are sequentially formed (see Plate 4 in the color insert).

blastozooid the individual resulting from asexual reproduction. *See:* **oozooid**.

blastula an early embryonic form of animals generally consisting of a layer of cells (blastomeres) surrounding a fluid-filled cavity, the blastocoel. In the sea urchin, the blastula is a single layer of 1000 to 2000 cells, surrounding the fluid-filled blastocoel. In the chick, the blastula consists of a discoidal cap of cells separated by a slit-like blastocoel from the underlying yolk (see Plate 10 in the color insert). *See:* **stereoblastula**.

blastulation formation of a blastula from a solid ball of cleaving cells. *See:* **morula**.

block to polyspermy any mechanism that prevents an egg from being fertilized by more than one spermatozoon; (1) fast block to polyspermy is a change in the electrical potential of the egg plasma membrane, as a result of increased sodium ion concentration inside the egg, resulting in the inability of sperm to attach to the egg plasma membrane; (2) slow block to polyspermy is the cortical reaction, as a result of exocytosis of the cortical granules of the egg, caused by an increase in calcium ion concentration inside the egg, which alters the egg cell surface and prevents the attachment of more than one fertilizing sperm.

blood islands aggregations of mesenchyme cells, derived from the splanchnic mesoderm of the yolk sac, which give rise to the yolk sac blood vessels (extra-embryonic) and the embryo's first blood cells. Fluid-filled spaces appear within the blood islands and separate central cells from peripheral cells. The peripheral cells become flattened, remain adherent, and completely enclose the central cells, forming the endothelial walls of primitive blood channels (extensions and anastomosis of these form networks of vessels). The central cells separate from each other and form primitive blood corpuscles. These processes begin in the peripheral part of the area opaca vasculosa and extend inward toward the body of the embryo.

blood plasma the liquid portion of the blood; that is, not including the blood cells or so-called formed elements.

blood-testis barrier formed by tight junctions between Sertoli cells, creates two compartments in the seminiferous tubules: the basal compartment extending from the basal lamina (basement membrane) to the blood-testis barrier, containing the spermatogonia and the earliest spermatocytes, and the adluminal compartment extending inward from the blood-testis barrier toward the lumen of the seminiferous tubule, containing the meiotically active cells and the differentiating spermatids. This barrier prevents the membrane antigens of differentiating sperm from escaping through the basal lamina, entering the bloodstream, and provoking an autoimmune response to one's own sperm, causing sterility (see Fig. 39).

Blumenbach, Johann Friedrich (1752–1840) a representative of German Naturphilosophie and professor at the University of Göttingen. In his essay, "On the formation force and generation affair" *(Über den Bildungstrieb und das Zeugungsgeschäft)*, published in three editions in 1771–1791, Blumenbach rejected any form of preformation, and as an experienced anatomist and anthropologist emphasized that the embryo formed from the very start by the action of a special "formation force." His natural history textbook (1830), published in 12 editions between 1780 and 1830, had a profound influence on his contemporaries. Among these was the Czech physiologist Jan Evangelista Purkyne. (In German and English papers, he spelled his name "Purkinje.")

BMP *See:* **bone morphogenetic proteins**.

BMP4 an anti-neuralizing protein secreted from the nonorganizer mesoderm. *See:* **bone morphogenetic proteins (BMPs)**, **Xwnt8**.

BMP4 and BMP7 these proteins can mimic the induction of neural crest, from cultured neural plate, by presumptive epidermis; induce the expression of the Slug protein and the RhoB protein in cells destined to become neural crest.

body cavity the coelom is the original, single-body cavity of the embryo. During development, it is subdivided into pericardial, peritoneal, and pleural cavities, which contain the viscera.

body contruction theory how the zygote gives rise to the body; i.e., embryology.

body folds folds in the germ layers of the developing embryo, which initially mark off the embryonic region from the extraembryonic region and, subsequently, form the sides and ventral surface of the developing embryo (see Fig. 6).

body stalk *See:* **connecting stalk**.

bone marrow hematopoietic (blood-forming) tissue found within certain bones.

bone morphogenetic proteins (BMPs) although originally discovered by their ability to induce bone growth, the BMPs regulate developmental processes as diverse as cell proliferation, apoptosis, cell migration, cell differentiation, and morphogenesis. The dorsal fates of the neural tube are established by the bone morphogenetic proteins; these proteins are expressed in the presumptive dorsal epidermis and have been shown to counteract the effect of Sonic hedgehog, permitting genes such as *msx1* and *Pax3* to be expressed in the

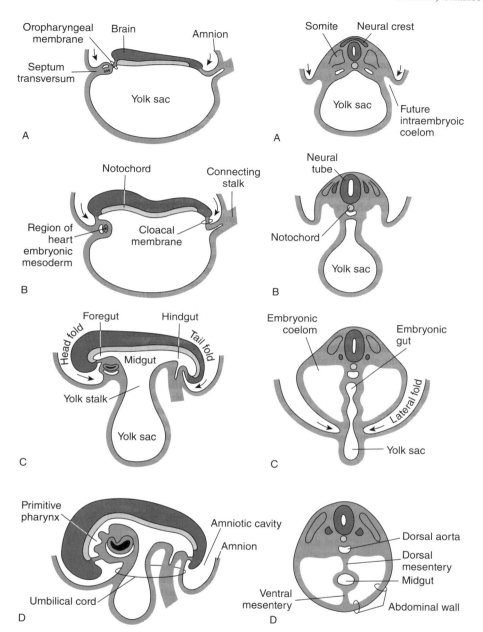

Figure 6. The body folds. (left) Diagrammatic sagittal sections of embryos during the fourth week, showing the contributions of the head and tail folds to the formation of the tubular embryonic body. (right) Diagrammatic transverse sections of embryos during the fourth week, showing the contributions of the two lateral folds to the formation of the tubular embryonic body. Reprinted from Frank J. Dye, *Human Life Before Birth*, Harwood Academic Publishers, 2000, fig. 8-12, p. 71.

dorsal portion of the neural tube. Both the insect and the vertebrate form their respective neural tubes by inhibiting a BMP signal to the ectoderm.

Bonellia viridis   a marine echiuroid worm (green spoonworm) showing extreme sexual dimorphism, where the environment determines sex.

Bonnet, Charles (1720–1793)   Swiss naturalist and philosopher; credited with the discovery of parthenogeneis while studying aphids; defended the incasement theory; endeavored to explain the phenomenon of regeneration on the basis of preformation. Claimed that within the first female of every species were contained the germs of all subsequent offspring, each generation within the previous one, like a nest of Chinese boxes; this doctrine was known as encapsulation or emboitement; when the last box had been opened, the human race would cease to be.

Born, Gustav (1851–1900) endorsed the approach of experimenting on embryos, but doubted Eduard Pflüger's interpretation, which placed so much importance on the efficacy of external factors. Born's work involving transplanting pieces of tissue from one organism to another, to determine the respective contributions of each of the two parts to the hybrid developing embryo, was influential and ultimately inspired Ross Harrison and Hans Spemann in their own successful work on tissue culture and embryonic transplantation.

bottle cells   so called because of their shape, resulting from changes in cell–cell attachment; in frog gastrulation, these presumptive endoderm cells initiate invagination and the beginning of the formation of the archenteron; however, experiments with *Xenopus* demonstrate that removal of the bottle cells does not affect the involution of the marginal cells.

boundary sequences   *See:* **insulator sequences**.

Bounoure, Louis (1934–) demonstrated that the vegetal region of fertilized frog eggs contains a material with staining properties similar to that of *Drosophila* pole plasm; he was able to trace this cortical cytoplasm into those few cells in the presumptive endoderm that would migrate into the genital ridge. He also showed (in 1939) that ultraviolet (UV) light treatment of the vegetal poles of fertilized frog eggs resulted in sterile adults. In 1966, L. D. Smith showed that when vegetal pole cytoplasm was transferred into the vegetal region of irradiated zygotes, primordial germ cells were found in the genital ridge. Thus, as with *Drosophila* pole plasm, frog zygotes contain a determinant for germ cell formation that is sensitive to UV irradiation and that can be transferred through the cytoplasm.

Boveri, Theodor (1862–1915)   German zoologist; sought to determine the relative contributions of nuclear and cytoplasmic material to development, as well as the relative contributions of the male and female parents; known for research in cytology, especially on fertilization in ascarids and sea urchin eggs; believed in the individuality of chromosomes. Boveri and Wilhelm August Oscar Hertwig (in 1887–1888) separately discovered the real nature of reduction division in the egg. Worked with a nematode, *Parascaris aequorin* (*Ascaris megalocephala*), demonstrated chromosome diminution in the somatoplasm, but not in the germplasm; i.e., the chromosomes are kept intact only in those cells destined to form the germ line; seemed to be consistent with Weissmann's concept of continuity of the germplasm; however, such instances of chromosome diminution and chromosome elimination are exceptions to the general rule that the nuclei of differentiated cells retain unused genes. By centrifuging *Ascaris* eggs (in 1910), concluded that the vegetal cytoplasm contained a factor (or factors) that protected nuclei from chromosomal diminution and determined them to be germ cells.

Bowman's capsule   the infolded, blind, end of the uriniferous tubule in which the glomerulus (tuft of capillaries) is found.

brachial   of or relating to the arms; not to be confused with branchial, which refers to gills.

brachy-   a word part meaning short, e.g., brachydactyly.

brachydactyly   abnormal shortness of the fingers or toes.

Brachyury   a *Xenopus* gene, turned on by moderate concentrations of the morphogen activin, responsible for instructing cells to become mesoderm. *See:* **goosecoid**.

brain-derived neurotrophic factor (BDNF)   a growth factor produced by the neural tube itself; the lack of this growth factor seems to inhibit production of dorsal root ganglia.

brain vesicles   subdivisions of the early developing brain; consisting first of three vesicles (prosencephalon, mesencephalon, and rhombencephalon) and then of five vesicles (telencephalon, diencephalon, mesencephalon, metencephalon, and myelencephalon) (see Figs. 7 and 8).

branchial   of or relating to the gills; not to be confused with brachial, which refers to the arms.

branchial arches   or gill arches; the successive bulges in the lateral walls of the pharynx found between sucessive branchial grooves and/or branchial pouches. Branchial arch I (the mandibular arch) is cephalic to the first postoral cleft; branchial arch II (the hyoid arch) is caudal to the first postoral cleft; branchial cleft I (the first postoral cleft) is the hyomandibular celft. Posterior to these, the arches and clefts are designated by their postoral numbers only. By 3 days, in the chick, the fourth visceral cleft has appeared and aortic

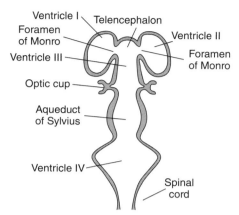

Figure 7. Lumen (cavity) of the brain. Four ventricles (I-IV) and the aqueduct of Sylvius make up the cerebrospinal fluid-filled lumen of the brain. Reprinted from Frank J. Dye, *Human Life Before Birth*, Harwood Academic Publishers, 2000, fig. 13-6, p. 109.

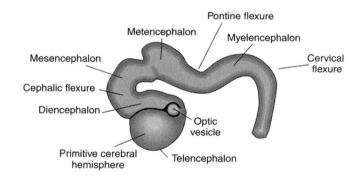

Figure 8. Flexures and vesicles of the brain. As the brain forms, it bends to give rise to three flexures: cephalic, pontine, and cervical. The five-vesicle brain—a highly diagrammatic representation of the five vesicles of the early brain (at 5 weeks). The prosencephalon gives rive to the telencephalon and diencephalon, the mesencephalon persists, and the rhombencephalon gives rise to the metencephalon and the myelencephalon. Reprinted from Frank J. Dye, *Human Life Before Birth*, Harwood Academic Publishers, 2000, fig. 13-5, p. 108.

arches can be observed running through the visceral arches. In the 4-day chick embryo, the arches are too thickened for the blood vessels to be visible. The mandibular arches make up the caudal boundary of the oral depression; maxillary processes develop on each side, as paired elevations in connection with the lateral part of the mandibular arches, forming the lateral boundaries of the mouth opening. Rostral to the mandibular arches, surrounding each nasal pit is a U-shaped elevation with its limbs directed toward the oral cavity; the lateral limb in each case is a nasolateral process, and the median limb in each case is a nasomedial process. The two nasomedial processes merge in the midline, and each fuses with the maxillary process of its own side to form the maxilla (upper jaw); the mandibular arches fuse to give rise to the mandible (lower jaw) (see Fig. 9).

branchial chamber   *See:* **gill chamber**.

branchial clefts   in some species, e.g., the frog, branchial furrows actually break through to form actual branchial clefts (between hatching and metamorphosis in the frog). The 55-hour chick embryo has branchial (gill) arches and clefts; the clefts form from a meeting of ectodermal gill furrows with endodermal pharyngeal pouches. The branchial arches and clefts are transitory, but they participate in the formation of some endocrine glands, the Eustachian tubes, face, and jaws. In birds, any open cleft is transitory; in the chick embryo,

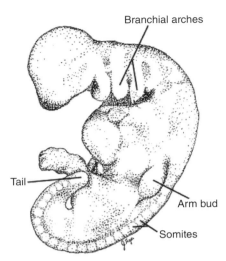

Branchial arches

Tail

Arm bud

Somites

Figure 9. The human embryo at the end of 4 weeks. Note the presence of branchial arches, tail, and arm buds; the head is off the chest. Reprinted from Frank J. Dye, *Human Life Before Birth*, Harwood Academic Publishers, 2000, fig. 8-13A, p. 72. Original artwork by John C. Dye.

the most posterior members of the series of clefts never become open, even though they are referred to as clefts. The first pair of postoral clefts is present at 46 hours; by 55 hours, there are three pairs of clefts.

branchial fistula   *See:* a fistula from the pharynx to an opening on the side of the neck; an example of a fistula between an internal structure and the external surface of the body.

branchial furrows   *See:* **branchial grooves (furrows)**.

branchial grooves (furrows)   the ectodermal invaginations of the lateral walls of the pharynx.

branchiomere   an embryonic metamere that will differentiate into a visceral arch and cleft; a branchial segment.

brassinosteroids   are animal-like hormones found in plants, e.g., brassinolide; demonstrates that plants, like animals, have steroid hormones; affect cell division and cell elongation.

BRCA1   a tumor-suppressor gene encoding BRCA1, which restricts cell division; in breast cancers becomes heavily methylated.

breech presentation   the relationship of the part(s) of the fetus to the birth canal at the beginning of labor where the buttocks and/or feet are closest to the birth canal.

Briggs, Robert (1911–1983)   American embryologist, who, with Thomas J. King, pioneered the technique of nuclear transplantation; Briggs and King discovered that as development proceeds, there is a progressive decrease in the percentage of transplanted nuclei that are capable of promoting normal development of enucleated eggs.

broad ligament of the uterus   a fold of peritoneum that extends laterally from the uterus to the pelvic wall on each side.

bronchioles   small subdivisions of the bronchi.

bronchus   (pl. bronchi) one of the two branches given off from the trachea at its bifurcation, one going to each lung.

brown fat cells   cells found in the late fetus and early newborn, containing mitochondria that convert food energy into heat rather than into adenosine triphosphate (ATP). Brown fat cells are an important source of warmth for the newborn.

bryophytes   literally, "moss plants"; includes those plants, such as the mosses, which exhibit more kinds of differentiated cells than the thallophytes, but lacking true vascular tissue.

bryostatin   a toxic chemical produced by the symbiotic bacteria of the bryozoan *Bugula neritina*; not present in adult tissues, it is transferred to the bryozoan eggs and embryos, which it renders unpalatable. *See:* **embryo defenses**.

buccal cavity   cavity of the mouth.

bud   something not yet mature or at full development, e.g., limb buds and tail buds of developing animals, embryonic parts of plants having meristematic activity; twigs have terminal buds (found at the ends of twigs) and lateral buds (found at the nodes of twigs). The terminal bud of a twig contains a terminal (apical) meristem, embryonic leaves, and axillary buds; a year's growth of a twig comes from its terminal bud. Axillary buds are also called lateral buds. An undeveloped shoot or flower. Also, to commence growth from buds.

budding   a type of asexual reproduction. The new individual is formed from a small out-growth on the surface of the parent. No organs of the parent are passed on as such to the offspring, but the bud is supported by the parental organism at least during the initial stages of its development; the most typical examples are found in coelenterates and tunicates.

bulb   an underground bud with thickened fleshy scales, as in the onion.

bulbourethral gland   a pair of male auxiliary sex glands that empty their secretions into the urethra; also called Cowper's glands (see Fig. 4). *See:* **Cowper's gland**.

bulbous arteriosus   the anterior division of the embryonic heart within the pericardial cavity; its proximal part is incorporated into the right ventricle; its distal part forms the aortic and pulmonary valve region of the heart.

bulge   a region of the hair follicle that serves as a niche for adult stem cells.

Bunker, Eng and Chang (1811–1874)   the conjoined twins who inspired the term "Siamese twins." *See:* **conjoined twins**.

bursa copulatrix   in female insects, an outgrowth of the median oviduct; will receive the spermatophore of the male during mating.

burst-forming unit, erythroid (BFU-E)   a lineage-restricted stem cell; it can form only one cell type, CFU-E (colony-forming unit, erythroid), in addition to itself.

bushy chorion   *See:* **chorion frondosum**.

C

C-cadherin also called EP-cadherin, is a cell adhesion molecule critical for maintaining cell adhesion between the blastomeres of the *Xenopus* blastula and is required for the normal movements of gastrulation.

C. elegans commonly used abbreviation for *Caenorhabditis elegans*.

C-value the amount of DNA per nucleus; the amount of DNA in picograms (trillionths of a gram) in a haploid genome of a given species. C-values during mitotic and meiotic cycles: (1) gamete = 1C, (2) zygote = 2C, (3) post S period of premeiotic mitosis = 4C, (4) mitotic anaphase = 2C, (5) post S period immediately preceding meiosis = 4C, (6) anaphase I = 2C, and (7) anaphase II = 1C.

C-value paradox the fact that the C value of a given species is not necessarily related to the genetic complexity of that species.

cadastral genes genes turned on by floral meristem identity genes that, in turn, initiate transcription of floral organ identity genes, e.g., *SUP* in *Arabidopsis*.

cadherins a large family of intercellular adhesion molecules; cell adhesion molecules whose cell-adhesive properties depend on calcium ions, e.g., N-cadherin (A-CAM), exhibited by nerve, kidney, lens, and heart cells; P-cadherin, exhibited by placental and epithelial cells; and E-cadherin (L-CAM or uvomorulin), exhibited by epithelial and mouse blastula cells. *See:* **cell adhesion molecules (CAMs)**.

caeno- a word part denoting new, fresh, recent, e.g., caenogenesis, caenomorphosis, and *Caenorhabditis*.

caenogenesis is a type of caenomorphosis where an addition of extra sequences or substitution of new sequences for old occurs; but not an actual omission of sequences.

caenomorphosis where shortcuts or alternative routes in preterminal developmental sequences are taken in arriving at adult structures essentially identical with those of ancestral species.

Caenorhabditis elegans a nematode worm, used extensively in developmental biology for cell lineage and aging studies. *See: ced* **genes**.

calcium phosphate a salt; crystals of which are deposited during bone formation.

callus a proliferation of parenchymatous cells produced in vascular plants to cover wounds and in regeneration; can be used to form *in vitro* cultures.

calvaria the flat bones of the skull, collectively, which make up the cranial vault containing the brain.

calyces plural of calyx.

calyx collectively, all of the petals of a given flower; a subdivision of the pelvis of the kidney.

CAM *See:* **chorio-allantoic membrane (CAM)**.

cambium a tissue composed of cells capable of active cell division, producing xylem to the inside of the plant and phloem to the outside; a lateral meristem.

Cambrian the geological period from 544 to 490 million years ago, during which the diversity of modern animal phyla expanded.

Cambrian explosion a relatively short period of approximately 15 million years near the beginning of the Cambrian period during which most of the invertebrate phyla appear.

Camerarius, Rudolph (1665–1721) Professor of Medicine at Tubingen, designed and carried out experiments on plant reproduction between 1691 and 1694; carried out experiments leading him to reason that anthers and stigmas must be sex organs.

CAMs *See:* **cell adhesion molecules (CAMs)**.

Dictionary of Developmental Biology and Embryology, Second Edition. Frank J. Dye.
© 2012 Wiley-Blackwell. Published 2012 by John Wiley & Sons, Inc.

canalicular period   a period during lung development when the lumens of the bronchi and terminal bronchioles enlarge.

canalization (developmental robustness)   developmental reactions are adjusted so as to bring about one definite end result regardless of minor variations in conditions during the course of the reactions; buffering of development so slight perturbations of genotype or slight perturbations of the environment will not lead to the formation of abnormal pheno-types, e.g., the role of Hsp90 as a developmental buffer.

cancer stem cell hypothesis   the hypothesis that the malignant part of a tumor is either an adult stem cell that has escaped its niche or a more differentiated cell that has regained stem cell properties.

candidate gene mapping   a method, starting with a correlation between the genetic mapping of a particuler syndrome and the genetic mapping for a particular gene, to find genes that are involved in normal human development and whose mutations cause con-genital developmental malformations; e.g., pedigree analysis was used to determine that Waardenburg sydrome type 2 is caused by a mutation in a gene on the small arm of chromosome 3, between bands 12.3 and 14.4; using a probe made from the mouse microph-thalmia gene, it was found that a human homologue (*MITF*) of the mouse microph-thalmia gene mapped to the exact same region as Waardenburg sydrome type 2. *See:* **positional gene cloning**.

canonical   authorized, recognized, accepted; applied to a mechanism underlying a develop-mental phenomen when it is the "classical" mechanism underlying the phenomenon; as opposed to a non-canonical mechanism, which underlies the same developmental phenom-enon, but is an alternative mechanism discovered after the "classical" mechanism.

capacitation   a change undergone by mammalian spermatozoa in the female reproductive tract, which has traditionally been regarded as necessary for development of the fertilizing ability of spermatozoa; probably involves removal of inhibitory factors from spermatozoa, accompanied by membrane protein and lipid modifications.

capillaries   the smallest blood vessels. Through the thin walls of these vessels, exchange of materials occurs between the blood and tissue fluids.

caput epididymis   the head (cephalic) end of the epididymis.

carapacial ridge   the structure in the early formation of the turtle carapace that is analogous to the apical ectodermal ridge.

carbohydrate   a class of organic molecules that carries out various functions in the cell; an especially important function is to serve as an energy source for cellular activities.

carcinoma   cancerous tumors that are derived from epithelial tissue.

cardiac bifida   the formation of a separate heart on each side of the body; reflecting the bilateral origin of the heart; may be experimentally created by surgically preventing the merger of the two cardiac tubes.

cardiac cells   heart cells.

cardiac jelly   the connective tissue found between the endocardium and the epimyocardium of the early developing heart. Continuous with interlaminar jelly, cardiac jelly is formed between the epimyocardium and the endocardium and has two functions: (1) gives mechan-ical unity to endocardium and myocardium during early stages of heart development and (2) serves as the substratum on which cells move in and knit together the two primordial layers of the heart (see Fig. 19).

cardiac muscle   the muscle of the heart; one of three general kinds of muscle tissue in the body.

cardiac myoblasts   myoblasts derived from splanchnic mesoderm that are the specific pre-cursors of cardiac muscle cells.

cardiac neural crest   located between the cranial and trunk neural crests; its cells develop into melanocytes, neurons, cartilage, and connective tissue (of the third, fourth, and sixth pharyngeal arches), the entire musculoconnective tissue wall of the large arteries as they originate from the heart, and contribute to the septum that separates the pulmonary cir-culation from the aorta.

cardiac prominence   the bulge of the chest region found on the surface of the embryo during the middle of the embryonic period.

cardinal veins   paired blood vessels, symmetrically placed on either side of the midline; there are two pairs, two anterior cardinal veins and two posterior cardinal veins. The anterior and posterior veins from both sides of the body become confluent dorsal to the heart as the two common cardinal veins (ducts of Cuvier), which enter the caudal end of the heart. The posterior cardinal veins lie just dorsal to the nephric tubules, which originate from the intermediate mesoderm in the angle between the somatic mesoderm and the somites. At the region of convergence of the omphalomesenteric veins, the common cardinal veins enter them dorsolaterally. Later in development, the proximal ends of the anterior cardinal veins become connected by formation of a new transverse vessel and empty together into the right atrium of the heart. The distal portions remain in the adult as the principle afferent vessels of the cephalic region, the internal jugular veins. The posterior cardinal veins become reduced and broken up, forming small vessels that aid in the formation of the inferior vena cava.

cardioblast any of certain early embryonic cells in insects from which the heart develops.

cardiogenic mesoderm that portion of the mesoderm that gives rise to the heart.

cardiomyocytes cardiac cells derived from cardiogenic mesoderm that form the muscular layers of the heart and its inflow and outflow tracts.

carotid arteries the pair of major arteries, found in the neck, which supplies the head with blood. In the chick embryo, late in the second day, plexiform channels extend from the first aortic arches toward the forebrain, foreshadowing the formation of the internal carotid arteries. With the regression of the first two pairs of aortic arches, the dorsal aortic roots become feeders to the original *internal carotid arteries*, making them appear to originate where the third pair of aortic arches merge with the dorsal aortic roots. After the first two pairs of aortic arches lose connections with the dorsal aortic roots, the parts of the ventral aortic roots that formerly fed them persist and become the main stems of the *external carotid arteries*. *See:* **aortic arches**.

carpel the specialized leaf of the female reproductive structure in flowering plants; a megasporophyll. Each carpel encloses one or more ovules.

carpellate flowers imperfect flowers lacking stamens; also called pistillate flowers.

carrier a person who carries a single recessive gene (allele) for a given characteristic and, therefore, does not exhibit the characteristic in his/her phenotype, although the gene (allele) may be passed onto the person's offspring.

cartilage a type of connective tissue; somewhat rigid, but less rigid than bone.

cascade of embryonic inductions a sequential series of embryonic inductions, where one induced structure becomes the inducer of the next structure in the series, etc.; e.g., the optic vesicle induces the lens of the eye, which in turn induces the cornea, or the Nieuwkoop center induces the cells above it to become the organizer that, in turn, induces the ectoderm above it to become the neural tube.

cascade of protein phosphorylations a frequent, important, component of intracellular signal tranduction pathways; this series of reactions, carried out by protein kinases, carries the signal received at the surface of the cell further into the interior of the cell.

caspases proteases activated by apoptotic signal transduction pathways, which cause the nucleus to condense, the cell to shrink, and the cell to display on its surface a signal for engulfment by phagocytic cells. Caspase-9 and caspase-3 are homologues of CED-3 of *Caenorhabditis elegans*. *See:* **Apaf-1**, *ced* **genes**.

catalyst a substance that increases the rate of a chemical reaction without itself becoming permanently changed. *See:* **enzyme**.

catalyze to increase the rate of a chemical reaction.

cataract a partial or complete opacity of the lens; although usually thought of as a result of aging, congenital cataracts may be caused by rubella virus during the embryonic period.

catenins a complex of proteins that anchors cadherins into the cell.

catheterization (see Fig. 11). *See:* **balloon catheterization**.

cauda epididymis the terminal portion of the epididymis, where it joins the vas deferens; the major storehouse of sperm in the male.

caudad toward the tail (cauda) end of the body.

caudal a term referring to the tail or tail end of an organism or structure (see Plate 5 in the color insert).

caudal intestinal portal (CIP) *See:* **posterior intestinal portal**.

caudal knob in fish embryos, a mass of cells, in and around the posterior portion of a thickened line of cells left behind by the closing blastoporal lips, from which the fish tail is largely formed by outgrowth. The thickened line of cells referred to here is, apparently, the homologue of the primitive streak of amphibians.

caudal **mRNA**   a crucial maternal product in early *Drosophila* development; it is uniformly distributed throughout the egg initially. A posterior-to-anterior gradient of the caudal protein is established by inhibition of caudal protein synthesis by the bicoid protein; caudal protein concentration is highest at the posterior end of the embryo.

caul   the membrane that sometimes covers the face of a newborn; consists of the amnion and chorion.

cavitation   the hollowing out of a fluid-filled space from a solid mass of cells, e.g., secondary neurulation; blastocyst cavity formation after the compaction stage of the early mammalian embryo.

CBFA1   a transcription factor that seems to be able to transform mesenchyme cells into osteoblasts; the *Cbfa1* gene is activated in mesenchymal cells by bone morphogenetic proteins.

Cbfa1 **gene**   *See:* **CBFA1, cleidocranial dysplasia**.

Cdc   an abbreviation for *cell division cycle*; as in cdc mutants (cell division cycle mutants). Cdc2 is the catalytic subunit of maturation promoting factor.

cdk   cyclin-dependent kinase; a protein kinase that has to be complexed with a cyclin protein in order to function. Various cdk-cyclin complexes trigger different steps in the cell cycle by phosphorylating specific proteins. Maturation promoting factor consists of the protein kinase Cdc 2 (a cdk) and Cyclin B (a regulatory subunit).

cDNA   a DNA copy of an RNA, made by reverse transcription.

cecum   the blind pouch of the intestine where the large intestine begins.

ced **genes**   genes in *Caenorhabditis elegans* that play a role in apoptosis; *ced-3* and *ced-4* genes encode proteins essential for apoptosis, and the product of the ced-9 gene turns off the *ced-3* and *ced-4* genes. *See:* **Apaf-1**, *Bcl-2* **genes, caspases**.

-cele   this word part refers to a ruptue or hernia; a number of significant birth defects involve ruptures, e.g., meningocele and omphalocele. *See:* **hernia**.

cell adhesion molecules (CAMs)   molecules that mediate cell adhesion. *See:* **strengths of adhesion**.

cell arrangements   there are two major types of cell arrangements in the embryo; epithelium and mesenchyme.

cell autonomous genes   the class of genes whose products affect only the differentiation or behavior of cells in which they are expressed. *See:* **non-cell autonomous genes**.

cell–cell adhesion   is specific for a given cell type; e.g., when dissociated cells are mixed together, the dissociated cell types sort out; i.e., cells have mechanisms for sensing other cells and expressing preferences for adhesion. Additionally, reconstituted tissues show specific spatial arrangements with respect to one another; i.e., the epidermis is always on the surface of the reaggregated mass and the mesoderm internal to it.

cell cycle   the series of stages a cell passes through between successive cell divisions; the classical cell cycle consists of four stages: G_1, S, G_2, and M (Gap 1, DNA synthesis, Gap 2, and Mitosis, respectively). The shortest eukaryotic cell cycles are the early embryonic cell cycles that occur in certain animal embryos immediately after fertilization; fly embryos have the shortest known cell cycles, each lasting as little as 8 minutes; on the other hand, the cell cycle of a mammalian liver cell can last longer than a year. Generally, the embryonic cell cycle is shortened by the elimination of G_1 and G_2, the resulting cell cycle consisting of only S and M, which underlies the rapidity of cleavage cell divisions. During cleavage of sea urchin embryos, the blastomeres may divide every 30 minutes; i.e., as fast as bacteria. Progression through the stages of the cell cycle is controlled by a conserved regulatory apparatus, which not only coordinates the different events of the cell cycle but also links the cell cycle with extracellular signals that control cell proliferation (see Fig. 10). *See:* **embryonic cell cycle**.

cell differentiation   the process by which a cell becomes specialized or differentiated; also called cytodifferentiation. The distinction between trophoblast and inner cell mass blastomeres represents the first differentiation event in mammalin development.

cell elongation   an important part of plant development, which contributes significantly to plant growth.

cell fusion   is an important phenomenon during development; two examples are the fusion of the spermatozoon and the egg at fertilization and fusion of myoblasts to form syncytial skeletal muscle fibers. Cells that normally do not fuse may be artificially made to fuse by the use of certain chemicals (polyethylene glycol) or inactivated Sendai virus; such agents are called fusogens.

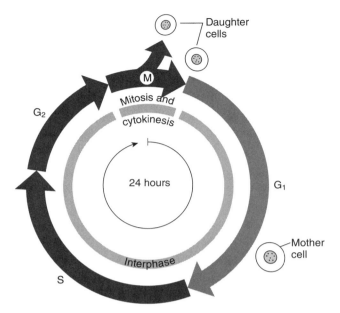

Figure 10. The cell cycle. Between the birth of a cell by cell division and its own division to form two daughter cells, the cell passes through the four stages that make up the cell cycle: G_1, S, G_2, and M. Reprinted from Frank J. Dye, *Human Life Before Birth*, Harwood Academic Publishers, 2000, fig. 3-1, p. 13.

cell hybridization   in a procedure called somatic cell hybridization, cells in culture are made to undergo fusion, by the use of chemicals or viruses; e.g., the fusion of human and mouse cells has been an important technique in localizing human genes to individual chromosomes or parts of chromosomes.

cell hypertrophy   increase in cell size.

cell junctions   specialized regions of contact between cells; generally falling into three categories: tight junctions, adhering junctions, and gap junctions.

cell lineage   the developmental history of individual blastomeres from their first cleavage division to their ultimate differentiation into cells of tissues and organs.

cell lineage restriction   when all the descendants of a particular group of cells remain within a boundary and never mix with an adjacent group of cells; e.g., in *Drosophila*, cells and their descendants from one parasegment (i.e., a compartment) never move into adjacent ones.

cell lineage studies   these studies are concerned with following the fate of individual cells; such studies have revealed that the same cells form essentially the same structures in many tubellarians, polychaete annelids, and mollusks (cephalopods being the main exception). This is a demonstration of the conservative nature of embryological development, as these groups have been separated from each other for approximately half a billion years. Classical cell lineage studies were carried out by Aleksandr O. Kowalewski (in 1883), Edmund B. Wilson (in 1892), and Edwin Grant Conklin (in 1897). Various methods have been used in cell lineage studies: in snails, cells could be identified and followed as a result of the determinate spiral cleavage pattern, in ascidians the segregation of identifiable cytoplasmic substances proved useful, and in insects recognizable genetic variants were induced randomly by radiation. *See:* **Whitman, C. O.**

cell-mediated immunity   that portion of immunity that directly involves cells (T-lymphocytes); as opposed to humoral immunity that involves antibodies produced by cells (B-lymphocytes).

cell proliferation   an increase in cell number by cell division.

cell-sorting   if cells with different adhesion systems are mixed, they will sort out into separate zones. *See:* **strengths of adhesion.**

cellular affinity   the phenomenon of selective adhesiveness observed among the cells of certain sponges, slime molds, and vertebrates.

cellular blastoderm   an early stage in *Drosophila* embryogenesis when all the cells of the embryo are arranged in a single-layered cover around the yolky core of the egg.

cellular ecosystem   reference to the morphogenetic field by Paul Weiss.

cellulose   a polysaccharide consisting of linear polymers of glucose molecules; orientation of the cellulose microfibrils is dictated by the microtubules within the plant cell; in turn, the orientation of the cellulose microfibrils determines the direction of plant cell expansion, before the cell wall becomes thick and the cellulose microfibrils cross-linked; as is common in newly forming cells of the meristem.

cement gland   a structure formed during early *Xenopus* development, from the anterior epidermis, ventral to the future mouth; this adhesive organ secretes a sticky mucus by which the recently hatched tadpole may adhere to vegetation, so that the tadpole may stabilize its position in the environment until its tail, muscular coordination, and mouth have developed to the point that the larva can swim and feed.

Cenozoic Era   started 65 million years ago; the Age of Mammals begins.

central cell   the largest and centrally located cell of the embryo sac; containing the two polar nuclei.

central dogma of molecular biology   basically holds that the flow of information is in the diection of: DNA→RNA→protein. *See:* **reverse transcriptase**.

central group *Hox* genes   *Hox* genes expressed in the region between the regions of expression of anterior and posterior group *Hox* genes; include the *Hox3* through *Hox8* genes in vertebrates.

central nervous system (CNS)   that portion of the nervous system derived from the neural tube; that is, the brain and spinal cord.

central zone   the center of a meristem that has a relatively low division rate; cell divisions in this zone replenish the meristem, similar to the way stem cells do in animals. *See:* **peripheral zone**.

centrioles   organelles found in animal cells that frequently are associated with microtubule-organizing centers, such as at the poles of mitotic and meiotic spindles. Centrioles themselves are made up of microtubules.

centrolecithal eggs   eggs with their yolk concentrated at the center of the egg; characteristic of insect eggs.

centromere   the part of a chromosome by which it attaches to the spindle, which is necessary for chromosome movement during cell division (see Fig. 29).

centrosome   the cell organelle that is found at the so-called cell center or at the poles of a cell division spindle. It apparently acts as a microtubule-organizing center (see Fig. 29).

cephalad   toward the head (cephalic) end of the body.

cephalic   a term referring to the head or head end of an organism or structure (see Plate 5 in the color insert).

cephalic neural crest   *See:* **cranial neural crest**.

cephalic precocity   because growth takes place from behind and pushes the cephalic region ahead of it, the cephalic end of an embryo will always be precocious in development; e.g., the cephalic somites will appear before the caudal somites.

cephalic presentation   the relationship of the part(s) of the fetus to the birth canal at the beginning of labor where the head is closest to the birth canal.

cephalochordates   animals placed in the subphylum, Cephalochordata, of the phylum Chordata; their notochord extends into their head; characterized by a notochord and small anterior brain vesicle. They constitute the nearest phylogenetic outgroup to the vertebrates. Also known, because of their body shape, as lancelets.

-cephaly   this word root refers to the head; a number of birth defects involve the abnormal development of the head, e.g., hydrocephaly.

Cerberus   a protein secreted by the amphibian organizer, promotes the formation of the cement gland, eyes, and olfactory placodes; the Cerberus protein can bind both bone morphogenetic proteins (BMPs) and Xwnt8.

cerci   a pair of segmented sensory appendages on the last abdominal segment of many insects and certain other arthropods.

cerebellum   that part of the brain that derives from the metencephalon and controls balance, posture, and movement.

cerebral hemispheres   the two major subdivisions of the cerebrum, derived from the telencephalon. The cerebrum is responsible for the "higher" brain functions, such as intellect and memory; in the frog, at the time of hatching, a vesicle, the cerebrum, grows out from the anterior end of the forebrain, then a medial longitudinal invagination of its anterior and dorsal wall creates the cerebral hemispheres (see Fig. 8).

cerebral vesicles   the early evaginations (outpocketings) of the telencephalon that give rise to the cerebral hemispheres.

cerebrospinal fluid (CSF)   the fluid, derived from the choroid plexuses, which circulates through the ventricles of the brain and the spinal canal of the spinal cord.

cerebrum   that portion of the brain derived from the telencephalon and responsible for the "higher" brain functions, such as intellect and memory in humans. In frogs, at metamorphosis, the cerebrum makes up half of the brain and masks the unchanged cranial flexure.

cervical canal   the lumen of the cervix of the uterus.

cervical cap   a contraceptive membrane that fits over the cervix of the uterus where it projects into the blind end of the vagina.

cervical flexure   a ventrally concave flexure of the embryonic brain occurring at the junction of the hindbrain and the spinal cord. In the chick embryo, the cervical flexure occurs further caudally (than the cranial flexure) in the future neck (myelencephalon-spinal cord region) region. The cervical flexure presses the pharyngeal region and the ventral surface of the head close together; the mandibular arch becomes the caudal boundary of the oral depression.

cervical mucus   the fluid found in the cervical canal, the consistency of which varies with the stages of the menstrual (uterine) cycle; the consistency of cervical mucus can be used as a fertility awareness method, as it becomes wetter at the time of ovulation.

cervical plug   highly viscous cervical mucus that fills the cervical canal during pregnancy; believed to prevent infections from ascending out of the vagina and into the uterus.

cervical sinus   a triangular depression caudal to the hyoid arch containing the posterior visceral arches and grooves; bilaterally, the deeply depressed region around the third and fourth clefts on the surface of 5-mm human embryos.

cervix uteri   the narrow lower portion of the uterus that projects into the vagina (see Figs. 11, 22, and 48).

cesarean section   the surgical removal of the fetus from the womb, as opposed to vaginal childbirth; usually carried out because vaginal childbirth is contraindicated.

CFU-E   colony-forming unit, erythroid; a cell capable of responding to the hormone erythropoietin to produce the proerythroblast, the first recognizable differentiated member of the erythrocyte lineage.

CFU-GM   the lineage-restricted stem cell, derived from a CFU-M,L, which gives rise to the granulocyte precursor cell and the monocyte.

CFU-Meg   the lineage-restricted stem cell, derived from the myeloid precursor cell; gives rise to megakaryocytes, which, in turn, give rise to platelets.

CFU-M,L   the colony-forming unit of the myeloid and lymphoid cells; a pluripotential hematopoietic stem cell, from which both the CFU-S and lymphocytes are derived. This cell gives rise to the CFU-S (blood cells) and the CFU-L (lymphocytes).

CFU-S   the colony-forming unit of the spleen; a pluripotential hematopoietic stem cell, capable of producing erythrocytes, granulocytes, and platelets. However, the immediate progeny of the CFU-S are lineage-restricted stem cells; e.g., the BFU-E (burst-forming unit, erythroid) can form only one cell type in addition to itself.

Chabry, Laurent (1855–1894)   a Frenchman, interested in teratology, worked with tunicate embryos; when he killed one cell at the two-cell stage, he essentially got Roux's results; however, he got the same results even if the two cells were separated; i.e., isolated tunicate blastomeres behave mosaically. It has been said that the tunicate embryo approximates "a mosaic of self-differentiating parts," constructed from information stored in the oocyte cytoplasm. Chabry's thesis (in 1887), "Normal and Teratological Embryology of Ascidia," is concerned primarily with the teratology of tunicates produced by experimental techniques; for his experimental techniques, he invented the microforge and the micromanipulator. He established a distinction between hereditary monstrosities, originating with the parents, and monstrosities. His observations on the consequences of the natural cell death of the tunicate *Ascidia aspersa* led to neopreformation, the rebirth of a theory of preformation; i.e., that each blastomere contains the potential of certain parts that are irremediably lost by its death; he believed this to be the case for the ascidia and for those animals whose blastomeres differentiate early. His production of predictable abnormal development by destruction of specific blastomeres demonstrated that, as in the experiments of Stanislas

Warynski and Hermann Fol, the kind of trauma that one inflicts allows one to predict which kind of monstrosity will be produced. *See:* **Roux, Wilhelm**.

chalaza(e)   as the chicken ovum rotates during its passage through the anterior glandular part of the oviduct, a twisted cord of albumin (chalaza) attaches on each side of the ovum; the chalazae keep the ovum placed in the center of the albuminous mass, as the albumin secreted further back in the oviduct is of a less firm consistency. The part of an ovule or seed where the integuments are connected to the nucellus, at the opposite end from the micropyle.

chancre   the initial lesion (ulcer) of syphilis.

chasmogamous   applied to flowers that open before fertilization and are usually cross-pollinated. *See:* **cleistogamous**.

checkpoints   temporal positions in the cell cycle at which controls of the cell cycle operate, e.g., G_1 checkpoint, G_2 checkpoint, and M checkpoint.

chemical teratogens   e.g., mercury, alcohol, retinoic acid, precocenes, methoprene, PCBs. *See:* **teratogen**.

chemoaffinity hypothesis   the complicated nerve fiber circuits of the brain grow, assemble, and organize themselves through the use of intricate chemical codes under genetic control; proposes that each retinal neuron carries a chemical label that enables it to connect reliably with an appropriately labeled cell in the optic tectum.

chemoattractant   in chemotaxis, the molecules toward which cells move. *See:* **chemotaxis**, **netrins**.

chemorepellent   in chemotaxis, the molecules away from which cells move. *See:* **chemotaxis**, **semaphorin proteins**.

chemotaxis   the movement of cells or organisms up a chemical concentration gradient toward the source of the chemical; in some species, plants and animals, chemotaxis plays a role in the spermatozoon finding the egg during fertilization. Chemotaxis and differential cell adhesion may aid primordial germ cells in homing in on the developing gonads; may play a role in guidance of migratory cells. *See:* **chemoattractant**, **chemorepellent**, **contact inhibition**, **fertilization**, **haptotaxis**, **netrins**, **slit protein**, **stereotaxis**.

chemotherapeutic drugs   drugs used for the treatment of cancer.

chiasmata   (sing., chiasma) visible, at the light microscope level, manifestations of crossing over.

chick   early chicken development has been extensively studied since at least the time of Aristotle. *Gallus ferrugineus* (East Indian Jungle Fowl) through domestication and mutation became "chickens."

Child, Charles Manning (1869–1954)   American zoologist; professor, University of Chicago. In 1900 began a series of experiments on regeneration in coelenterates and flatworms; his gradient theory, of 1911, maintained that each part of an organism dominates the region behind and is dominated by that in front. In 1915 showed that axial gradients can be followed in development.

childbirth   the vaginal delivery of a baby.

chimera   an organism consisting of parts derived from more than one pair of parents. The chimera of Greek mythology consisted of a lion's head, a goat's body, and a serpent's tail. Chimeras may be made in the laboratory by, in the case of mammals, removing the zona pellucida of each embryo and pushing the sticky embryos together, which merge together to form a single chimeric embryo; which, if transferred into the uterus of a maternal host, may develop into a chimeric organism. Such constructs have been created with, for example, different strains of mice. A plant that contains genetically distinct cell types is a chimera.

Chironomus   gall midges, a genus of the insect order, Diptera; the larval salivary glands are an excellent source of polytene chromosomes.

chlamydia   a sexually transmitted disease caused by the bacterium *Chlamydia trachomatis*.

chloasma   the mask of pregnancy; patchy hyperpigmentation located chiefly on the forehead, temples, and cheeks.

cholinergic neurons   the parasympathetic neurons derived from neural crest cells of the neck that produce the neurotransmitter acetylcholine. *See:* **adrenergic neurons**.

Chondrichthyes   the class of vertebrates composed of cartilaginous fishes; sharks, skates, rays, and chimeras.

chondrocranium   the cartilaginous, embryonic cranium of vertebrates.

chondrocytes   cartilage cells; those cells, derived from mesodermal mesenchyme, which lay done the connective tissue, cartilage. Growth hormone stimulates the production of IGF-I by epiphyseal growth plate chondrocytes; the combination of growth hormone and IGF-I apparently provides a strong signal to divide.

chondrogenesis   formation of cartilage.

chondroskeleton   the parts of the bony skeleton formed from cartilage.

chordamesoderm   that portion of the mesoderm in the chordate embryo that gives rise to the notochord.

Chordata   one of the major phyla (classification divisions) of animals; it includes the vertebrates (including humans) and the protochordates.

chordate   an animal placed into the phylum Chordata because it possesses a notochord sometime during its development. The ability of the mesoderm to form a notochord and its overlying ectoderm to become a neural tube separated the chordates from the remaining invertebrates. *See:* **chordate characteristics**.

chordate characteristics   (1) a notochord, at least in the embryo; (2) presence of a pharynx with pouches or slits in its walls, at least in the embryo; and (3) a dorsal, tubular, nervous system. Vertebrates are a subgroup of the chordates possessing a vertebral column in addition to the foregoing three characteristics.

chordin   the second organizer protein found; capable of inducing a secondary neural tube; secreted by the organizer and prevents the bone morphogenetic protein (BMP) from binding to the ectoderm and mesoderm near the organizer. *See:* **follistatin**, **noggin**.

chordoneural hinge   the growing region at the tip of the dorsal blastopore lip of the frog embryo that, instead of involuting into the embryo, keeps growing ventrad; contains precursors for both the posteriormost portion of the neural plate and the posterior portion of the notochord.

chorio-allantoic membrane (CAM)   the mesodermal layer of the allantois (splanchnic mesoderm) fuses with the mesodermal layer of the serosa (somatic mesoderm); in the double layer of mesoderm, a rich vascular network develops, connected with the embryonic circulation by allantoic arteries and veins; this highly vascular fusion membrane is the chorio-allantoic membrane. The serosa is also referred to as the chorion. *See:* **amnion**.

choriocarcinoma   a form of cancer originating in chorionic tissue (derived from trophoblast).

chorion   (1) one of the four extraembryonic membranes formed during the development of higher vertebrates, including humans. It provides the fetal contribution to the formation of the placenta. The chorion is made up of trophoblast and extraembryonic somatic mesoderm; it develops outgrowths called villi. Initially the chorion is uniformly villated (up to approximately the eighth week), but later, it is only partially villated, consisting of the smooth chorion laeve and the bushy chorion frondosum; the chorion frondosum becomes the placenta fetalis. (2) A tough egg coat found around the eggs of some species, e.g., insects and fish. Insect eggs are typically covered by a chorion that is secreted by the follicular epithelium (see Plate 3 in the color insert and Fig. 35).

chorion frondosum   or bushy chorion, the portion of the chorion that retains its villi and becomes the placenta fetalis or fetal contribution to the placenta (see Fig. 11).

chorion laeve   or smooth chorion, the portion of the chorion that does not retain its villi and does not contribute to the placenta.

chorionic plate   the chorionic membrane of the placental region (see Fig. 35).

chorionic sac   that sac-like portion of the conceptus, which contains the embryo, amniotic cavity, and amnion.

chorionic villi   feather-like extensions of the chorion. Those that make up part of the placenta are the sites at which actual exchanges of materials occur between fetus and mother. During early human development, the trophoblastic masses of the trophoblastic syncytium increase rapidly in extent and give rise to sprawling, anastomosing strands enclosing irregular spaces called trophoblastic lacunae; at this stage the embryo is said to be previllous. Toward the end of the second week, the trophoblast becomes molded into masses more suggestive of villi. At first, young villi consist entirely of epithelium with no connective tissue core; these are primary villi. Two types of cells appear: (1) outer cells that are large and lose their intercellular boundaries and that make up the syncytiotrophoblast and (2) deeper cells that remain smaller and maintain distinct boundaries and that make up the cytotrophoblast. While the primary villi have been forming, the inner face of the blastocoel has been receiving an ingrowth of allantoic vessels and mesoderm; early in the third week, mesoderm pushes into the primary villi so that the trophoblastic cells become an epithelial

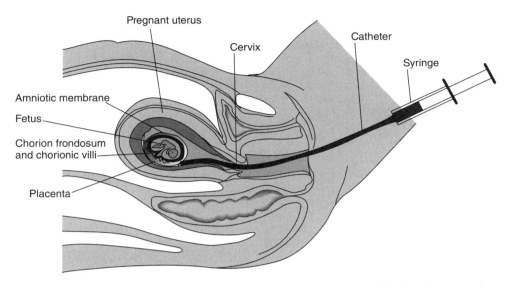

Figure 11. Diagram shows the setup for the procedure of chorionic villus sampling. Not shown are other instruments used, such as the ultrasound probe (for guiding the catheter) and the speculum (for making the uterine cervix accessible). Reprinted from Frank J. Dye, *Human Life Before Birth*, Harwood Academic Publishers, 2000, fig. 19-16, p. 174.

layer over a framework of connective tissue (secondary villi). Blood vessels soon appear; villi with a vascular connective tissue core are tertiary or "true" chorionic villi (around the close of the third week). The mature villus is exposed to maternal blood and has three components: epithelial, connective tissue, and vascular. Cytotrophoblast cells at the ends of villi proliferate, penetrate the syncytiotrophoblast, and become firmly fixed to the endometrial stroma as anchoring (stem) villi versus their branches, which are called free villi. The cytotrophoblastic columns of these villi become joined with one another to form the cytotrophoblastic shell, by the end of the third week (see Fig. 11). *See:* **chorionic villus sampling (CVS)**.

chorionic villus sampling (CVS)   a very early prenatal diagnostic procedure, made possible because the chorionic villi contain dividing cells derived from the conceptus (see Fig. 11).

choroid fissure   the break in the continuity of the optic cup, resulting from the eccentric invagination of the optic vesicle that gives rise to the optic cup.

choroid plexus   one of several thin, highly vascular membranes that hang down into the ventricles of the brain; sites of cerebrospinal fluid formation and of the blood-brain barrier. (1) The anterior choroid plexus hangs down into ventricle III, and (2) the posterior choroid plexus hangs down into ventricle IV.

chromaffin cell   A cell that stains readily with chromium salts, especially a cell of the adrenal medulla; chromaffin cells are differentiated secretory cells derived from neural crest cells during embryogenesis, and the term usually refers to groups of cells found in the adrenal medulla; the chromaffin cells of the adrenal medulla secrete two catecholamines, epinephrine and norepinephrine, which affect smooth muscle, cardiac muscle, and glands in the same way as sympathetic stimulation.

chromatid   from anaphase to S-phase of the cell cycle, a chromatid is a chromosome; from G_2 of the cell cycle to metaphase, a chromatid is a longitudinal half of a chromosome; and during the S-phase of the cell cycle, chromatids are replicated (see Figs. 26 and 29).

chromatin   a dispersed state of the chromosomes found during the interphase stage of the cell cycle; the physical basis, composed of DNA and protein, of chromosomes; when chromosomes are dispersed (as during the interphase of the cell cycle), they are referred to as chromatin (see Fig. 29). *See:* **euchromatin**, **heterochromatin**.

chromatin diminution   *See:* **chromosome diminution**.

chromatophores   pigment cells (see Plate 14 in the color insert).

Chromosomal Theory of Heredity   the concept that genes are found on chromosomes.

chromosome bands   (1) striations *induced* to appear on chromosomes, including human chromosomes, by various chemical and physical treatments; these banding procedures, introduced around 1969, made it possible to identify unequivocally different chromosomes and different chromosome parts; (2) *naturally occuring* striations appearing along the lengths of polytene chromosomes; these bands are quite visible without chemical or physical treatments of the chromosomes.

chromosome diminution   the elimination of whole chromosomes or heterochromatic chromosome segments in mitosis, meiosis, or interphase, e.g., as occurs in the blastomeres (somatic cell lineage) of early cleavage-stage *Ascaris* embryos, except in those blastomeres that are in the germ line lineage; i.e., in *Ascaris megalocephala*; this leads to differences in the chromosome complements of the germ line and the somatic cells.

chromosome elimination   the loss of chromosomes from nuclei during certain early cleavage divisions in Dipteran and Hemipteran insects; only the germ line cells retain the full chromosome complement; all somatic cells have a reduced number of chromosomes; a mechanism of somatic versus germ line differentiation.

chromosome puffs   enlarged regions along the lengths of polytene chromosomes, which are manifestations of actively transcribing genes; each puff corresponds to decondensed chromatin of the polytene chromosome, at which RNA is being transcribed from DNA.

chromosomes   (1) a condensed state of the chromatin of the cell nucleus (DNA and protein), these "colored bodies" are visible, through the microscope, during mitosis or meiosis; (2) chromosomes found in eukaryotic organelles (mitochondria and chloroplasts) and prokaryotic cells are simpler, circular organelles, visible at the electron microscope level (see Fig. 29).

cilia   hair-like organelles that project from the surfaces of some cells; may move cells through a liquid medium, as in the movement of the conceptus by ciliated cells lining the fallopian tube, or may move a liquid medium over cells, as in the movement of fluid by ciliated cells of the respiratory tract.

circadian clocks   internal "clocks" that are approximately 1 day long; phytochromes and blue-light receptors are strongly implicated in entraining (setting) circadian clocks in plants.

circuit   a regulatory pathway that includes particular target genes; two circuits are different if they employ the same pathway but regulate different target genes.

circulation   in the chick the heart begins to pulsate feebly before actual circulation of the blood; the first contractions occur at 29 hours in the ventricular myocardium (only this region has fused); by 33 hours, pulsation has increased, the atrium has been established, and the atrial myocardium serves as the pacemaker; at 40 hours the arterial channels out to the yolk are complete and the shuttling back and forth of blood corpuscles gives way to a jerky progression.

circulatory arcs   in the chick embryo, three circulatory arcs develop: (1) the intra-embryonic arc, (2) the vitelline arc (primarily extra-embryonic), and (3) the allantoic arc (extra-embryonic; develops late in the third day). Intra-embryonic blood vessels originate from mesodermal cells *in situ*. The courses of circulation through the circulatory arcs are as follows: (1) intra-embryonic circulation: heart-ventral aortic roots-aortic arches-dorsal aortae-body of embryo-anterior and posterior cardinal veins-common cardinal veins-heart, (2) vitelline circulation: heart-ventral aortic roots-aortic arches-dorsal aortae-omphalomesenteric arteries-vitelline vascular plexus-sinus terminalis-vitelline veins-omphalomesenteric veins-heart, and (3) allantoic circulation: allantoic arteries originate from the aorta at the level of the allantoic stalk and a rich plexus of vessels is found in the mesoderm of the allantois; blood is carried from the allantois to the heart by the allantoic veins (the two main vessels enter the embryonic body through the allantoic stalk, extend cephalad, and enter the sinus venosus on either side of the entrance of the omphalomesenteric veins).

***cis* regulation**   *cis*-acting; a genetic element that is physically on the same chromosome and located near the region it affects; it is usually defined by a specific nucleotide sequence. *See: **trans**-regulation*.

cis*-regulatory element**   a discrete region of DNA that affects transcription of a gene; *cis*-regulatory elements are promoters and enhancers. *See: **trans**-regulatory **factor.

cistron   for all practical purposes, synonymous with the word *gene*.

Types of Cleavage Patterns

Yolk Amont (Terminology)	Yolk Distribution (Terminology)	Cleavage Pattern	Vertebrate Examples
Large (macrolecithal)	Nonuniform (telolecithal)	Meroblastic, discoidal	Fish, reptiles, birds
Moderate (mesolecithal)	Nonuniform (telolecithal)	Holoblastic, unequal	Amphibians
Small (microlecithal)	Uniform (isolecithal)	Holoblastic, equal	Mammals

Figure 12. Types of cleavage patterns. The type of cleavage pattern is determined by yolk and genes. Note the effects of the amount and distribution of yolk on cleavage patterns. Note also that genes have an effect on cleavage patterns. Reprinted from Frank J. Dye, *Human Life Before Birth*, Harwood Academic Publishers, 2000, table 8-3, p. 61.

c-Kit protein   a membrane RTK encoded by the *W* gene in mice; the receptor for stem cell factor; the receptor for Mgf (encoded by the *steel* gene); mutations in either c-Kit or Mgf lead to deficiencies in the directed migration of primordial germ cells (PGCs), as well as defective migration of melanoblasts originating from the neural crest.

clade   a group of species descended from a common ancestral species; also known as a monophyletic group.

class II promoter   a promoter recognized by RNA polymerase II.

class switching   during B-cell development, many B cells switch from making one class of antibody to making another; this process is called class switching.

cleavage   a process universally undergone by embryos after formation of the zygote at fertilization; characterized by mitotic divisions without intervening growth. During cleavage, the size of blastomeres decreases even as their number increases; this is attributed to a lack of net increase in the volume of the embryo during cleavage, accomplished by elimination of interphasic growth between the cleavage divisions. In amphibian embryos, there is a 1000-fold decrease in the cytoplasmic volume to nuclear volume. Because of the elimination of growth between divisions of the blastomeres, cleavage divisions may be quite rapid; in *Drosophila* cleavage divisions occur every 10 minutes for more than 2 hours. The concomitant rapid rate of increase in cell number during cleavage sharply contrasts with the much lower rate during gastrulation and subsequent stages of development. Mammals exhibit among the slowest cleavage rates in the animal kingdom, approximately 12 to 24 hours apart (see Fig. 12).

cleavage embryogenesis   the production, in gymnosperms, of two or more embryos by a single zygote, after cell division begins.

cleavage furrow   created by a contractile ring of actin microfilaments, this furrow, characteristic of dividing blastomeres of early cleavage, eventually bisects the cytoplasm of the diving cell (see Fig. 29).

cleavage nucleus   the nucleus of a zygote formed by fusion of male and female pronuclei; also known as segmentation nucleus.

cleavage patterns   embryos of different animal species, because of differences in amount and distribution of yolk in their eggs, and genetic makeup, undergo different modes of cleavage called cleavage patterns. For example, human embryos undergo holoblastic, equal, rotational cleavage.

cleft lip   sometimes referred to as harelip, this is a congenital defect of the upper lip, resulting from the lack of fusion of the maxillary and nasomedial processes.

cleft palate   a fissure through the palate; a congenital defect resulting from failure of fusion of embryonic palatine processes resulting in a fissure through the palate.

cleidocranial dysplasia (CCD)   a human syndrome in which the skull sutures fail to close; DNA from such patients have either deletions or point mutations in the *CBFA1* gene; apparently CCD is caused by heterozygosity of the *CBFA1* gene.

cleidoic egg   an egg laid on land; constructed so as to prevent loss of its water supply, rid the body of metabolic wastes, and permit respiration to go on; the so-called locked up egg that released the reptiles from an aquatic environment; the common chicken egg is a cleidoic egg.

cleistogamous   flowers that self-fertilize without opening. *See:* **chasmogamous**.

cloaca   the common chamber at the caudal end of the embryo/fetus into which the digestive, urinary, and reproductive systems open. In the frog, the cloaca is the general region where the endoderm of the rectum joins the ectoderm of the proctodaeum; the dorsal wall of the cloacal chamber receives the urogenital ducts; at metamorphosis, the ventral part of the cloaca gives rise to an anteriorly directed outgrowth, the urinary bladder. In the chick, the cloaca first appears, during the fourth day, as a dilated terminal portion of the gut; this establishes at the outset the relationship of the cloaca and the intestine that persists in the adult bird. In the adult bird, the cloaca is a common chamber into which the intestinal contents, the urine, and the products of the reproductive organs are received for discharge. The proximal portion of the allantoic stalk, the homologue of the urinary bladder of mammals (there is no urinary bladder in birds), opens directly into the cloaca. The ducts that drain the developing excretory organs open into the cloacal region on either side of the allantoic stalk.

cloacal membrane   the membrane, in the embryo, between the rectum and the proctodeum; future site of the anus (see Fig. 6).

cloacal plate   a double layer of (1) endoderm, from the anal evagination of the foregut, and (2) ectoderm, from the proctodaeal invagination; when the cloacal plate ruptures, the cloacal opening of the embryo is created.

clonal analysis   is a form of fate mapping in which a single cell is labeled and the position and cell types of its progeny are identified at a later stage.

clonal restriction   found in insect compartments defined by clonal retriction lines, which lines are not transgressed by cell clones generated after a certain stage of development.

clonal selection theory   the concept that each animal first randomly generates a vast diversity of lymphocytes and then those cells that react against the foreign antigens that the animal actually encounters are specifically selected for action.

clone   (1) a population of cells derived from a single cell by mitosis; (2) a group of organisms produced by asexual reproduction, e.g., vegetative propagation in plants, parthenogenesis in animals, and nuclear transfer techniques involving the transfer of somatic cell nuclei of identical genetic constitution into enucleated eggs.

cloning   as applied to organisms, the process (natural or artificial) of making several, genetically identical, independent organisms from a single initial organism. Amphibian cloning experiments suggested: (1) a general restriction of potency occurred with progressive development and (2) the differentiated cell genome was remarkably potent. Debate persisted about the totipotency of such nuclei, but there was little doubt that they were extremely pluripotent. Cloning may be applied at the molecular, nuclear, and cellular levels, as well as at the organismal level.

clubfoot   a congenital malformation in which the forefoot is inverted and rotated.

c-mos   *See:* **cytostatic factor**.

Cnidaria   the equivalent name for Coelenterata.

cnidarians   animals belonging to the phylum Cnidaria.

CNS   central nervous system.

co-activator   factor that has no transcription-activation ability of its own, but helps other proteins stimulate transcription; a protein that interacts with the activator in control of transcription; some co-activators possess activities that modify local chromatin conformation.

cochlea   the auditory part of the inner ear; a cone-shaped tube.

codominance   the form of inheritance wherein both alleles of genes are expressed.

codons   the triplets of nucleotides in messenger RNA molecules that specify (through the intermediary of transfer RNAs with their anticodons) the position of a specific amino acid.

Coelenterata   a phylum of animals exhibiting primary radial symmetry; the solid body wall, which encloses a gastrovascular cavity, consists of two well-defined epithelial layers, epidermis and gastrodermis, and an intermediate, mesoglea, layer. *See:* **Ctenophora**.

coelenterates   animals belonging to the phylum Coelenterata.

coeloblastula   a simple, hollow blastula with a single cell-layered wall.

coelom   the original body cavity that appears between layers of somatic and splanchnic mesoderm. The coelom originates as bilaterally symmetrical chambers; later in development, it gives rise to the unpaired body cavity. In the embryos of higher vertebrates, body folds will divide the coelom into intraembryonic and extra-embryonic regions (see Fig. 6).

coelomate   an animal exhibiting a true coelom, i.e., one completely lined with mesoderm.

coenobium   a colony of motile cells formed by some green algae, e.g., the Volvocales, with the form of the colony being characteristic of the species. In some species, some differentiation may be found as a polarized colony with reproductive cells at only one pole of the colony.

coenocyte   a multinucleate mass of cytoplasm formed by repeated nuclear divisions without division of the cytoplasm.

cohesion proteins   protein rings that encircle the sister chromatids during meiosis, provide a scaffold for the assembly of the meiotic recombination complex, resist the pulling forces of the spindle microtubules, and thereby keep the sister chromatids attached together and promote recombination.

coitus   the act of copulation; also known as intercourse.

coitus interruptus   a form of birth control that entails the withdrawal of the penis from the vagina before ejaculation occurs.

colcemid   an antimitotic drug, derived from and closely related to colchicine (the $COCH_3$ group of colchicine is replaced by a methyl group); often used in place of colchicine, especially with animal cells; its binding to tubulin, unlike that of colchicine, is readily reversible.

colchicine   a chemical derived from the autumn crocus (*Crocus autumnale*), a plant indigenous to the eastern Mediterranean, which inhibits the maintenance and formation of microtubules. An antimitotic drug, this chemical is used to inhibit cellular and developmental activities that depend on the integrity of microtubules, e.g., the ingression and migration of primary mesenchyme cells during sea urchin gastrulation. Also, used extensively for making spreads of mitotic chromosomes preparatory to making karyotypes.

coleoptile   the sheath-like structure that encloses the plumule of grass embryos; it is the first part of the embryo to break through the ground and protects the apical meristem of the shoot.

coleorhizae   the structure that ensheaths the root of grass embryos; a mirror image of the coleoptile.

colinear   having corresponding parts arranged in the same linear order. *See:* **colinearity**.

colinearity   the correspondence between the sequence of DNA nucleotide triplets in structural genes and the sequence of amino acids in the polypeptide chains encoded in those genes; the correlation between the order of *Hox* genes on a chromosome and the rostrocaudal order of gene deployment in the embryo.

collagen   a fibrous protein (actually a family of proteins) and a major component of the extracellular matrix; some receptor tyrosine kinases (RTKs) are activated by collagen.

collagen, type IV   a major component of the type of extracellular matrix called the basal lamina.

collecting tubules   the tubules of the kidney that collect urine from the uriniferous tubules and transfer it to the calyces of the kidney.

colliculi   mounds, as in mounds of tissue.

colors   traditionally, embryologists have used specific colors to designate specific germ layers: blue for ectoderm, red for mesoderm, and yellow for endoderm.

combinatorial   of, relating to, or involving combinations. *See:* **combinatorial regulation**.

combinatorial regulation   control of gene transcription by two or more transcription factors; the spatial patterns of gene expression are often delimited by the combined action of transcription factors.

commensalism   a symbiotic relationship that is beneficial to one partner and neither beneficial nor harmful to the other partner.

commisure   a seam; the union-line between two parts, e.g., a connecting band of nerve tissue.

commitment   a state in which a cell's developmental fate has become restricted even though it is not yet displaying overt changes in cellular biochemistry and function.

committed stems cells   multipotent and unipotent stem cells that have the potential to become relatively few cell types.

community effect induction of cell differentiation in some tissues depends on there having to be a sufficient number of responding cells present for differentiation to occur.

compaction early during cleavage in mammalian development, the blastomeres become so intimately associate that their individual boundaries become obscured; this is compaction. *See:* **cavitation**.

comparative embryology comparison of developmental processes in different species, providing greater insight into developmental processes.

compartments domains of cell lineage restriction; a compartment can be defined as a region in the embryo that contains all the descendents of the cells present when the compartment is set up and no others. In a compartment, the cells may all be under common genetic control. The compartment boundary is remarkably sharp and straight and does not correspond to any structural feature in the wing of *Drosophila* (also, the pattern of the wing is not dependent on cell lineage, as demonstrated by using the *Minute* technique). The specification of cells as the posterior compartment of a segment (the anterior of a parasegment) initially occurs when the parasegments are set up and is attributed to the *engrailed* gene. Complex intercellular circuitry consolidates the compartment boundaries, involving, for example, the segment polarity genes, *wingless* and *hedgehog*; compartments tend to act as discrete developmental units. *See:* **cell lineage restriction**.

compensatory mutation a mutation that restores or maintains the function of a gene or *cis*-regulatory element by counteracting the effects of one or more different mutations.

compensatory regeneration regeneration in which cells divide, but maintain their differentiated function; they produce cells similar to themselves and do not form a mass of undifferentiated tissue; characteristic of mammalian liver regeneration. *See:* **epimorphic regeneration**, **mophallactic regeneration**.

competence the ability to respond to an inductive signal; reactive state permitting development in response to a stimulus, as part of an embryo in response to an inductive signal; also responsiveness to the signals that turn on the relevant combination of transcription factors.

competence factor a factor that makes a responder competent to respond to the inductive signal from the inducer; e.g., the PAX6 protein seems to be important in making the ectoderm competent to respond to the inductive signal from the optic vesicle. *See:* **induction**.

competence map identification of what a group of cells may form when transplanted to different locations of a host embryo. *See:* **fate map**, **specification map**.

competence transfer competent tissue may transfer its responsiveness from one inducing agent to another; the embryologist Conrad Hal Waddington called the transfer of competence "genetic assimilation." *See:* **genetic assimilation**.

complete flower a flower that possesses all four floral organs; i.e., possesses: sepals, petals, stamens, and carpels.

conceptacle cavity or chamber in which gametangia are borne.

conception fertilization.

conceptus that which results from conception (fertilization), namely, the embryo or fetus and its associated membranes.

condensation as it pertains to chromosomes; to become compact and visible (as opposed to dispersed chromatin).

conditional development development exhibit by cells whose ultimate fates depend on their location in the embryo.

conditional mutation any of a class of mutations whose viability is dependent on a set of permissive conditions; they perish under nonpermissive or restrictive conditions; a special class of these mutations is the temperature-sensitive mutation.

conditional specification specification by interactions between cells, relative positions are important; the mode of commitment where the fate of a cell depends on the conditions in which the cell finds itself, characteristic of all vertebrates and few invertebrates. *See:* **autonomous specification**.

cone reproductive structures found on conifers; of two general varieties: pistillate (female, egg producing) and staminate (male, pollen producing). *See:* **strobilus**.

congenital with birth; present from the time of birth, as with congenital birth defects.

congenital anomalies the modern medical term for birth defects.

congenital dislocation of the hip a potentially crippling abnormality, commonly involving one or both hip joints and, although present at birth, is often discovered only after the child starts to walk.

congenital syphilis   *See:* **syphilis**.

conjoined twins   twins that have, during their development, fused to a greater or lesser degree; also called Siamese twins. *See:* **Bunker, Eng and Chang**.

conjugation   a phenomenon exhibit by paramecia whereby sex (the combining of genes from two different individuals), but not reproduction, is accomplished.

Conklin, Edwin Grant (1863–1952)   professor of biology at Princeton University; conducted lineage studies of ascidian embryogenesis; published work on *Styela partita* and extended the principle of organ-forming, germinal, areas to the chordate embryo; concluded that each colored plasm (of the egg cytoplasm) delineates a specific embryonic fate and there occurs a segregation of cytoplasmic determinants at fertilization. *See:* **Vogt, Walter**.

connecting stalk   the connection between the trophoblast and the inner cell mass as it develops into the embryo. It gives rise to part of the umbilical cord; also called the body stalk (see Fig. 6).

conservation   (1) as applied to proteins, little change in the primary sequence of a given protein over long stretches of evolutionary time; e.g., the histone proteins are more highly conserved than cytochrome c throughout eukaryotes; (2) as applied to genes, little change in the sequence of deoxyribonucleotides of a given gene over long stretches of evolutionary time; (3) evolutionary conservation, as in that of homeotic gene organization and transcriptional expression in flies and mice.

constitutive   in a genetic context, gene products that are active all the time or genes that are expressed all the time.

constitutive heterochromatin   the common form of heterochromatin that usually does not change its nature; e.g., heterochromatin present in the proximity of centromeres and telomeres and in the nucleolus organizer region.

constitutive mutation   a mutation that causes a gene to be expressed at all times, regardless of normal controls.

constraints   *See:* **developmental constraints**.

contact guidance   the concept that the specificity of growth of nerve cell axons depends on physical cues, such as the orientation of fibers in the substratum.

contact inhibition   motile cells contact each other and become paralyzed, after a brief period motile activity starts again leading to movement in a different direction and separation of the cells that had made contact; may play a role in guidance of migratory cells.

contact placenta   a placenta where the chorion may be very closely apposed to maternal tissues while still being readily separable, as in the pig. *See:* **deciduous placenta**.

continuity of the germplasm   the concept, attributed to August Weismann, that cells of the germplasm are immortal while somatic cells are destined to die.

contraception   prevention of conception (fertilization).

contractile proteins   proteins, such as actin and myosin, that are directly involved in muscle contraction.

contractile ring   an array of actin microfilaments found beneath the plasma membrane of a blastomere just below, and only, where the cleavage furrow will form; the contractile ring is believed to be the physical basis of cytokinesis.

contraction   as applied to muscles, the shortening of the length of muscles fibers.

contragestion   prevention of the establishment of a pregnancy; i.e., implantation.

conus arteriosus   cardiac outflow tract; precursor of both ventricles.

convergence   a type of morphogenetic movement wherein cells move toward (converge on) a specific region, such as the dorsal midline of the embryo.

convergent extension   a basic morphogenetic movement; the phenomenon whereby cells intercalate to narrow a tissue and at the same time elongate it.

cooperativity   a phenomenon discovered in the hemoglobin protein molecule, where the binding of a single oxygen molecule by a hemoglobin molecule increases its affinity for additional oxygen molecules, even though the oxygen-binding sites are too far apart to interact directly; the phenomenon depends on conformational (shape) changes in the molecule. This phenomenon is characteristic of allosteric enzymes where the binding of ligands at the regulatory (allosteric) site affects the activity of the catalytic (active) site at a remote location. Also, the enhanced binding of a protein to a *cis*-regulatory DNA sequence resulting from the interaction with other bound proteins. *See:* **gene expression, ligand**.

co-option   the recruitment of genes into new developmental or biochemical functions; the recruitment of preexisting units for new functions, e.g., a protein that functions as an enolase or alcohol dehydrogenase enzyme in the liver can function as a structural crystallin protein in the lens; can be observed on the molecular, pathway of development, and morphological levels.

copulation sexual intercourse.

cord blood blood obtained from the umbilical cord at the time of delivery; a rich source of stem cells.

co-repressor a protein that interacts with a repressor in the control of transcription; co-repressors may modify the local conformation of chromatin.

cork cambium also called phellogen; a plant lateral meristem, the activity of which results in the formation of cork tissue and phelloderm. Cork cambium is part of the bark and produces the outer bark of woody plants.

cornea the portion of the eye in front of the lens, responsible for most of the refraction (bending) of light that focuses the image on the retina; induced from surface ectoderm during early development by the lens of the eye.

corolla collectively, all of the petals of a given flower.

corona dispersing enzyme the acrosome of some mammals releases corona-dispersing enzyme that separates follicle cells from one another (in guinea pig).

corona penetrating enzyme (CPE) mammalian acrosomal enzyme; an enzyme apparently responsible for allowing sperm to penetrate between the tightly bound corona radiata cells.

corona radiata collectively, those follicle cells that remain adherent to the outer surface of the zona pellucida after ovulation.

corpora allata the source of juvenile hormone, in insect larvae, which inhibits metamorphosis. Secretory cells of the corpora allata are active during larval molts but inactive during the metamorphic molt. *See:* **prothoracic gland**.

corpora bigemina *See:* **optic lobes**.

corpora quadrigemina four eminences of tissue; the paired superior colliculi, concerned with vision, and the paired inferior colliculi, concerned with hearing; derived from the roof of the mesencephalon.

corpus body; (1) the major portion of the uterus, which contains the uterine cavity; (2) meristems in flowering plants are organized into layers that encompass cells in both the central and peipheral zones; the inner layer is referred to as the corpus (see Fig. 48). *See:* **tunica**.

corpus albicans the body formed, in the ovary, when the corpus luteum undergoes involution and its site becomes infiltrated with connective tissue; literally, "white body" (see Figs. 27 and 32).

corpus epididymis the body (corpus) or main portion of the epididymis, found between the caput epididymis and cauda epididymis.

corpus hemorrhagicum a transition body, in the ovary, between the mature graafian follicle and the corpus luteum; after ovulation a slight amount of blood may be deposited within the antrum of the follicle, giving rise to this body.

corpus luteum formed from the mature (graafian) follicle in the ovary during the second half of the menstrual (uterine) cycle. It is an important source of the sex steroids, estrogen and progesterone; literally, "yellow body" (see Figs. 27 and 32).

corpuscles of Meissner ovoid corpuscles attached to nerve fibers; found at the tips of fingers and toes; a corpuscle is a small, round body.

correlated progression an evolutionary consequence of the modular nature of development; changes in one part of the embryo induce changes in another; skeletal cartilage informs the placement of muscles; muscles induce the placement of nerve axons.

cortex the outer portion of something, as in the ovarian cortex, cortex of the adrenal gland, or outer cytoplasm of a cell.

cortical granules abundant, membrane-bounded, organelles found in the periphery (cortex) of many animal eggs; derived from the Golgi apparatus and homologous to the acrosome of spermatozoa. At the time of fertilization, the contents of the cortical granules are released and result in a number of consequences as part of the fertilization process; an important consequence is the slow block to polyspermy.

cortical reaction the slow block to polyspermy; the release of the contents of cortical granules at the time of fertilization, which alters the surface of the egg and makes it refractory to supernumerary sperm. During the cortical reaction, the vitelline membrane, which surrounds the plasma membrane of many eggs, is chemically altered to form the fertilization membrane and the underlying, fluid-filled, perivitelline space, within which early development occurs. In sea urchins, where the cortical reaction has been particularly well studied, the consequences, in addition to the formation of the fertilization membrane and the perivitelline space, included the formation of the hyaline membrane and an altered plasma membrane composition. The plasma membrane of the fertilized egg is quite different from that of the unfertilized egg because the membranes of thousands of cortical granules are incorporated into the original membrane of the egg. *See:* **zona reaction**.

cortical rotation a cytoplasmic rearrangement, especially in amphibian eggs; rotation of the egg cortex relative to the egg interior, initiated by sperm penetration, leading to a reduction in the pigmentation of the animal hemisphere opposite the sperm entry point. *See:* **gray crescent**.

cortices layers into which neurons of the brain are organized.

cotyledons (1) divisions of the fetal part of the placenta resulting from the presence of the placental septa of the maternal part of the placenta; (2) seed leaves; found in the seeds of flowering plants, there are two in the seeds of dicots (where they are food-storing organs), and there is one in the seeds of monocots (where they are food-absorbing organs, the endosperm is the food-storing tissue in monocot seeds).

covalent bond the type of chemical bond resulting from a sharing of a pair of electrons.

Cowper's glands (see Fig. 4). *See:* **bulbourethral gland**.

CpG islands regions of DNA containing many unmethylated CpG sequences; usually associated with active genes; clusters of CGs, in the vertebrate genome, in approximately 40,000 regions near the 5′ ends of the genes; predominantly unmethylated and thought to indicate the positions of "housekeeping" genes.

cranial ectodermal placodes a series of ectodermal thickenings from which the major sensory organs of the head develop by interacting with the neural tube, e.g., olfactory placodes and auditory placodes.

cranial flexure a flexure of the embryonic brain; a ventrad bending of the head end of the neural tube. In the 55-hour chick embryo, cranial flexure causes the brain to be bent nearly double on itself, as a result: the mesencephalon is the most anteriorly located part of the brain, and the optic vesicles and auditory vesicles are brought opposite each other at nearly the same anteroposterior level.

cranial ganglia ganglia associated with some of the cranial nerves.

cranial nerves twelve pairs of nerves associated with the human brain, each pair of which may be one of three types of nerves: sensory, motor, or mixed.

cranial neural crest that portion of the neural crest found in the head; its cells produce the craniofacial mesenchyme that differentiates into the cartilage, bone, cranial neurons, glia, and connective tissues of the face; also its cells give rise to thymic cells, odontoblasts, and bones of the middle ear and jaw.

cranial vault *See:* **calvaria**.

craniobuccal pouch a diverticulum from the buccal cavity in the embryo from which the anterior lobe of the pituitary is developed. Also known as Rathke's pouch.

craniorachischisis a birth defect resulting from the failure of the entire neural tube to close over the entire body axis.

cranium refers to the skull, specifically the part that encloses the brain.

Cretaceous-Paleocene discontinuity 65 million years ago, an asteroid collided with earth; dust obscures the sun, and a massive extinction occurs, nearly all dinosaurs and 70% of all species become extinct.

Cretaceous Period 65–135 million years ago; rise of angiosperms, dinosaurs reach peak and rapidly decline, and birds persist.

cri du chat cry of the cat; name of a syndrome resulting in early death of the child, resulting from a deletion of the short (p) arm of chromosome 5.

-crine this word part means to separate, e.g., used in autocrine, endocrine, juxtacrine, and paracrine.

critical day length the point in the relative lengths of day and night at which 50% of the population entered diapause; in some species, a period of diapause has become an obligatory part of the life cycle, e.g., often observed in temperate-zone insects that overwinter, where diapause is induced by changes in the photoperiod.

critical periods in development the theory of critical periods in development originated in the early 1900s with the discovery that various chemical, surgical, or environmental insults to embryos produced drastically different effects depending on the stage in development at which the insult was applied; e.g., thalidomide causes phocomelia when taken by the

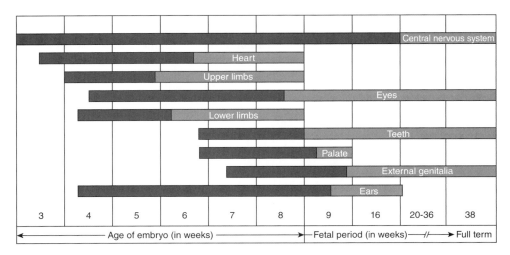

Figure 13. Critical periods in development. An important concept in trying to understand birth defects is that of "critical periods in development." Different organs and structures are most vulnerable to teratogens during specific periods in development. Most of the embryonic period, 3 weeks to 8 weeks, is very critical for most organs and structures because this is the time when most organs and structures are forming. The darker shading indicates periods more sensitive to birth-defect causing agents (teratogens). Adapted from Moore KL, Persaud TVN. Before We Are Born, 4th ed. Philadelphia: WB Saunders, 1993. Reprinted from Frank J. Dye, *Human Life Before Birth*, Harwood Academic Publishers, 2000, fig. 19-12, p. 171.

pregnant women during the latter part of the embryonic period when the limbs begin their early development (see Fig. 13).

crossing over reciprocal exchange of homologous segments of homologous chromosomes during prophase I of meiosis; increases genetic diversity of gametes and, therefore, of the species (see Fig. 26). *See:* **random segregation of chromosomes**, **spermatocytes**.

crossopterygians fossil fish, ancestral to the amphibians.

cross-talk interactions between seemingly independent intracellular signal transduction pathways, e.g., cooperation between two pathways resulting in formation of the AP-1 transcription factor, which turns on interleukin 2 transcription during lymphocyte differentiation.

crura cerebri in the frog, nerve tracts that originate as a pair of ventrolateral thickenings of the mesencephalon.

cryogenic storage storage at low temperature, as with sperm and embryos.

cryptic genetic variation the genetic variation in components of developmental programs that exists in the absence of phenotypic variation; pertubations to developmental programs may unmask this variation and lead to phenotypic variation.

cryptochrome *See:* **daylength**.

cryptogam literally, "hidden gametes"; includes the mosses and ferns, plants that reproduce by means of spores and in which the production of gametes is not obvious; a plant that does not produce seeds. *See:* **phanerogam**.

cryptorchidism hidden testis; refers to an undescended testis that has not appeared in the scrotum.

crystallins proteins made by the cells of the lens of the eye; gives the lens its crystal-clear property.

CSF cerebrospinal fluid.

Ctenophora the comb jellies; a phylum of animals whose members are diploblasts.

culmination the process undergone by the mound of aggregated cells in the life cycle of *Dictyostelium discoideum* to give rise to the sorocarp. *See:* **sorocarp**.

cumulus in mice, the follicle cells, collectively, that surround those oocytes ovulated from a given ovary (see Plate 8 in the color insert).

cumulus oophorus   a mound of follicle (granulosa) cells entending from the membrana granulosa into the antrum of the Graafian follicle, it contains the nearly mature oocyte; cumulus oophorus means "egg-bearing cloud."

cutis plate   dermatome, the outer layer of each somite is called the cutis plate or dermatome; the inner larger mass is called the myotome. *See:* **dermatome**.

cyclic adenosine monophosphate (cAMP)   an example of a group of naturally occurring chemicals known as cyclic nucleotides. cAMP is an important signaling molecule; e.g., (1) it functions as an intracellular "second messenger" in human cells and (2) it functions as an extracellular chemoattractant in the aggregation phase of the life cycle of *Dictyostelium discoideum*.

cyclin-dependent kinase   *See:* **cdk**.

cyclins   proteins that vary in amounts with different stages of the cell cycle that complex with cdks to produce active kinases that phosphorylate-specific proteins, which results in the passage of the cell through the cell cycle. Cyclin B is the regulatory subunit of the maturation (mitosis) promoting factor.

cyclopamine   a teratogen, found in the plant *Veratrum californicum*, known to cause cyclopia in vertebrates; blocks the synthesis of cholesterol. Cyclopamine, an alkaloid, can block the functions of the Hedgehog family of paracrine factors; blocking the *sonic hedgehog* gene signal leads to the failure of the optical field to separate. *See:* **cyclopia**, **jervine**, *Veratrum californicum*.

cyclopia   a birth defect characterized by fusion of the cerebral hemispheres, formation of only one central eye, and no pituitary gland.

cyclostomes   the most primitive of all living vertebrates.

cystic duct   the duct leading from the gall bladder; it joins the hepatic duct to form the common bile duct.

cystocytes   the cells of the clone of cells derived from the oogonium by meroistic oogenesis in insects.

cytochalasin B   a member of a class (cytochalasins) of fungal metabolites that disrupts the integrity of actin microfilaments and, therefore, cellular and developmental activities that depend on their activity; e.g., cytochalasin B has been used to uncouple karyokinesis (nuclear division, mitosis) from cytokinesis (division of the cytoplasm, which depends on the activity of a contractile ring of actin microfilaments) of eggs undergoing cleavage. Depolymerization of actin microfilaments will cause sea urchin eggs to lose their spherical shape.

cytodifferentiation   *See:* **cell differentiation**.

cytogenetics   the marriage of the disciplines of cytology (the study of cells) and genetics (the study of heredity).

cytokines   growth factors that regulate blood cells and lymphocytes.

cytokinesis   division of the cytoplasm as a part of mitosis or meiosis (see Figs. 10 and 29).

cytokinins   a class of plant hormones; can release buds from apical dominance.

cytology   the study of cells.

cytomegalovirus   a teratogenic virus.

cytonemes   fine, actin-based, intercellular processes, similar to thin filopodia, that may transport inductive signals through extracellular spaces.

cytoplasm   that portion of the cell between the plasma membrane and the nuclear membrane.

cytoplasmic bridges   thin strands of cytoplasm linking cells, e.g., between nurse cells and developing eggs, and between developing spermatozoa; in higher plants, cytoplasmic bridges are called plasmodesmata. These cytoplasmic connections are larger than those of gap junctions and allow the transfer of large macromolecules and, perhaps, organelles. *See:* **nurse cells**, **spermatogenesis**.

cytoplasmic determinant   a substance or substances, located in part of an egg or blastomere, that guarantees the assumption of a particular commitment by the cells, which inherit it during development; are often mRNAs that are localized to a part of the cell in association with the cytoskeleton, e.g., the *bicoid* and *nanos* mRNAs in the *Drosophila* egg.

cytoplasmic incompatibility   embryos resulting from crosses between uninfected males and infected females are viable, but those resulting from crosses between infected males and uninfected females are not, e.g., a common effect of *Wolbachia* infection of arthropod populations.

cytoplasmic localization the nonhomogeneous distribution of substances in the cytoplasm of a cell, especially an egg or oocyte, e.g., the spatial localization of morphogenetic determinants within the egg cytoplasm.

cytoplasmic matrix that portion of the cytoplasm found between the organelles and containing the cytoskeleton of the cell; also called cytosol.

cytoplasmic organelles subcellular structures specialized for specific functions, e.g., mitochondria, lysosomes, and Golgi apparatus.

cytoplasmic rearrangements the eggs of a number of species, e.g, *Styela partita*, and frogs, when fertilized, undergo rearrangements of their cytoplasm such that shortly after fertilization the arrangement of cytoplasmic components is quite different from that in the unfertilized egg. The purpose of these cytoplasmic rearrangements is to put specific morphogens in specific parts of the fertilized egg, so that cleavage will partition the morphogens into different subsets of cells; these morphogens, in turn, will activate different subsets of genes in the different subsets of cells. In some cases, these cytoplasmic rearrangements are quite visible, as in the formation of the yellow crescent in *Styela partita* and in the formation of the gray crescent in frogs. The beginnings of gastrulation and induction are to be found in the cytoplasmic rearrangements that attend fertilization; several lines of evidence lead to this model: (1) prevent cytoplasmic rearrangements, (2) artificially mix cytoplasmic regions, and (3) remove the equatorial cells of the midblastula frog and salamander embryo.

cytosine methylation appears to be a major mechanism of transcriptional regulation in vertebrates. *See:* **DNA methylation**.

cytoskeletal proteins those proteins, such as tubulin, actin, and keratin, that make up the microtubules, microfilaments, and intermediate filaments, respectively, of the cytoskeleton.

cytoskeleton that portion of the cytoplasm of the cell made up of fibrillar, protein, components—namely, microtubules, microfilaments, and intermediate filaments—which is largely responsible for cell shape and cell movement.

cytostatic factor a complex of c-mos (an oncoprotein) with cdk2, that inhibits cyclin breakdown, which is responsible for arrest of the secondary oocyte ("unfertilized egg") in second meiotic metaphase.

cytotrophoblast the inner cellular component of the trophoblast, covering the chorion and the chorionic villi during the first half of pregnancy (see Figs. 5 and 44).

cytotrophoblastic shell that portion of the cytotrophoblast, which penetrates the overlying syncytiotrophoblast, to the maternal endometrium, and attaches the chorionic sac firmly to the maternal endometrial tissue.

D

-dactyly a part of a word referring to the fingers or toes; there are a number of specific kinds of birth defects involving dactyly depending on what happens to the digits during development. Note that during development the digits are "sculpted" out of a hand plate or foot plate by specific, genetically programmed, cell death.

Dareste, Camille (1822–1899) influeneced by Etienne Geoffroy Saint-Hilaire, he conducted experiments with the goal of obtaining experimental transformations of embryos; he subjected chicken eggs to varnishing of the shell, refrigeration of the eggs, and concussion of the eggs before incubation, for example. Although he was able to produce different types of abnormalities, he could not obtain them specifically. Dareste, as did E. Geoffroy Saint-Hilaire, employed the so-called indirect method, which involved exposing an entire egg to a teratogenic factor; as opposed to the direct method, which involved placing an instrument directly on a precise point of the embryo. He has gone down in history as one of the first true experimental teratologists. *See:* **Fol, Hermann**; **Warynski, Stanislas**.

Darwin, Charles (1809–1882) resuscitated the theory of pangenesis in his *The Variation of Animals and Plants under Domestication*, Volume 2 (1868), to account for the origin of variation. He theorized that body cells secrete minute corpuscles or "gemmules" that record growth patterns for the area they represent. These gemmules were believed to be carried by body fluids and the bloodstream to the reproductive organs and there packed into the eggs or sperm. In the new individual, they were considered to determine the visible characteristics and the growth pattern.

Darwin and embryology Charles Darwin recognized that embryonic resemblances were a very strong argument in favor of the common ancestry of different animal groups; "Community of embryonic structure reveals community of descent," Darwin said in *Origin of Species*; he looked to embryonic and larval stages for homologies that would be obscured in the adult. *See:* **descent with modification**.

dauer German, continuance.

dauer larva a nonfeeding, metabolically dormant stage that may be induced in the life cycle of *Caenorhabditis elegans* by overcrowding or insufficient food.

dauer larva pathway starvation of *Caenorhabditis elegans* larvae creates a long-lived dauer larva by repressing the pathway to adult development and by increasing the reactive oxygen species (ROS)-degrading enzymes. *See:* **reactive oxygen species (ROS) theory of aging**.

dauer stage a nonfeeding, metabolically dormant stage that *Caenorhabditis elegans* larvae may enter when overcrowded or starved.

dauerblastula a hollow ball of ciliated cells derived from the isolated animal hemisphere of an early sea urchin embryo.

daughter cells those cells resulting from the division (mitotic or meiotic) of preexisting cells.

DAX1 **gene** apparently a gene involved in ovary determination. *See: SOX9* **gene**, *SRY* **gene**.

daylength many plants have a daylength requirement for flowering, either a shortened day (8 hours) or a long day (12 to 16 hours); the ability to sense the length of day, as well as the quality of light, is under the control of two families of photoreceptors: phytochromes (which respond to red light) and cryptochromes (one of two known proteins that respond to blue light). *See:* **cryptochrome**, **long-day plants**, **phototropin**, **short-day plants**.

de novo **methylases** DNA methyl transferase enzymes that methylate previously unmethylated cytosine guanines (CGs). *See:* **maintenance methylases**.

deacetylated histones histones not enzymatically modified by the addition of acetyl groups; deacetylated histones form stable nucleosomes. *See:* **acetylated histones**.

Dictionary of Developmental Biology and Embryology, Second Edition. Frank J. Dye.
© 2012 Wiley-Blackwell. Published 2012 by John Wiley & Sons, Inc.

Decapentaplegic (Dpp) gene   a zygotic gene in *Drosophila*, produces a key morphogen necessary for the correct patterning of the fifteen imaginal discs. *See:* **imaginal discs**.

decapentaplegic protein   the more dorsal part of the dorsoventral axis in the *Drosophila* embryo is thought to be determined by a gradient in the activity of this protein. The graded activity of decapentaplegic protein along the dorsoventral axis is attributed to its interaction with a secreted protein, such as short gastrulation, which is expressed in the ventral region and diffuses into the dorsal region. The decapentaplegic protein is a secreted signaling protein, which is involved in a variety of signaling processes throughout *Drosophila* development; is the *Drosophila* homologue of the vertebrate bone morphogenetic proteins (BMPs).

decidua   the endometrium during pregnancy. The decidua has three parts depending on their positions relative to the position of the conceptus: (1) decidua basalis, (2) decidua capsularis, and (3) decidua perietalis (see Fig. 44).

decidua basalis   that portion of the decidua underlying the conceptus; gives rise to the placenta materna or maternal contribution to the placenta (see Fig. 35).

decidua capsularis   that portion of the decidua overlying the conceptus.

decidua parietalis   that portion of the decidua that neither underlies nor overlies the conceptus.

decidua reaction   alteration of the endometrium in response to elevated progesterone levels and in preparation for implantation of the blastocyst.

deciduous placenta   a placenta where the chorion and the maternal tissues are so intimately integrated that the two tissues cannot be readily separated: as is the case with most mammals, including humans. *See:* **contact placenta**.

dedifferentiation   the loss, by differentiated cells, of their differentiated characteristics, e.g., as occurs in a regeneration blastema of a salamander limb after amputation.

deep homology   when homologous developmental pathways are used for the same function in both protostomes and deuterostomes they are said to have deep homology. *See:* **homologous developmental pathways**.

deep layer (cells)   that portion of the zebrafish blastula, covered by the enveloping layer and underlaid by the yolk syncytial layer, which forms the embryo proper. *See:* **enveloping layer**, **yolk syncytial layer**.

defect experiment   involves destroying a part of an embryo and observing the development of the defective embryo.

defensins   antibacterial peptides, found in shrimp eggs, and tobacco hornworm embryos. *See:* **embryo defenses**.

delamination   the splitting of a single layer of cells into two layers, as in the formation of somatic mesoderm and splanchnic mesoderm from lateral plate mesoderm.

deletion   as applied to chromosomes, a missing portion of a chromosome.

Delta-Notch system   in this inducing factor family, important in neurogenesis, somitogenesis, and imaginal disc development, both the ligand (Delta) and the receptor (Notch) are integral membrane proteins, so their interaction can only take place if the cells making them are in contact.

denatured DNA   DNA not in its natural state, which is that of a two-stranded double helix; rather, the weak hydrogen bonds of the DNA molecules have been broken allowing the two strands of the double-stranded molecule to separate. *See:* **native DNA**.

denatured protein   protein not in its natural state; where protein molecules have had their weaker chemical bonds broken and therefore no longer have their natural shape (e.g., the albumin protein in cooked egg white) and are no longer able to carry out their normal functions. *See:* **egg chemistry**, **native protein**.

dendrites   nerve cell processes, generally more numerous and shorter than axons, which carry nerve impulses toward the cell body of the nerve cell.

dental papilla   that portion of the embryonic tooth, located inside the enamel organ of a developing tooth, and forming the dentin and the dental pulp of the tooth that together with the enamel organ makes up the tooth germ.

Dentalium   genus of molluscs; E. B. Wilson studied cleavage and polar lobe extrusion and reincorporation in *Dentalium*; his research led him to conclude that the polar lobe cytoplasm contained the mesodermal determinants.

denticles   outgrowths of the chitinous cuticle of the larvae of *Drosophila*.

dentin   a bonelike connective tissue composing the bulk of a vertebrate tooth.

dentinoblast   a mesenchymal cell that forms dentin. *See:* **odontoblast**.

deoxyribonucleic acid　*See:* **DNA**.

deoxyribonucleotides　monomers (building blocks) of DNA.

dependent development　also called conditional development. *See:* **conditional development**.

derivative　that which is derived or comes from something else; e.g., chromatophores are neural crest derivatives; i.e., chromatophores come from neural crest cells.

dermal papilla　a small node of dermal fibroblasts formed beneath the hair germ; stimulates more rapid division of the hair germ basal cells; the basal cells respond by producing post-mitotic cells that will differentiate into the keratinized hair shaft.

dermal tissue　one of three tissue systems that emerge at the globular stage of angiosperm embryogenesis; the dermal tissue will form the protoderm. *See:* **ground tissue**, **procambium**.

dermatome　that portion of a somite that gives rise to the dermis of the skin of the back; in frogs, the main part of each dermatome (cutis plate) breaks up and some cells form the dermal layer of the dorsal region, while other cells form connective tissue; it appears that dermis of the ventral regions is derived from part of the somatopleure.

dermis　the inner layer of the skin, derived from mesoderm. *See:* **epidermis**.

dermomyotome　that portion of the somite, not including sclerotome, which gives rise to dermis of the back and muscles (see Fig. 41).

descent　(1) as applied to the gonads, the normal movement of the testes from their site of origin into the scrotum or the normal movement of the ovaries from their site of origin to their final location in the pelvic cavity; (2) one of the seven cardinal movements of labor (see Fig. 40).

descent with modification　Charles Darwin's theory, he stated, would explain "unity of type" as descent from a common ancestor, and the changes creating the marvelous adaptations to the "conditions of existence" would be explained by natural selection; is critical for developmental regulatory genes as well as for the organisms they help construct; one of the major contributions of developmental biology to the study of evolutionary mechanisms has been to underline the importance of descent with modification during evolution; e.g., descent with modification may involve changes in regulation of the *Hox* gene with attendant alterations in time or domain of expression, or changes in the target genes of *Hox*.

descriptive embryology　study of normal developmental processes in organisms.

desiccation　drying out; loss of water.

desmosomes　a type of cell junction, joining one epithelial cell to another and giving structural integrity to an epithelium.

determinants　are cytoplasmic factors (e.g., proteins and RNAs) in the egg and in embryonic cells that can be asymmetrically distributed at cell division and so influence how the daughter cells develop.

determinate cleavage　a type of cleavage that separates portions of the zygote with specific and distinct potencies for development as specific parts of the body. *See:* **indeterminate cleavage**.

determination　the second part of commitment to cell fate; the fate of a cell has become cell-intinsic and is irreversible; the cell's commitment to eventually differentiate into a specific cell type and no other. Blastomeres are determined by cytoplasmic segregation or by interactions among blastomeres. In a molecular sense, determination means that the cells have lost their responsiveness, or competence, to the signals that originally turned on the relevant combination of transcription factors. *See:* **differentiation**, **specification**.

determinative development　*See:* **mosaic development**.

deuterencephalic　the hindbrain, in the context of the regional specificity of the neural structures that are produced during development of the central nervous system.

deuterostomes　those animals in which the blastopore gives rise to the anus and mouth formation is secondary; includes the phyla Echinodermata and Chordata.

Deuterostomia　a "so-called," superphylum containing coelomate phyla characterized by radial cleavage, formation of the anus from the blastopore, and enterocoely. *See:* **Protostomia**.

deutocerebrum　*See:* **neuromeres**.

deutoplasm　yolk.

development　the complex, organized changes that an individual undergoes from fertilization until death. The three major aspects of development are as follows: cell proliferation, cytodifferentiation, and morphogenesis. Development has a time axis, and one can predict that the changes will go from the simpler to the more complex.

developmental anatomy   the study of the dynamic (changing) anatomy of the developing organism.

developmental biologist   a biologist who specializes in the study of embryology or developmental biology.

developmental biology   the study of development throughout the life cycles of organisms; including such postembryonic development as flower development and regeneration.

developmental constraints   interactions of developmental modules that limit the possible phenotypes that can be created. *See:* **morphogenetic constraints, phyletic constraints, physical constraints**.

developmental domains   domains characterized by a unique profile of zygotic gene activity; made by the beginning of zygotic transcription and visualized by *in situ* hybridization for specific transcription factors; e.g., the entire mesoderm can be visualized by the expression of *brachyury. See:* **insulator sequences**.

developmental estrogen syndrome   fertility problems, cancers, and obesity later in life resulting from exposure to estrogenic compounds early in development in both men and woman.

developmental evolutionary synthesis   expands evolutionary theory to include the population genetic approach (microevolution) as well as to explain macroevolutionary phenomena.

developmental genes   genes that control development; many of the genes that control development of *Drosophila* are similar to those that control development of vertebrates and other animals. Many of the key genes in vertebrate development were originally identified as developmental genes in *Drosophila*. Developmental genes act in a strict temporal sequence, forming a hierarchy of gene activity.

developmental geneticist   a geneticist who specializes in the genetics of development; especially concerned with the relationship between gene expression and development.

developmental genetics   the discipline that examines how the genotype is transformed into the phenotype. *See:* **transmission genetics**.

developmental modularity   divergence through dissociation; development occurs through discrete and interacting modules; such modules include morphogenetic fields, imaginal discs, cell lineages, insect parasegments, and vertebrate organ rudiments. Modular units allow different parts of the body to change without interfering with other functions; modularity allows three processes to alter development: dissociation, duplication and divergence, and co-option.

developmental Northern blot   a Northern blot in which RNAs from several different stages or tissues are compared simultaneously. *See:* **Northern blot**.

developmental origin of health and disease (DOHD) hypothesis   expansion of the *fetal origin of adult disease (FOAD) hypothesis* for two reasons: (1) the plasticity of development occurs in both prenatal *and early postnatal periods* and (2) the change in terminology emphasizes not only the causes of disease, but also *disease prevention and adult health promotion.*

developmental plasticity   phenotypic plasticity when observed in embryonic or larval stages of animals or plants: phenotype is not merely the unrolling of genotype.

developmental program   the dynamic ways in which genes are connected to other genes by the complex regulatory mechanisms that, in their interactions, determine when and where a particular gene will be expressed.

developmental robustness   *See:* **canalization**.

developmental switches   the regulatory sequences in those genes (e.g., the *dorsal* and the *hunchback* genes in *Drosophila*) whose products are morphogens, which when "thrown" by the binding of transcription factors, activate genes and set cells off along new developmental pathways.

developmental symbiosis   symbiosis in which at least one of the symbionts has its development facilitated by the symbiotic relationship, e.g., the symbiosis between photosynthetic algae and the egg masses of the spotted salamander, *Ambystoma maculatum*. The algae provide oxygen for eggs within the masses. *See:* **autopoietic**.

developmental systems theory (DST)   a school of philosophy that takes developmental biology very seriously and that incorporated epigenetic ideas.

Devonian the geological period from approximately 409 to 362 million years ago, during which terrestrial forms invaded the land.

dextrotropic cleavage a clockwise spiral cleavage pattern. *See:* **levotropic cleavage**.

diacylglycerol (DAG) *See:* **phosphoinositol pathway**.

diakinesis the fifth and last stage of prophase I of meiosis, during which the chromosomes undergo terminalization of chiasmata, i.e., the chiasmata tend to lose their original position and move toward the ends of the chromosomes. Also, during diakinesis, the nucleolus becomes detached from its special bivalent and disappears, and the bivalents are considerably more contracted than previously.

diapause an arrestation of growth that may take place at any stage of development. *See:* **critical day length**.

diaphragm the sheet of muscle and tendon found between the thoracic and abdominal cavities, the contraction of which is essential for breathing.

diaphragmatic hernia is a hernia that passes through the diaphragm into the thoracic cavity; it may be congenital or acquired.

diaphysis shaft of a long bone.

dickkopf a protein that interacts directly with the Wnt receptors, preventing Wnt signaling.

dicots *See:* **dicotyledonous plants**.

dicotyledonous plants plants whose seeds contain two seed leaves or cotyledons.

dictyate the prolonged diplotene stage of the first meiotic prophase.

Dictyostelium discoideum a cellular slime mold that exhibits a multicellular plasmodium during its life cycle. Unlike typical multicellular organisms that develop from a fertilized egg, *D.d.* develops from the aggregation of many separate cells; also, unlike typical multicellular organisms, there is no input of maternal information; rather, a self-organization, relying on signaling between cells, occurs. *D.d.* has been extensively studied by developmental biologists interested in (1) cell communication and (2) cell differentiation. *See:* **cyclic adenosine monophosphate (cAMP)**, **differentiation inducing factor (DIF)**.

diencephalon the second division, from the rostral end, of the five-vesicle early brain; derives from the forebrain (prosencephalon) and gives rise to the optic vesicles, pineal gland, and neurohypophyseal portion of the pituitary gland; gives rise to the retinas (vision), epithalamus (pineal gland), thalamus (relay center for optic and auditory neurons), and hypothalamus (regulates temperature, sleep, and breathing) (see Fig. 8).

diestrus the long, quiescent period after ovulation in the estrous cycle in mammals; the stage in which the uterus prepares for the reception of a fertilized ovum.

diethylstilbesterol (DES) a synthetic estrogen that is teratogenic; can cause infertility by changing the patterning of the Müllerian duct.

differential adhesion hypothesis a model that explains patterns of cell sorting based on thermodynamic principles. *See:* **Moscona**.

differential cell affinities the selective recognition of cells and extracellular matrices, by cells, which seems to play a major role in morphogenesis.

differential gene expression a concept incorporating three postulates: (1) every cell nucleus contains the complete genome established in the fertilized egg, (2) the unused genes in differentiated cells retain their potential for being expressed, and (3) only a small percentage of the genome is expressed in each cell and a portion of the RNA synthesized in the cell is specific for that cell type.

differential RNA splicing *See:* **alternative RNA splicing**.

differential substrate affinity hypothesis postulates that different cells recognize different molecules in various extracellular matrices.

differentiation the process by which cells reach the final stage of development in their lineage; the process by which the cells become mature and specialized. *See:* **determination**, **specification**.

differentiation inducing factor a chemical produced by aggregated cells of *Dictyostelium discoideum* that causes some of the cells to differentiate into prestalk cells and some of the cells to differentiate into prespore cells.

diffuse placenta a placenta having villi diffusely scattered over most of the surface of the chorion; found in whales, horses, and other mammals.

digestive proteins those proteins whose function is the digestion (hydrolysis; breakdown) of large molecules.

digestive system   in the chick, immediately caudal to the glottis, the narrowed region of the foregut is the esophagus while further caudally the slightly dilated region is the stomach; mesenchymal cells around the endoderm provide the muscular and connective tissue coats. The liver develops as a diverticulum from the ventral wall of the gut immediately caudal to the stomach. First recognized at approximately the 22 somite stage, the liver appears just as the part of the gut from which it originates is acquiring a floor; therefore, the hepatic evagination is located for a short time on the lip of the anterior intestinal portal; it grows cephalad toward the folk where the omphalomesenteric veins enter into the sinus venosus. The proximal part of the original evagination remains open to the intestine as the duct of the liver; later its regional differentiation gives rise to the common bile duct, the hepatic duct, the cystic duct, and the gallbladder; cellular cords, which bud off from the diverticulum, become the secretory units of the liver, the hepatic tubules. As the anterior intestinal portal moves caudad in the lengthening of the foregut, the proximal portions of the omphalomesenteric veins are brought together and fuse. In its growth, the liver surrounds the fused portion of the omphalomesenteric veins; this foreshadows the way in which the proximal part of the returning vitelline circulation is to be involved in the establishment of the hepatic-portal circulation of the adult. The pancreas is derived from evaginations appearing in the walls of the intestine at the same level as the liver diverticulum. In birds, three pancreatic buds appear, a single medial-dorsal (third day) and a pair of ventrolateral (end of fourth day) buds. The dorsal pancreatic bud originates directly opposite the liver diverticulum. The ventrolateral buds develop close to the point where the duct of the liver connects with the intestine, so that the duct of the liver and the ventral pancreatic ducts open into the intestine by a common duct, the ductus choledochus or common bile duct. Later in development, the masses of cellular cords derived from the three pancreatic primordia grow together and fuse into a single glandular mass. In mammals, only two pancreatic primordia are formed; the dorsal duct of Santorini and the ventral duct of Wirsung. By the fourth day, the midgut has been practically replaced by the extension of the foregut and hindgut.

digestive tube   the tubular portion of the digestive system found between the mouth and the anus.

digit   a finger or toe.

dihydrotestosterone   the hormone derived from testosterone by action of the enzyme 5-alpha-ketosteroid reductase 2; appears to control the formation of the external genitalia.

dilatation and evacuation (D & E)   a method of abortion used early in pregnancy.

dilatation stage of labor   the dilatation stage begins with the first true contractions of labor; contractions that begin to cause dilatation (enlargement) of the cervix.

dimer   made of two parts, as proteins made of two polypeptides.

dimorphic   with two forms.

diocoel   the cavity of the diencephalon, which becomes the third ventricle of the brain.

dioecious   literally, "two houses"; having male and female reproductive organs on different individuals; refers to those flowering plants where male and female flowers are found on different plants. *See:* **monoecious**.

dipleurula   a hypothetical, ancestral, bilaterally symmetrical, larva of which existing larvae may be modifications; also, any bilaterally symmetrical, ciliated echinoderm larva.

diplobiont   an organism characterized by alternating, morphologically dissimilar, haploid and diploid generations.

diploblastic   having two germ layers, referring to embryos and certain lower invertebrates.

diploblasts   animals possessing two cell layers; the outer ectoderm layer and the inner endoderm layer; includes cnidarians and ctenophores.

diploid   the normal and characteristic number of chromosomes for a given species; twice the haploid number of chromosomes; generally regarded as the species-specific number of chromosomes; 46 for humans.

diploid merogony   development of a part of an egg in which the nucleus is the normal diploid fusion product of egg and sperm nuclei.

diplont   an organism with diploid somatic cells and haploid gametes.

diplontic life cycle   life cycles exhibited by most animals, in which only the gametes are in the haploid state; i.e., undergo mitosis only in the diploid generation. *See:* **haplodiplontic life cycle, haplont**.

diplospory   development of gametophytes from somatic cells (megaspore mother cells), without the production of spores; the resulting gametophyte is diploid. *See:* **apospory**.

diplotene stage   the fourth stage of prophase I of meiosis, during which pairs of nonsister chromatids of each bivalent repel each other and are kept from falling apart by chiasmata; human oocytes may remain in this stage for decades before continuing on to produce "eggs" capable of fertilization.

Diptera   flies; insects with a single pair of membranous wings, includes *Drosophila*.

direct development   occurs when the embryo abandons the stages of larval development entirely and proceeds to construct a small adult; typified by frog species that lack tadpoles (e.g., *Eleutherodactylus coqui*, a small frog found in Puerto Rico) and by sea urchins that lack pluteus larvae (e.g., *Heliocidaris erythrogramma*, an Australian sea urchin).

disc   also spelled disk.

discoblastula   a blastula formed by cleavage of a meroblastic egg; the blastoderm is disc-shaped.

discogastrula   a gastrula formed from a discoblastula.

discoid   disc-like.

discoidal cleavage   a type of cleavage producing a disc of cells at the animal pole (see Fig. 12).

Discontinuity, Theory of   put forward by William Bateson in his book *Material for the Study of Variation* (1894); to explain the long process of evolution; i.e., species do not develop in a predictable sequence of very gradual changes but instead evolve in a series of discontinuous jumps.

disjunction   separation of chromatids or homologous chromosomes during anaphase. *See:* **nondisjunction**.

disk   also spelled disc.

disomy   two bodies; the normal condition where human chromosomes exist as pairs. *See:* **monosomy**, **trisomy**.

disparity   the variety of designs in body plans.

disruptions   congenital abnormalities caused by exogenous agents (chemicals, viruses, radiation, hyperthermia), e.g., cyclopia and fetal alcohol syndrome. *See:* **malformations**.

dissociation   by means of mutation or environmental perturbation one part of the embryo can change without the other parts changing; the ability of one module to develop differently from other modules. *See:* **allometry**, **heterochrony**.

distal   an anatomic term referring to a structure that is distant from an anatomic reference point; toward the tip, or the end of the organ, opposite the end of attachment (see Fig. 38). *See:* **proximal**.

Distal-less   a gene found as a single copy in both insects and amphioxus, and in both this gene is expressed in the epidermis and central nervous system. Vertebrates, however, have five or six closely related copies of *Distal-less*, each of which probably originated from a single ancestral gene that resembles the one in amphioxus. One vertebrate homologue of *Distal-less* is expressed in the neural crest cells. Amphioxus lacks neural crest cells. It has been speculated that the new type of *Distal-less* gene could cause the migratory ectodermal cells of amphioxus to evolve into neural crest cells. Also, because vertebrate *Distal-less* homologues are expressed in the forebrain, with an expression pattern similar to that observed in the anterior of the amphioxus neural tube, it has been suggested that the vertebrate forebrain is homologous to the anterior neural tube of amphioxus.

distal tip cell   a single nondividing cell, at the blind end of the *Caenorhabditis elegans* gonad, which controls whether germ cells enter into mitotic or meiotic divisions.

divergence   a process that allows redundant structures to assume new roles; as in gene divergence. *See:* **duplication**.

diversity   the number of species in a clade; the divergence of morphology.

diverticulum   a blind-ending pouch or sac resulting from an outpocketing of a cavity; e.g., the liver diverticulum is, initially, a pouch resulting from an outpocketing of the foregut.

dividing egg   an expression sometimes used for the early embryo during the cleavage stage of development.

dizygotic twins   *See:* **fraternal twins**.

DNA   (deoxyribonucleic acid) the chemical substance that actually makes up the genes or hereditary material of almost all organisms, with the exception of some RNA viruses.

DNA-binding domain   that domain of a transcription factor that recognizes a particular DNA sequence, e.g., homeodomain or zinc-finger domain.

DNA methylation   a process that seems to be able to stabilize a pattern of gene transcription; the promoters of inactive genes become methylated at certain cytosine residues and the resulting methylcytosine stabilizes nucleosomes and prevents trascription factors from binding. *See:* **cytosine methylation**.

DNA microarrays (DNA chips)   a chip containing many tiny spots of DNA, used as a dot blot to measure the expression of many genes at once; are fabricated by high-speed robotics, generally on glass but sometimes on nylon substrates, for which probes with known identity are used to determine complementary binding, thus allowing massively parallel gene expression and gene discovery studies. An experiment with a single DNA chip can provide researchers information on thousands of genes simultaneously—a dramatic increase in throughput. There are two major application forms for the DNA microarray technology: (1) identification of sequence (gene/gene mutation); and (2) determination of expression level (abundance) of genes. A "probe" is the tethered nucleic acid with known sequence, whereas a "target" is the free nucleic acid sample whose identity/abundance is being detected.

DNA microchip   a chip with many oligonucleotides built in separate tiny spaces on the chip, used as a dot blot to measure the expression of many genes at once.

DNA replication   the process by which a DNA molecule produces two identical daughter DNA molecules; each strand of the original DNA molecule acts as a template for the formation of a new, complementary strand (semiconservative replication). The basis of this complementarity is specific hydrogen bond formation between the nitrogenous bases of nucleotides according to the base-pairing rules.

DNA synthesis   the formation of DNA.

DNase footprinting   a method of detecting the binding site for a protein on DNA by observing the DNA region that this protein protects from degradation by DNase.

DNase hypersensitive sites   regions of chromatin that are approximately 100 times more susceptible to attack by DNase I than bulk chromatin; these usually lie in the 5′-flanking regions of active or potentially active genes.

DNase protection assay   an assay used to confirm the results of a gel mobility shift assay; if a DNA-binding protein binds to its DNA target sequence and the complex is incubated in a solution of DNase I, the DNA-binding protein (e.g., transcription factor) will protect the DNA target sequence from being cleaved by the enzyme. *See:* **gel mobility shift assay**.

Dolly   Dolly the sheep is the first mammal cloned from a cell from an adult animal; she was a result of a collaborative experiment between the Roslin Institute and PPL Therapeutics to test the suitability of different sources of cells for nuclear transfer. She was derived from cells that had been taken from the udder of a 6-year-old Finn Dorset ewe and then cultured in the laboratory; individual cells were then fused with unfertilized eggs from which the genetic material had been removed, and 29 of these "reconstructed eggs," each now with a nucleus from the adult animal, were then implanted in surrogate Blackface ewes. One gave rise to a live lamb, Dolly, some 148 days later. Other cloned lambs were derived in the same way from cells taken from embryonic and fetal tissue.

domain   a specific region of a larger structure with a specific function, e.g., the DNA-binding domain of a steroid hormone receptor (a protein), the homeobox domain of a homeotic gene (DNA), and the apical domain of the plasma membrane of an epithelial cell. Developmental domains are each characterized by a unique profile of zygotic gene activity. Also, the eukaryotic genome is, apparently, divided into relatively independent developmental regions called domains. *See:* **maternal effect genes**.

domains of commitment   made possible by the beginning of zygotic transcription and visualized by *in situ* hybridization for specific transcription factors; e.g., the entire mesoderm can be visualized by the expression of *brachyury* (needed to activate later mesoderm genes and to control gastrulation movements); the organizer region (Spemann's organizer) is characterized by the expession of a variety of genes including *siamois, goosecoid, not,* and *lim 1*; the ventral mesoderm expresses the homeobox genes *vent1* and *vent 2, vent 1 + vent2* specify lateral plate mesoderm, and *vent2* only specifies somitic mesoderm; the endoderm can be visualized as the domain of transcription factors including *mix1* and *sox17*.

dominant   descriptive of one allele, of a nonidentical pair of alleles, which is expressed.

dominant negative mutation   a mechanism of dominance; the mutant form of the gene product itself has no function, but it interferes with the function of the wild-type gene product, as in dimer formation; e.g., mutation of the gene for fibrillin (a secreted glycoprotein that forms multimeric microfibrils in elastic connective tissue), which causes Marfan syndrome. *See:* **gain-of-function mutation, haploinsufficiency**.

dominant negative protein the protein, resulting from the mutant form of the gene, which itself has no function, but interferes with the function of the wild-type gene product. *See:* **dominant negative mutation**.

dominant-recessive inheritance the form of inheritance wherein dominant (expressed) and recessive (not expressed) forms (alleles) of genes are involved.

dormancy the embryos and larvae of many animals can curtail development when conditions are not favorable for growth and survival; e.g., brine shrimp embryos can undergo dormancy under dehydrating conditions and come "back to life" when water is available. *See:* **seed dormancy**.

dorsal an adjective referring to the dorsum or back of an animal.

dorsal a maternal gene in *Drosophila* that encodes dorsal protein, which is a transcription factor with a vital role in organizing the dorsoventral axis of the embryo. The embryo initially becomes divided into four regions along the dorsoventral axis, amnioserosa, dorsal ectoderm, ventral ectoderm, and mesoderm; patterning along this dorsoventral axis is controlled by the distribution of dorsal protein. Initially uniformly distributed in the cytoplasm, dorsal protein, under the influence of signals from the ventrally activated Toll receptors, enters the nuclei in a graded fashion, with the highest concentration in the ventral nuclei. Embryos lacking the Toll receptor are "dorsalized," no ventral structures develop; in the absence of another maternal gene product, the cactus protein, almost all of the dorsal protein is found in the nuclei and the embryos are "ventralized," no dorsal structures develop.

dorsal aortae these originate as paired blood vessels throughout their lengths; with development, they fuse in the midline to form the unpaired dorsal aorta (this occurs first where the common cardinal veins cross the course of the aorta, then progresses cephalad and caudad; cephalically this fusion never extends to the pharyngeal region, but caudally fusion involves the whole length of the aorta) (see Fig. 19).

dorsal axial mesoderm notochord and somites (see Fig. 19). *See:* **mesodermal mantle**.

dorsal closure a process that brings together the two parts of the epidermis of the *Drosophila* embryo at the dorsal surface.

dorsal determinant one of two determinants found in the fertilized *Xenopus* egg; initially localized at the vegetal pole, it shifts to the dorsal side during cortical rotation; causes the formation of the organizer in the region that will become the dorsal lip of the blastopore. *See:* **ventral determinant**.

dorsal lip of the blastopore that portion of the blastopore of amphibians over which cells pass that will sequentially become the roof of the archenteron and the notochord; the "organizer" of classical, experimental embryology. In fish embryos, the dorsal lip of the blastopore is a small, crescent-shaped, opening at the caudal end of the embryonic shield.

dorsal mesocardium a membrane, derived from a double layer of splanchnic mesoderm, which for a time suspends the developing heart in the pericardial region of the coelom.

dorsal mesoderm also called the paraxial mesoderm, somitic mesoderm, segmental mesoderm, and epimeric mesoderm; the mesoderm that becomes segmented to form somites.

dorsal protein the product of the *dorsal* gene in *Drosophila*; it is a transcription factor, the gradient of which acts as a morphogen along the dorsoventral axis of the embryo.

dorsalization induction of dorsal-type mesoderm. *See:* **vegetalization**.

dorsalizing factors those molecules that cause dorsalization, e.g., Xwnt8, activin, Vg1, and Siamois.

dorsoventral axis the dorsal (back, top) to ventral (belly, bottom) axis. *See:* **axes**.

dorsum the back of an animal.

dosage compensation the equalization of specific genes products in male and female cells, e.g., inactivation of an X chromosome in female mammalian cells so that both male and female cells will have equal amounts of X chromosome gene products.

double assurance hypothesis an engineering term, used by Hans Spemann (in 1938) in the sense that the embryo uses more than one mechanism to accomplish its ends; he applied it to vertebrate eye development, specifically lens induction. More recently, it has been applied to the embryo's possible use of both planar and vertical signals to induce its nervous system.

double fertilization in angiosperms, the pollen tube brings two spermatozoa to the embryo sac; one fertilizes the egg to form the diploid zygote, which gives rise to the embryo, while the other "fertilizes" the binucleated central cell to form the triploid endosperm. Generally restricted to angiosperms, but also found in the gymnosperms *Ephedra* and *Gnetum*, where no endosperm is formed.

double-negative gate   activation of the genes of specification by repression of a repressor; activation by the repression of a repressor.

down-mutation   a mutation, usually in a promoter, that results in less expression of a gene. *See:* **up-mutation**.

downregulation   reduction in responsiveness of a cell to a stimulus after the first exposure, this is often produced by a reduction in the number of available receptors expressed on the surface that can result from internalization of the ligand receptor complex or from decreased expression of the receptor; e.g., the cell surface receptors for several growth factors are internalized after growth factor binding and eventually degraded in lysosomes, the effect of this process is to remove the receptor-ligand complexes from the plasma membrane terminating the response of the cell to growth factor stimulation. Process that decreases ligand and receptor interactions or reduces the responsiveness of a cell to a stimulus following first exposure. Classically the concept referred to hormone receptors but contemporary usage includes other cell surface receptors. *See:* **upregulation**.

downstream   (1) in terms of a signal transduction pathway (cascade), downstream would be further from the cell surface/receptor; e.g., in the MAK-kinase pathway, Raf is downstream from Ras; and (2) in terms of a gene, other genes regulated by the gene in question; HOM genes regulate a diverse and large number of downstream genes. *See:* **upstream**.

Down's syndrome   name of a syndrome, including mental retardation, resulting from an extra (trisomy) chromosome 21 (see Fig. 24).

Driesch, Hans (1867–1941)   his goal was to reduce embryology to the laws of physics and mathematics; he carried out *isolation* experiments (in 1892) with sea urchin embryos; he isolated blastomeres from the 2-cell, 4-cell, and 8-cell stages and found that individual isolated blastomeres developed into a complete larvae; this result was not expected from the early experimental work of Wilhelm Roux. Nor was it consistent with August Weismann's concept of the continuity of the germplasm. Driesch had discovered regulative development; the momentous consequences of his experiments included (1) the demonstration that the prospective potency of an isolated blastomere was greater than its prospective fate; (2) the conclusion that the sea urchin embryo is an "equipotential harmonious system"; and (3) the conclusion that the fate of a nucleus depended solely on its location in the embryo. Driesch eventually came to believe that development could not be explained by physical forces; rather, he invoked a vital force, entelechy; he renounced the study of developmental physiology and became a professor of philosophy, proclaiming vitalism until his death in 1941. Like Roux he was a student of Ernst Haeckel. *See:* **Horstadius, Sven**; **Roux, Wilhelm**; **Weismann, August**.

Drosophila   a genus of dipteran insects, widely used in pioneering studies of genetics; more recently, it has been used to elucidate mechanisms of developmental genetics; the control of development by genes is better understood in the fruit fly, *Drosophila*, than in any other organism.

dry sperm   collected sperm (actually semen, a liquid), as from a sea urchin by the induced shedding of gametes, which has not yet been mixed with sea water; once mixed with sea water, the spermatozoa remain viable for only a short time.

dual gradient model   the assumption that in the sea urchin egg two factors or principles exist that are mutually antagonistic and yet interact with each other at the same time, and that normal development is dependent on a certain equilibrium between the two principles. This theory, according to Boris Balinsky, was originally suggested by Theodor Boveri in 1910 and developed in application to the sea urchin egg by J. Runnstrom (in 1928) and Sven Horstadius (in 1928).

duct of Cuvier   either of the paired common cardinal veins in a vertebrate embryo. *See:* **cardinal veins**.

duct of Santorini   the dorsal pancreatic duct in a vertebrate embryo. *See:* **digestive system**.

duct of Wirsung   the ventral pancreatic duct in a vertebrate embryo. *See:* **digestive system**.

ductus arteriosus   a blood vessel shunt between the pulmonary trunk and arch of the aorta of the mammalian embryo, which shunts blood past the undeveloped lungs into the systemic circulation, so that the right ventricle may get the exercise required for its normal development; the ductus arteriosus is obliterated between the first and fifth days after birth, becoming the ligamentum teres.

ductus choledochus   the common bile duct.

ductus deferens   *See:* **vas deferens**.

ductus venosus   a small branch of the umbilical vein, in the fetus, that bypasses the liver and enters the inferior vena cava; the ductus venosus is obliterated between the first and fifth days after birth, becoming the ligamentum venosum of the liver.

duodenum   the first (most cephalic) portion of the small intestine.

duplex uterus   in many marsupials, there is no fusion of the embryo's mullerian ducts and the female tract is double all the way to the urogenital sinus; consequently, marsupials are said to have a duplex uterus; and they also have paired vaginas.

duplication   a process that allows for the formation of redundant structures; as in gene duplication or chromosome duplication. *See:* **divergence**.

dyad   a group of two.

dynamic anatomy   dramatically changing anatomy, as in the embryo or fetus, as opposed to the adult.

dynein   an ATPase that powers the beating of flagella, including those of spermatozoa. *See:* **motor proteins**.

dysgenesis   defective development.

dysplasia   abnormal development or growth, e.g., chondroectodermal dysplasia, a dysplastic state of skin and cartilage.

E

E-cadherin   also called epithelial cadherin, a cell adhesion molecule expressed on all early mammalian embryonic cells, even at the one-cell stage; later restricted to epithelial tissues of embryos and adults.

ecdysis   *See:* **molting**.

ecdysone   the insect molting hormone; a steroid hormone synthesized by the prothoracic gland; actually, ecdysone is a prohormone that is converted in other tissues to its active form, 20-hydroxyecdysone. Each molt is triggered by a surge of ecdysone in the hemolymph of the insect (the character of the molt is determined by the concentration of juvenile hormone).

ecdysone receptors   the receptors specifically binding 20-hydroxyecdysone.

Ecdysozoa   animals whose bodies are covered by an exoskeleton; a great clade of protostomes, including the arthropods, nematodes, onychophora, and priapulids; named for the shared characteristic of molting.

Echinoderm Theory   of the origin of vertebrates; holds that echinoderms, hemichordates, and chordates diverged from a common ancestral type more than 600 million years ago. Also called Hemichordate Theory.

eclosion   the hatching of the adult insect from its pupal case.

ECM   *See:* **extracellular matrix**.

eco-evo-devo   *See:* **ecological evolutionary developmental biology**.

eco-devo   *See:* **ecological developmental biology**.

ecological developmental biology   an approach to embryonic development that studies the interactions between a developing organism and its environment; postulates that we are defined, in part, by the "other"; it depicts our identities as *becoming with* the "other."

ecological evolutionary developmental biology   the integration of ecological developmental biology into evolutionary biology.

ecomorph   *See:* **morph**.

ecotype   a subspecies whose phenotype is characteristic of the species in one particular geographic range. *See:* **ecotype**, **genetic assimilation**, **phenocopy**.

ectoderm   the outer of the three primary germ layers; literally, "outer skin." In vertebrate embryos, the original ectoderm gives rise to three sets of cells: (1) neural tube, (2) epidermis, and (3) neural crest cells.

ectodermal placodes   circumscribed thickenings of the ectoderm of the head that contribute to sensory structures and cranial ganglia.

ectodysplasin (EDA) cascade   a gene cascade specific for cutaneous appendage formation.

ectogenesis   development of an embryo or of embryonic tissue outside the body in an artificial environment, e.g., preimplantation mouse embryo culture.

ectogony   in plants, the influence of pollination and fertilization on structures outside the embryo and endosperm; effect may be on color, chemical composition, ripening, or abscission.

ectomere   a blastomere that will differentiate into ectoderm.

ectomesenchyme   mesenchyme produced from the ectoderm.

ectopic embryos   the plant embryos that originate from fertilization or from apomixis are found inside the ovary; however, plant embryos can also form from somatic tissue that has received the right balance of hormones in tissue culture.

ectopic kidney   a congenital anomaly in which the kidney is held in an abnormal position.

ectopic pregnancy   a pregnancy that becomes established in an abnormal location, such as the fallopian tube or abdominal cavity, as opposed to the normal uterine location.

Dictionary of Developmental Biology and Embryology, Second Edition. Frank J. Dye.
© 2012 Wiley-Blackwell. Published 2012 by John Wiley & Sons, Inc.

Types of Eggs

Yolk Amount (Terminology)	Yolk Distribution (Terminology)	Vertebrate Examples*
Large (macrolecithal)	Nonuniform (telolecithal)	Fish, reptiles, birds
Moderate (mesolecithal)	Nonuniform (telolecithal)	Amphibians
Small (microlecithal)	Uniform (isolecithal)	Mammals

*Eggs are classified according to their amount and distribution of yolk.

Figure 14. Types of eggs. Eggs are classified according to their amount and distribution of yolk. Reprinted from Frank J. Dye, *Human Life Before Birth*, Harwood Academic Publishers, 2000, table 8-2, p. 61.

ectoplacental cone   a cone-shaped mass of small, mitotically active, cells that originates from cells overlying the inner cell mass of the mouse blastocyst; the proliferative center of the placental trophoblast, which is most conveniently transplanted to produce extrauterine growths of pure trophoblast.

edema   swelling of part(s) of the body as a result of retention of water.

Ediacara   the fauna that lived 575–544 million years ago; named for their fossil deposits in the hills of South Australia.

Edwards, Robert (1925–)   (Britain) recipient of the 2010 Nobel Prize for Physiology or Medicine for the development of human *in vitro* fertilization (IVF) therapy.

effacement   loss of form; as of the uterine cervix during labor.

efferent ductules   as applied to the epididymis, the tiny tubules emerging from the testis and joining the duct of the epididymis (see Fig. 15).

EGF   *See:* **epidermal growth factor**.

EGF receptors   epidermal growth factor receptors; are tyrosine kinases and activate the MAP kinase pathway.

egg   or ovum, the female gamete (see Figs. 14 and 22).

egg activation   stimulation of the egg, by natural or artificial means, to begin development. *See:* **fertilization**, **parthenogenesis**.

egg capsules   containers of eggs/embryos/larvae providing a protective physical barrier, especially found protecting developing gastropods.

egg chamber   the ovary in which the *Drosophila* oocyte will develop, containing 15 interconnected nurse cells and a single oocyte.

egg chemistry   in hens' eggs, the egg white is almost all water (88%) and protein (11%); the difference between the semifluid raw egg and the semirigid cooked one is that heating drives the natural (native) globular protein molecules into a network (denatured); salt or acid can change the electrical environment of the globular protein molecules so that they become denatured spontaneously and the protein cooks a lot faster; if the egg cracks, the egg white is cooked at the crack and seals it right away; salting the water is a sort of first aid for a ruptured egg. As the egg is warmed, during cooking, a small portion of the protein in the egg white decomposes and its sulfur and hydrogen unite to form hydrogen sulfide gas ("the smell of rotten eggs"); this gas generated in the egg white collects in the coolest part of the egg, which during cooking is in the center where the yolk is found; yolks contain iron, which has a terrific attraction for sulfur; the iron kicks out the hydrogen and forms solid iron sulfide, which results in a dark iron sulfide deposit at the surface of the yolk (the dreaded green yolk). The cold water rinse; by chilling the shell in cold water, the hydrogen sulfide is forced there, pulling it away from the iron-laden yolk (Grosser, 1983).

egg coats   (1) coverings surrounding an egg when it is ovulated from the mammalian ovary. Around human eggs, two such egg coats are found: the outer, cellular, corona radiata, and the inner, noncellular, zona pellucida. (2) Such coats are also found surrounding the eggs of nonmammalian species, e.g., the vitelline membranes and jelly coats of frog and sea urchin eggs.

egg cylinder   a stage in mouse development (6 days); also, at this time, the cylindrical structure containing both the epiblast and the extraembryonic tissue derived from the polar trophectoderm.

egg follicle   the complex consisting of an oocyte and its surrounding layer of follicle cells.

egg grooming   a type of parental behavior that protects embryos against pathogens, e.g., prevents fungal growth over the developing eggs. *See:* **embryo defenses**.

egg jelly   a glycoprotein meshwork outside the vitelline envelope in many species; most commonly it is used to attract and/or to activate sperm.

egg physics   if there is a question about whether hens' eggs are cooked, all one has to do is twirl the eggs on a table, those that spin faster are the hard-cooked eggs. The yolk of a hard-cooked egg is spinning quite close to the vertical axis of rotation; however, when a raw egg is spun, the yolk, much denser than the egg white, moves toward the shell, away from the axis of rotation, increasing the moment of inertia; as the mass moves away from the axis of rotation, increasing the average radius, the rotational frequency must get smaller (Grosser, 1983).

egg shell   in the chicken, consists of calcite (a form of calcium carbonate) and a sparse matrix of protein; the main part of the shell is the palisade layer (once called the spongy layer), which consists of columns of tightly packed calcite crystals, from the inner mammillary knobs to the shell cuticle; occasional pores are found through the palisade layer. The outer shell membrane has anchoring fibers that attach to the mammillary knobs. A hen may lay five to seven eggs per week; the average shell weighs 5 grams; 40% of which (2 grams) is calcium laid down during the final 16 hours of calcification = 125 milligrams/hour; the total calcium in the hen's blood is 25 milligrams = 12 minute supply; the immediate source of the calcium is the blood, and the ultimate source is food; however, because calcium is not absorbed from the intestine fast enough, the deficit is made good by the skeleton. A hen can mobilize 8–10% of the total amount of calcium in its skeleton in 15–16 hours; this is in association with a system of secondary (medullary) bone in marrow cavities, produced in hen bones by the influence of estrogens and androgens. A suggested control mechanism is a drop in the level of plasma calcium causing a release of parathyroid hormone, leading to resorption of bone tissue by osteoclasts and enlarged osteocytes. A couple of reasons why egg shells crack: (1) if the air space is heated faster than the air can escape through the shell, a terrific internal pressure builds up and the shell cracks, and (2) if the egg goes from the refrigerator into hot water, there is a thermal shock; different regions of the shell cannot expand in unison and the shell cracks (Grosser, 1983).

egg tube   *See:* **ovariole**.

egg water   sea water in which the jelly coats of sea urchins were allowed to dissolve; this egg water seemed to be able to agglutinate the spermatozoa of the same species; played an important role in the fertilizin theory of fertilization.

eggs   chicken eggs: brown-shelled eggs come from brown-feathered hens (porphyrins); two "yolks" are attributed to double ovulation and cause abnormal development because there is not enough food or space; a speck of blood in an egg is from a blood clot at the time of ovulation; the air space increases in size with the age of the egg.

einsteckung   a procedure for combining competent ectoderm with organizer tissue, where bits of amphibian organizer tissue are inserted into the blastocoel of an early gastrula and become pressed against the ventral ectoderm by gastrulation movements; Hans Spemann and Otto Mangold devised the *einsteck* method of transplantation, wherein a transplant could be slipped into the blastocoel of the gastrula through a small slit in the upper hemisphere. During the ensuing gastrulation movements, the blastocoel narrows, and the transplant adheres to the inner layer of the ventrolateral ectoderm. If it has neural inducing ability, it should induce a secondary neural plate.

ejaculate   a portion of semen ejected from the penis during one ejaculation event. The usual ejaculate in a fertile man is approximately 2–6 mL, with a concentration of anything from 40 to 250 million spermatozoa/mL. Men may be fertile even though their sperm counts may be as low as 1 million/mL; however, the presence of many immotile or abnormally shaped spermatozoa in the ejaculate is usually a good indication of male infertility. The quantity of ejaculate varies greatly from 0.1 mL in the mouse to more than 100 mL in the boar. The numbers of spermatozoa ejaculated and sites of deposition for a number of species are mouse, 50 million (uterus); cattle, 3 billion (vagina); and man, 280 million (vagina).

ejaculation the process of discharging one portion of semen from the penis.

ejaculatory ducts the pair of highly muscularized ducts that pierce the substance of the prostate gland, thereby providing a passageway between the vasa deferentia and the urethra. In male insects, the paired vasa deferentia join the anterior end of the ejaculatory duct, which opens onto the tip of the penis.

electron microscope a type of microscope that uses a beam of electrons rather than light; provides great detail in images.

electroporation a method for incorporating substances (e.g., genes) directly into cells by using a high-voltage pulse to "push" the substance into cells. *See:* **microinjection**, **transfection**.

emboitement or encapsulation, the doctrine that within the first female of every species were contained the germs of all subsequent offspring, each generation within the previous one, like a nest of Chinese boxes. *See:* **Bonnet, Charles**; **encapsulation**.

emboly formation of a gastrula by the process of invagination.

embryo the early developing animal before it begins to look like the adult of the species; the young plant within a seed (see Fig. 22).

embryo bank a collection of embryos kept in cryogenic storage.

embryo body plan the following embryonic basic body plans are approximately equivalent: pig embryo of 5 mm (the pig embryo reaches 5 mm at approximately 17–18 days), human embryo of just over 1 month, and the chick embryo of 4–4.5 days. In birds, the eyes and midbrain (which will contain the optic centers) are relatively larger and further developed. The heart, liver, and mesonephros are mopre conspicuous and advanced in mammalian embryos.

embryo defenses *See:* **antifungal symbiotic bacteria**, **bryostatin**, **defensins**, **egg grooming**, **lysozyme**, **protective cloning**, **orphan embryos**.

embryo sac the female gametophyte (megagametophyte) of angiosperms; a structure, found in the ovule of the ovary, consisting of seven cells and eight nuclei: one egg, two synergids, three antipodal cells, and one binucleate central cell. The end of the embryo sac, where the egg is located and next to the micropyle, through which the pollen tube grows, is the micropylar end of the embryo sac; the opposite end, where the antipodal cells are found, is the chalazal end of the embryo sac.

embryo transfer (ET) the artificial movement of an embryo, usually from an IVF site, into the uterine cavity, with the expectation of implantaion of the embryo into the endometrium (see Fig. 22). *See:* *in vitro* **fertilization (IVF)**.

embryoblast the inner cell mass; that portion of the blastocyst that actually gives rise to the embryo (see Plate 7 in the color insert).

embryogenesis the formation of the embryo; development from the ovum, as opposed to blastogenesis. In plants, covers development from the time of fertilization until dormancy occurs.

embryogeny the origin and growth of the embryo.

embryologist a biologist who specializes in the study of embryology or developmental biology.

embryology the sudy of development; traditionally concerned with the phase of development before metamorphosis in amphibians, hatching in birds, and birth in mammals. According to E. J. Gardner (1965), embryology did not come into its own as a science until the 19th century.

embryonal carcinoma (EC) *See:* **teratocarcinoma**.

embryonal region *See:* **embryonic shield**.

embryonate containing an embryo; as an embryonated hen's egg.

embryonic an adjective referring to the embryo.

embryonic-abembryonic axis in mammalian blastocysts runs from the site of attachment of the inner cell mass (the embryonic pole) to the opposite pole.

embryonic axis (1) collectively, the set of mostly dorsal organ rudiments (neural tube, notochord, laterally adjacent somites and other mesodermal structures) in vertebrates; either formed or induced by the dorsal lip of the blastopore in the early gastrula. (2) A line of orientation such as the anteroposterior, or animal-vegetal, axis of an embryo.

embryonic cell cycle an abbreviated cell cycle, consisting of only S and M phases, as opposed to the canonical cell cycle of eukaryotic cells, consisting of G_1, S, G_2, and M. *See:* **cell cycle**.

embryonic disc   the early embryo when it consists of only the two or three germ layers.

embryonic ectoderm   the outermost germ layer of the embryo; the name given to the mouse epiblast once it has developed into an epithelial sheet.

embryonic endoderm   the innermost germ layer of the embryo.

embryonic germ (EG) cells   primordial germ cell (PGC)-derived cells; pluripotent embryonic stem cells, produced in culture by treating PGCs with stem cell factor, leukemia inhibition factor, and basic fibroblast growth factor (FGF2). EG cells have the potential to differentiate into all the cell types of the body; they are often considered to be embryonic stem (ES) cells, ignoring the distinctiveness of their origin.

embryonic induction   a developmental phenomenon wherein one part of the embryo determines the fate of another part of the embryo. As an example, inductions are involved in the early development of *Xenopus*, as follows: (1) the animal-vegetal axis develops during oogenesis; (2) fertilization causes cytoplasmic rearrangements that subdivide the vegetal region into dorsovegetal (DV) and ventrovegetal (VV) areas; (3) during cleavage, mesodermal *induction* occurs such that the DV region (Nieuwkoop center) induces the organizer activity (O) in the dorsal marginal cells above it, whereas the VV induces the cells above it to become ventral mesoderm (M4); (4) a signal from the organizer *induces* the ventral mesoderm near it into lateral mesoderm (M2, M3, M4); (5) during gastrulation, the ventral and lateral mesoderm go to the sides of the gastrula, while the dorsal mesoderm expands and *induces* polarity in the ectodermal cells, this causes the ectodermal cells to become the different regions of the neural tube (N1, N2, N3, N4); and (6) the uninduced ectoderm becomes epidermis. *See:* **mesoderm induction**, **Nieuwkoop center**, **organizer**.

embryonic mesoderm   the germ layer of the embryo found between the outer ectoderm and the inner endoderm.

embryonic period   the first 8 weeks of human development during which the developing organism does not resemble a human.

embryonic pole   that region of the blastocyst where the inner cell mass is located.

embryonic shield   a shield-shaped, thickened region of the fish blastoderm anterior to the dorsal lip of the blastopore. In chick embryos, the thickened part of the area pellucida adjacent to the primitive streak is called the embryonic shield or embryonal region.

embryonic stem cells (ES cells)   stem cells derived from the embryo, e.g., derived from normal mouse inner cell mass cells and cultured *in vitro*. Use of these cells to make chimeras allows investigators to introduce an engineered gene with a known mutation into ES cells and then to breed mice that have the same mutation in all their cells.

embryopathy   any abnormal development of an embryo, either morphological or biochemical.

embryophytes   (1) literally, "embryo plants"; an older term used in plant classification to encompass the Bryophyta, Pteridophyta, and Spermatophyta; In these plants, the zygote produces an embryo, which remains attached to and secures food from the haploid plant (gemetophyte) that produced the zygote. Embryophytes are grouped together on the basis of at least temporarily producing a dependent multicellular structure, the embryo. Encompassing the bryophytes and tracheophytes; plants in which the zygote gives rise to an embryo within protective tissues of the female structure, e.g., archegonium or embryo sac. (2) A synonym for plants.

embryotomy   any mutilation of the fetus in the uterus to aid in its removal when natural delivery is not possible.

embryotroph   as primary villi grow out of the chorion, they destroy adjacent maternal tissue, producing a liquefied material, embryotroph, which may nourish the embryo. *See:* **implantation**.

emulsify   to bring a lipid (e.g., a phospholipid) into aqueous (watery) suspension by means of molecules that are partially soluble in oil and partially soluble in water.

enamel hypoplasia   underproduction of the enamel of the teeth that may be caused by tetracycline (antibiotic) use, as well as by other factors.

enamel knot   a group of epithelial cells that makes up the signaling center for tooth development; first observed at the beginning of the cap stage of tooth germ development, as a nondividing population of cells in the center of the growing cusps. The enamel knot is the source of sonic hedgehog, FGF4, BMP7, BMP4, and BMP2 secretion.

enamel organ   the epithelial portion of a tooth germ that secretes enamel.

encapsulated nerve endings   to have the nerve terminations wrapped in connective tissue. *See:* **free nerve terminations**.

encapsulation   emboitement. *See:* **Bonnet, Charles**; **emboitement**.

-encephaly   this word part refers to the brain; in addition to the mental retardation associated with many chromosome abnormalites, structural malformations of the brain are represented in human development.

end bud   *See:* **tail bud**.

endoblast   *See:* **hypoblast**.

endocardial cushion tissue   a type of embryonic connective tissue that plays an important role in heart development. It divides the atrioventricular canal into two channels, it contributes to the formation of the interventricular septum, and it provides the valves of the heart.

endocardial primordia   loosely associated cells detach from the medial face of the splanchnic mesoderm, along the margins of the anterior intestinal portal, and give rise to a pair of tubular structures, the endocardial primordia. *See:* **endocardial tube**.

endocardial tube   the first cardiac tube formed. It originates from mesenchyme cells, derived from the medial faces of splanchnic mesoderm in the region of the amniocardiac vesicles, which first organize themselves into a pair of tubes, which then fuse into the single endocardial tube that becomes the lining of the heart (see Fig. 19).

endocardium   the endothelial lining of the heart.

endochondral ossification   ossification that occurs within cartilage.

endocrine cells   cells that produce hormones.

endocrine disruptors   exogenous chemicals that interfere with the normal function of hormones, e.g., diethylstilbesterol.

endocrine factors   molecules (hormones) that function in endocrine signaling, e.g., estrogens, testosterone, and progesterone.

endocrine mimics (endocrine disruptors)   exogenous substances that act like hormones in the endocrine system and disrupt the functions of endogenous hormones, e.g., DDT, PCBs, bisphenol A, and phthalates.

endocrine signaling   a mode of cell-cell communication in which signaling molecules (endocrine factors) are released into the circulatory system that may affect cells (target cells) that are some distance from the signaling cell, e.g., the effect of the anterior lobe of the pituitary gland hormones on the gonads. *See:* **autocrine signaling**, **juxtacrine signaling**, **paracrine signaling**.

endoderm   also known as entoderm; the inner of the three primary germ layers; literally, "inner skin."

endoderm determinant   *vegT* mRNA appears to act as a determinant for the endoderm in *Xenopus*: (1) it is localized to the prospective endoderm, (2) it will induce endodermal markers if injected into other parts of the embryo, and (3) antisense ablation of maternal *vegT* mRNA in oocytes, followed by maturation and fertilization, will produce embryos lacking endoderm.

endolymphatic duct   the dorsal evagination of the otocyst, in the embryo, which gives rise to the endolymphatic sac.

endomere   a blastomere that forms endoderm.

endomesoderm   pharyngeal endoderm and head (prechordal) mesoderm; constitutes the leading edge of the amphibian dorsal blastopore lip (the most anterior portion of the organizer); endomesoderm induces the most anterior head structures, by blocking the Wnt pathway and BMP4.

endometriosis   the disease resulting from the migration of portions of the endometrium into abnormal locations, e.g., pelvic cavity or fallopian tubes.

endometrium   lining of the uterus (see Figs. 27, 44, and 48).

Endopterygota   also Holometabola; insects passing through a complex metamorphosis, always accompanied by a pupal instar. The wings develop internally, and the larvae are usually specialized. *See:* **Holometabola**.

endoskeleton   the skeleton found inside the body, as in humans; as opposed to on the surface of the body as in insects (exoskeleton).

endosperm   nutritive tissue found in angiosperms, resulting from the fertilization of the binucleated central cell, of the embryo sac, by a spermatozoan; this triploid tissue provides nourishment for the developing embryo. It is digested by the growing sporophyte either before or after the maturation of the seed. It has been suggested that endosperm may be regarded as a derivative of a second embryo, modified during the course of evolution to perform a new function. There is evidence that the endosperm genome influences the zygote. *See:* **double fertilization**.

endosperm cell   the triploid cell that gives rise, by growth and cell division, to the nutritive endosperm tissue; which provides nourishment for the developing angiosperm embryo.

endosteal osteoblasts   osteoblasts that line the bone marrow and are responsible for providing the niche that attracts hematopoietic stem cells (HSCs), prevents apoptosis, and keeps the HSCs in a state of plasticity.

endosymbiosis   symbiosis where one partner (a cell) lives inside another partner (also a cell); examples of developmental endosymbiosis involve bacteria of the genus *Wolbachia* and arthropods. *See also:* **cytoplasmic incompatibility**.

endothelial cells   the cells lining the heart, blood vessels, and lymph vessels; in heart development, endothelial cells first become arranged in the form of two parallel tubes, which soon become fused into a single tube, which becomes the lining of the heart, the endocardium.

endotheliochorial placenta   a type of placenta in which the maternal blood is separated from the chorion by the maternal capillary endothelium; occurs in dogs.

endothelins   small peptides secreted by blood vessels that have a role in vasoconstriction and can direct the extension of certain sympathetic axons that have endothelin receptors, e.g., targeting of neurons from the superior cervical ganglia to the carotid artery.

energids   the nuclei and their associated cytoplasmic islands of the syncytial blastoderm of the early *Drosophila* embryo.

engagement   one of the seven cardinal movements of labor; the entrance of the presenting part of the fetus into the upper pelvic passage (see Fig. 40).

engrailed   a segment polarity gene, in *Drosophila*, activated by pair-rule genes; it encodes a transcription factor, which is expressed in the anterior region of every parasegment. The expression of this gene delimits a boundary of cell lineage restriction. *Engrailed* is also a selector gene. The *engrailed* gene plays a key role in segmentation and is expressed throughout the life of the fly; its activity first appears at the time of cellularization as a series of 14 transverse stripes. The *engrailed* gene needs to be expressed continuously throughout larval and pupal stages and into the adult to maintain the character of the posterior compartment of the segment; thus, *engrailed* is also a selector gene, i.e., a gene whose activity is sufficient to cause cells to adopt a particular fate. *Engrailed* is turned on by the pair-rule genes, and its expression is maintained by the segment polarity genes, *wingless* and *hedgehog*, which stabilize the compartment boundary.

enhanceosome   the complex formed by enhancers coupled to their activators.

enhancer   a regulatory sequence of nucleotides in DNA to which gene regulatory proteins bind and influence the rate of transcription, by maximizing the levels of mRNA transcription, of a structural gene that may be thousands of base pairs away. Enhancers act on promoters and may be located upstream or downstream of the promoter; they are used in tissue specific gene expression.

enhancer modularity   the DNA regions that form the enhancers of genes are modular; i.e., not only anatomical units are modular; indeed the modularity of development is determined by the modularity of the gene enhancers; each enhancer element allows the gene to be expressed in a different tissue; e.g., the *Pax6* gene is expressed in the pancreas, the neural tube, the retina, the cornea, and the lens; in each of these tissues, there is an enhancer element in the regulatory region of the *Pax6* gene so it can be independently activated in each separate tissue. *See:* **developmental modularity**.

enhancer trap   consists of a reporter gene fused to a relatively weak eukaryotic promoter (the weak promoter will not initiate the transcription of the reporter gene without the help of an enhancer); if this recombinant enhancer trap is introduced into an egg or oocyte and the reporter gene becomes expressed, this means that it has come within the domain of an active enhancer; by isolating this activated region of the genome in wild-type organisms, the normal gene activated by this enhancer can be discovered.

entelechy   an internal goal-directed force.

enterocoel   a coelom formed by enterocoely. *See:* **enterocoely**.

enterocoely   formation of the coelom by budding of the mesodermal rudiment from the gut; true enterocoely is not found in vertebrates, but it is found in some protochordates and echinoderms.

entoblast   a blastomere that differentiates into endoderm.

entoderm   also endoderm. *See:* **endoderm**.

Entwicklungsmechanik   the physiological approach to embryology (developmental mechanics).

enveloping layer   that portion of the zebrafish blastula that covers the deep cells. *See:* **deep layer**, **yolk syncytial layer**.

environmental integration   the influence of cues from the environment surrounding the embryo, fetus, or larva on their development.

environmental mismatch hypothesis   if the predicted and future environments do not match, health risks or other disadvantages may develop; e.g., if the thermal environment (cooler conditions) triggers a minimal number of active sweat glands in human infants, this becomes maladaptive in later life when the adults need to survive in a hot and humid climate. *See:* **predictive adaptive responses**.

enzyme-linked receptors   cell-surface receptors found in plasma membranes, which upon binding their ligand (e.g., a paracrine factor) activate enzyme activity. These receptors are important in developmentally significant intracellular signal transduction pathways. *See:* **serine/threonine kinases**, **tyrosine kinases**.

enzymes   protein (sometimes RNA molecules, but then generally called ribozymes) molecules that increase the rates of biochemical reactions; organic (carbon-containing) catalysts.

EP-cadherin   *See:* **C-cadherin**.

epaxial muscles   muscles formed from those myoblasts, of the dermamyotome, closest to the neural tube; the deep muscles of the back.

ependyma   the cellular membrane that lines the ventricles of the brain and the spinal canal of the spinal cord (see Fig. 43).

ependymal cells   the cells derived from the lining of the neural tube, which make up the ependyma (see Fig. 18).

Eph receptors   cell-surface receptors, used by neural crest cells to recognize the ephrin proteins; binding of ephrin proteins activates the tyrosine kinase domains of the Eph receptors in neural crest cells; these kinases probably phosphorylate proteins that interfere with the actin cytoskeleton, which is critical for cell migration. *See:* **ephrin protein**.

ephrin proteins   the main proteins involved in restriction of neural crest cell migration. *See:* **Eph receptors**.

epialleles   altered chromatin configurations that act just as genetic alleles might act; it does not matter to the developing system whether a gene has been inactivated by a mutation or by an altered chromatin configuration, the effect is the same.

epiblast   the early human embryo is composed of two layers; the upper layer, the epiblast, gives rise to the three germ layers. The fate map for bird and mammalian embryos is restricted to the epiblast; the hypoblast does not contribute any cells to the embryo. The epiblast gives rise to the embryonic ectoderm, embryonic mesoderm, and embryonic endoderm, as well as to the amniotic ectoderm, of human embryos (see Fig. 44). *See:* **hypoblast**.

epiboly   a basic morphogenetic movement; the spreading of cells on a surface, as in gastrulation movements of fish resulting in the envelopment of yolk by cells. In amphibian gastrulation, the major mechanism of epiboly seems to be an increase in cell number coupled with a concurrent integration of several deep layers into one layer. In chick gastrulation, epiboly occurs by migration of presumptive ectodermal cells along the underside of the vitelline membrane.

epicardium   the covering of the heart; the visceral pericardium.

epicotyl   that portion of the angiosperm embryonic axis, above the attachments of the cotyledons (seed leaves).

epidermal cells   cells of the epidermis, which are derived from the ectoderm.

epidermal ectoderm   the portion of the ectoderm that gives rise to the epidermis (see Plate 5 in the color insert). *See:* **neuroectoderm**.

epidermal growth factor (EGF)   a protein growth factor; it stimulates the proliferation of many cell types and acts as an inductive signal in embryonic development.

epidermis   the outer layer of the skin, derived from ectoderm and itself composed of multiple cell layers. *See:* **dermis**.

epididymis   [epi "on" + didymos "testicle"] the proximal portion of the male reproductive duct consisting of several efferent ductules of the epididymis and the epididymal duct. The

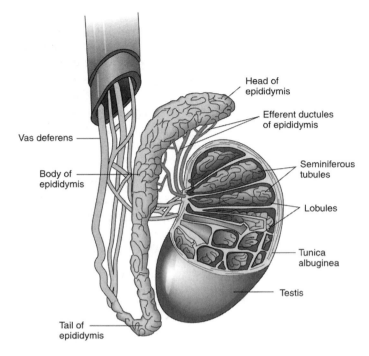

Figure 15. The epididymis. The efferent ductules of the epididymis convey sperm from the testis to the duct of the epididymis. Reprinted from Frank J. Dye, *Human Life Before Birth*, Harwood Academic Publishers, 2000, fig. 5-4, p. 36.

epididymal duct is highly contorted and consists of three parts: caput (head), corpus (body), and cauda (tail) (see Figs. 4 and 15).

epigaster   the portion of the intestine in vertebrate embryos that gives rise to the colon.

epigenesis   each organism is entirely new; the doctrine that the new individual is not already preformed in the spermatozoon or egg, but that development involves the gradual formation of structures from unorganized or undifferentiated material in the egg; historically, it seemed to best explain the variation between generations and the direct observations of organ formation. Aristotle probably originated the epigenesis theory. William Harvey is believed to have coined the word "epigenesis" to refer to the process that drives the egg. According to E. J. Gardner (1965), epigenesis found no 17th-century supporters. Friedrich Wolff, a century after Harvey, became the most celebrated proponent of epigenesis; Wolff was able to observe microscopically the building up of the chick embryo, and he saw no evidence of an encapsulated chick in the egg. *See:* **Harvey, William**; **preformationism**.

epigenetic assimilation   (epigenetic) fixation of an adaptive response even without the presence of genetic variation in the population.

epigenetic change   a change in the expression of a gene, not in the structure of the gene itself.

epigenetic inheritance   heritable phenotypes that are not encoded in the genome; the idea that inherited traits can be passed on through changes in chromatin structure as well as through changes in DNA sequence; a contribution of eco-devo.

epigenetic methylation   *See:* **DNA methylation, epigenetic change**.

epigenetic origin of adult diseases   *See:* **fetal origins of adult disease (FOAD) hypothesis**.

epigenetic progenitor model   an epigenetic hypothesis. Alterations of DNA methylation can prevent the normal functioning of DNA repair genes; without the DNA repair proteins, mutations accumulate; if these mutations prevent the normal functioning of growth regulatory genes, cancers may develop. *See:* **somatic mutation hypothesis, tissue organization field**.

epigenetic theory according to Conrad Hal Waddington, discrete and separate entities of classical genetics are displaced by collections of genes that could "lock in" development through their interactions.

epimere the dorsal part of the mesoderm in chordate embryos.

epimorphic regeneration a type of regeneration in which the new individual is produced by the addition of parts to the remaining piece, as in salamnder limb regeneration. *See:* **compensatory regeneration, morphallactic regeneration.**

epimorphosis See **epimorphic regeneration.**

epimyocardial primordium thickened splanchnic mesoderm, applied to the lateral aspects of the endocardial tubes. *See:* **epimyocardial tube.**

epimyocardial tube derived from the coming together of two layers of splanchnic mesoderm around the endocardial tube; gives rise to the muscular wall of the heart, myocardium, and the covering of the heart, epicardium.

epimyocardium the layer of the embryonic heart from which both the myocardium and the epimyocardium develop.

epinephrine the hormone, produced by the adrenal glands, associated with the "fight or flight" reaction of animals.

epineural sinus in insects, the body cavity begins as the epineural sinus or space that is mainly formed by the separation of the yolk from the embryo in the midventral region; the epineural sinus and most of the coelom sacs becoming confluent form the permanent body cavity.

epiphyseal fusion fusion of an epiphysis with the diaphysis of a long bone, accompanied by ossification of the cartilage of the epiphysis and bringing an end to growth in length of the bone.

epiphyseal growth plates cartilaginous regions of long bone growth in the epiphyses of long bones; as long as the epiphyseal growth plates are able to produce chondrocytes, the bone continues to grow. *See:* **growth plate closure.**

epiphyses the ends of a long bone.

epistasis the phenomenon where the effects of one gene are modified by one or several other genes, which are sometimes called modifier genes; the gene whose phenotype is expressed is said to be epistatic, whereas the phenotype altered or suppressed is said to be hypostatic. Epistasis can be contrasted with dominance, which is an interaction between alleles at the same gene locus.

epithelial cadherin *See:* **E-cadherin.**

epithelial cells cells that are tightly connected to one another in sheets or tubes; one of the two major types of cell arrangements in the embryo. *See:* **cell arrangements.**

epithelialization conversion from mesenchymal tissue into an epithelial tissue.

epithelial-mesenchymal interactions interactions between an epithelium and its underlying mesenchyme, which play a major role in organ formation; as in the interactions between the epithelial enamel organ and the underlying mesenchymal dental papilla in tooth germ morphogenesis. The epithelium may come from ectoderm, mesoderm, or endoderm, and the mesenchyme is generally of mesodermal or neural crest origin.

epithelial-to-mesenchymal transitions occur when cells leave an epithelium and move off as individual cells or as a mesenchymal mass, e.g., formation of the neural crest from the dorsal neural tube. *See:* **mesenchyme-to-epithelium transition.**

epitheliochorial placenta a type of placenta in which the maternal epithelium and fetal epithelium are in contact, exhibited by pigs; also known as villous placenta.

epithelium a layer of cells that lines a cavity or covers a surface.

epoophoron [Etymology: Gk, *epi + oophoron,* ovary] a blind longitudinal duct (Gartner's) and several transverse ductules, in the mesosalpinx near the ovary, which represent remnants of the reproductive part of the mesonephros in the female; homologue of the head of the epididymis in the male.

equator as applied to the spindle of a dividing cell, the imaginary plane located midway between the poles of the spindle; when the centromeres of all the chromosomes of the cell are aligned on the equator, the cell is said to be in metaphase of mitosis or meiosis.

equivalence group in the context of early mammalian development, the inner cell mass (ICM) cells constitute an equivalence group; i.e., each ICM cell has the same potency and their fates will be determined by interactions among their descendants.

ERK extracellular signal regulated kinase; also called MAP kinase.

erythroblast   a member of the erythrocyte lineage, between the earlier proerythroblast and the later reticulocyte; synthesizes enormous amounts of hemoglobin.

erythroblastosis fetalis   destruction of red blood cells by maternal antibodies produced in response to Rh incompatibility.

erythrocyte   the final, fully differentiated, member of the erythrocyte lineage; does not divide, synthesize RNA, or synthesize protein. Leaves the bone marrow and delivers oxygen to the cells of the body.

erythroid progenitor cell (BFU-E)   a committed stem cell that can form only red blood cells.

erythropoiesis   the process by which red blood cells (erythrocytes) are produced.

erythropoietin   a protein growth factor; promotes proliferation, differentiation, and survival of red blood cell precursors.

ES cells   *See:* **embryonic stem cells**.

esophageal atresia   is an example of an abnormal closure of a canal, in this case, the esophagus; as this prevents the fetus from swallowing amniotic fluid, polyhydramnios results.

esophagus   the portion of the digestive tube between the pharynx and the stomach. In the frog, shortly after hatching, a portion of the foregut, between the future glottis and the opening of the bile duct, elongates; the anterior part of this elongation gives rise to the esophagus; for a brief time, there is a closure between the esophagus and the pharynx.

estrogen   female sex hormone; actually a small family of closely related female sex steroids. It is produced by ovarian follicles, as well as by other cells, and is responsible for the development and maintenance of female secondary sex characteristics. Estrogen enables the development of the Müllerian ducts into the uterus, oviducts, and upper end of the vagina.

estrous cycle   the physiological changes that take place between periods of estrus in the female mammal.

estrus   the period in female mammals during which ovulation occurs and the animal is receptive to mating.

ethylene   a volatile gas and the plant hormone with the simplest structure, C_2H_4; effects include fruit ripening, root hair growth, abscission, and senescence.

eucaryotes   *See:* **eukaryotes**.

euchromatin   dispersed and, generally, active chromatin; "normal" chromatin as opposed to heterochromation.

eukaryotes   are organisms that are made up of eukaryotic cells; i.e., cells with true (membrane-bound) nuclei. Also called eucaryotes.

euploid   an exact multiple of the haploid number of chromosomes for a given species.

eustachian tubes   the pair of tubes that extends between the pharynx and the cavity of the middle ears.

Eutheria   placental mammals; an infraclass of the Theria (live-bearing mammals), composed of 16 extant orders.

euviviparity   the supplementation of the food supply of the egg (yolk) by food from the mother (uterine). The maternal supply varies tremendously from one extreme of very slight dependence (as in some fishes, like the dogfish shark) to the other extreme of complete and absolute dependence (as in eutherian mammals).

euviviparous   *See:* **euviviparity**.

evagination   an outpocketing; e.g., the optic vesicles develop as evaginations of the forebrain.

evanescent   unstable; tending to vanish quickly.

even-skipped   a pair-rule gene in *Drosophila*. In the expression of the second *even-skipped* stripe, bicoid and hunchback proteins are required to activate the *even-skipped* gene, but they do not define the boundaries of the stripe of expression; these are defined by the Krüppel and giant proteins, by a mechanism based on repression of *even-skipped*. The anterior edge of the stripe is localized at the point of threshold concentration of giant protein, whereas the posterior border is similarly specified by the Krüppel protein.

"evo-devo"   the developmental genetic approach to evolution. *See:* **evolutionary developmental biology**.

evolution   the change in the gene frequency of a population over time. The result of hereditary changes affecting development. The control of development by ecology.

evolutionarily conserved regions   regions of the genome that are identical (or nearly so) between certain species, indicating that they encode genes whose functions are extremely important; such conserved regions often contain regulatory sequences that control the expression of transcription factors. *See:* **conservation**, *hedgehog*, **homologous regulatory genes**, **human accelerated regions**, **polycomb proteins**.

evolutionary developmental biology (evo-devo)   much of this approach to developmental biology focuses on variation produced by differences in the regulatory regions of genes active in constructing the embryo.

evolutionary embryology   the study of how changes in development may cause evolutionary changes and of how an organism's ancestry may constrain possible types of change.

Evolutionary Synthesis   *See:* **Modern Synthesis**.

evolutionary tree   *See:* **phylogenetic tree**.

exalbuminous seed   a seed without albumen; the nutritive tissue in a seed.

excretion   the discharge of waste.

exocoelom   the extension of the coelom (body cavity) between the extraembryonic regions of the mesoderm. *See:* **extraembryonic coelom**.

exocytosis   secretion; release of substances from a cell by fusion of a vesicle with the plasma membrane.

exogastrula   an artificially produced, abnormal gastrula that has not internalized its mesodermal and endodermal components such that the gut appears to be everted out of the blastopore. An amphibian exogastrula will not undergo neurulation because the "organizer" is not adjacent to ectoderm; however, some neural markers (e.g., N-CAM) are expressed. Exogastrulas have been experimentally produced in sea urchin embryos as well as in amphibian embryos.

exogastrulation   an abnormal process of gastrulation such that the gastrula's mesoderm and endoderm are not internalized; results in the formation of an exogastrula. Curt Herbst made the discovery (in 1893) that sea urchin embryos would exogastrulate if lithium salts were added to the sea water in which they were kept, thereby introducing the phenomenon of exogastrulation to experimental embryology. Treatment with lithium ions also causes exogastrulation in amphibian embryos.

exon   the portion of a gene that is expressed as part of a protein molecule; also, a region of a gene that is ultimately represented in that gene's mature transcript; the word refers to both the DNA and its RNA product. In eukaryotic genes, exons are interspersed with introns. *See:* **intron**.

exon shuffling   the process by which new genes are generated by recombination of exons of other genes.

Exopterygota   also Hemimetabola; insects passing through a simple and sometimes slight metamorphosis, rarely accompanied by a pupal instar. The wings develop externally, and the young are generalized nymphs. *See:* **Hemimetabola**.

exoskeleton   the skeleton found on the surface of the body, as in insects, as opposed to inside the body (endoskeleton), as in humans.

expansin   a protein, unique to plants, that causes plant cell walls to expand; plant cells may expand their volumes by nearly 20,000-fold. *See:* **morphogenesis**.

experimental embryology   comparison of developmental processes in different species, under normal and manipulated conditions, providing the greatest insight into developmental processes. According to Viktor Hamburger, experimental embryology, built on an organismic, holistic view of embryos and their development, attained a commanding position in the field of biology during the first half of the 20th century, with Hans Spemann's organizer experiment of 1924 being the crowning achievement of this period; furthermore, the radical shift of emphasis to the cellular and subcellular levels and, from the 1950s on, to the molecualr level, transformed experimental embryology to developmental biology.

explant   to transfer from the body to an artificial medium for growth.

explantation   removal of living tissue from its original site in the body for tissue culture.

expressivity   the phenotypic expression of a gene or genotype, which may be slight, intermediate, or severe; the degree to which a gene or genotype is expressed. *See:* **penetrance**.

expulsion   one of the seven cardinal movements of labor; the forcing of the fetus out of the birth canal (see Fig. 40).

exstrophy   a term referring to the eversion or turning inside out of a part.

exstrophy of the bladder   is the condition where part of the abdominal wall and the anterior wall of the bladder are missing and the posterior wall of the bladder bulges through the opening.

extant   related to living species.

extension   one of the seven cardinal movements of labor; the straightening out of the fetal head (see Fig. 40).

external auditory meatus   external ear canal.

external os   the opening of the cervical canal of the uterus into the vagina.

external rotation   one of the seven cardinal movements of labor; rotation of the fetal head after it has emerged from the birth canal (see Fig. 40).

extinct   related to species no longer living.

extracellular matrix (ECM)   the biochemically complex material outside of and secreted by cells; knowledge of the molecular basis of morphogenesis will be better understood as the, yet unknown, complete details of the composition of the ECM are understood.

extraembryonic   those portions of the conceptus not incorporated into the developing embryo.

extraembryonic coelom   the extension of the coelom (body cavity) between the extraembryonic regions of the mesoderm. *See:* **exocoelom**.

extraembryonic endoderm   formed by delamination of the hypoblast cells from the inner cell mass to line the blastocoel cavity; will form the yolk sac.

extraembryonic membranes   the four membranes, amnion, chorion, yolk sac, and allantois, developed from the conceptus that will not be incorporated into the embryo. Although only temporary structures, these membranes are essential for mammalian development. Also called fetal membranes.

extraembryonic mesoderm   derived from the trophoblast, this mesoderm does not become part of the embryo.

extrauterine fetus   a premature baby may be thought of as an extrauterine fetus, and in terms of brain development, human babies have been considered, in general, to be extrauterine fetuses.

exuviae   the cast-off skin of an insect resulting from a molt.

F

Fabricius, Hieronymus (1537–1619) wrote two treatises on embryology: *On the Form of the Foetus* (1600) and *On the Formation of the Egg and the Chick* (1621); he was the first author to propound a reasoned scheme for the mechanics of generation. He did not believe that the spermatic fluid reached the genital organs or formed any part of generation in the animal.

facultative heterochromatin euchromatin that has been heterochromatinized, e.g., the chromatin of the Barr body X-chromosome.

fallopian tubes also called oviducts or uterine tubes, these tubes convey eggs or embryos from the site of ovulation at the ovary into the uterine cavity. The human ovum takes 3 to 4 days to pass through the fallopian tube. Spermatozoa have been found in the fallopian tube within 5 minutes of ejaculation in several species including the human (see Figs. 16, 22, and 34).

fascicles independent bundles of axons.

fasciclins cell adhesion molecules (CAMs) belonging to the immunoglobulin superfamily of CAMs; in insects, they aid the migration of axons.

fat bodies accumulations of fatty tissue at specific locations in the body; e.g., gonads in some animals have associated fat bodies; e.g., in mice, the gonads are embedded in fat bodies; in frogs, the gonads have attached fat bodies.

fate map a diagram that shows what will become of each region of the embryo in the course of normal development; traditionally created with the use of vital stains, e.g.; mark the surface of regions of amphibian embryos with vital stains and follow the fates of those regions into later developmental stages; this allows one to project back (map) onto the surface of amphibian embryos what their various regions will give rise to (fates) under normal conditions. Fate maps demonstrated that different amphibians gastrulate differently; e.g., W. Vogt's fate map (from 1929) for *Rana fusca* demonstrated that the presumptive mesoderm of its late blastula is on the surface of the embryo, whereas Raymond E. Keller's fate map (from the 1970s) for *Xenopus laevis* demonstrated that the presumptive mesoderm of its late blastula is beneath the surface of the embryo. Recent fate maps use methods employing chick-quail chimeras, fluorescent dyes, green fluorescent protein, and retroviral transfection, for example. *See:* **competence map**, **specification map**.

fecundation the act of fertilizing; fertilization.

fecundity the innate potential reproductive capacity of the individual organism as denoted by its ability to produce offspring; fruitful in offspring or vegetation.

fertile capable of reproducing; directly concerned with sexual reproduction. The fertile parts of flowering plants are flowers, as opposed to roots, stems, and leaves that are vegetative parts of flowering plants. Capable of bearing seeds; capable of bearing pollen. *See:* **vegetative**.

fertilin proteins proteins in the sperm plasma membrane of mammals essential for sperm membrane-egg membrane fusion. *See:* **meltrins**.

fertility awareness methods methods that identify the fertile days of the month; these methods are the rhythm (calendar) method, the basal body temperature (BBT) method, and the cervical mucus method.

fertilization the interaction between gametes culminating in the formation of the zygote; generally refers to the fusion of a spermatozoon with an ovum, but fertilization is a multistep process that begins with interaction between the gametes before they physically meet and that ends with the fusion of the male and female pronuclei. The phenomenon of penetration by a single sperm, the fate of the sperm within the egg, and the equal participation

Dictionary of Developmental Biology and Embryology, Second Edition. Frank J. Dye.
© 2012 Wiley-Blackwell. Published 2012 by John Wiley & Sons, Inc.

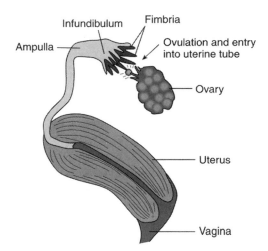

Figure 16. A fallopian tube and its relationship to the ovary and uterus. Reprinted from Frank J. Dye, *Human Life Before Birth*, Harwood Academic Publishers, 2000, fig. 5-8, p. 40.

of the egg and sperm nuclei in fertilization were completely demonstrated in animals by the research of Wilhelm August Oscar Hertwig, August Weismann, and Hermann Fol between the years 1875 and 1879. *See: **in vitro** fertilization (**IVF**).*

fertilization cone an upwelling of the surface of the egg where it is being penetrated by the spermatozoon; visible through the light microscope, as well as through the electron microscope, and resulting from the coalescence of egg plasma membrane microvilli.

fertilization consequences (1) activation of the egg, (2) amphimixis, and (3) displacements of cytoplasmic substances.

fertilization envelope *See:* **fertilization membrane or fertilization envelope**.

fertilization membrane or fertilization envelope the altered vitelline membrane resulting from fertilization; e.g., during the cortical reaction of sea urchin eggs, the vitelline membrane, which surrounds the plasma membrane of many eggs, is chemically altered to form the fertilization membrane (see Plate 10 in the color insert).

fertilizin a substance found in the egg water of sea urchins, which appeared to agglutinate the sperm of the same species (this was thought to be analogous to the agglutination of cells by antibodies) and which was once thought to play a role in fertilization. *See:* **egg water**.

Fertilizin Theory of fertilization held that the binding of fertilizin, from jelly coats of sea urchins, with antifertilizin, fertilizin receptors on the heads of sea urchin sperm, was the basis of species specificity in sea urchin fertilization.

fetal alcohol syndrome (FAS) a syndrome of birth defects in the children of alcoholic mothers; babies with FAS are characterized as having a small head size, an indistinct philtrum, a narrow upper lip, and a low nose bridge; the brain of such a child may be dramatically smaller than normal and often shows defects in neuronal and glial migration.

fetal hemoglobin the hemoglobin found in fetal red blood cells, made up of two alpha polypeptide chains and two gamma polypeptide chains; has a higher affinity for oxygen than adult hemoglobin and binds oxygen given up by adult hemoglobin in the placenta. *See:* **adult hemoglobin, hemoglobin**.

fetal membranes *See:* **extraembryonic membranes**.

fetal origins of adult disease (FOAD) hypothesis broadening of the David J. P. Barker hypothesis to encompass a large range of prenatally influenced illnesses. *See:* **developmental origin of health and disease (DOHD) hypothesis**.

fetal period in human development, from the end of the eighth week until birth.

fetal plasticity or fetal programming, refers to the processes by which a stimulus given during an intrauterine period of development can have a lasting or lifelong effect, e.g., the "Barker hypothesis" (named after David J. P. Barker).

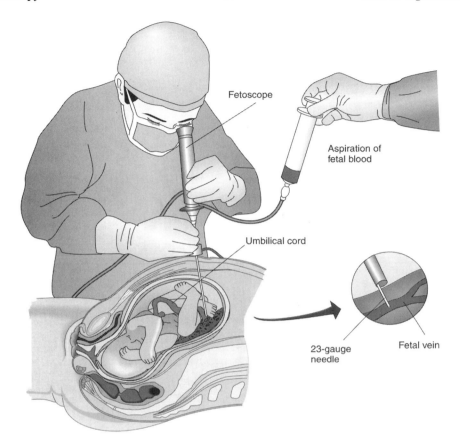

Figure 17. This drawing shows what fetoscopy involves. Using fiberoptics and a fetoscope, the obstetrician is able to directly observe the fetus in its amniotic cavity. Reprinted from Frank J. Dye, *Human Life Before Birth*, Harwood Academic Publishers, 2000, fig. 19-14, p. 172.

fetoscopy a prenatal diagnostic technique allowing direct, visual observation of the fetus in the womb (see Fig. 17).

fetus when the developing mammal begins to resemble the adult, it is called a fetus; until then it is called an embryo. In human development, from the beginning of the ninth week after fertilization through the fortieth week of gestation or until birth.

fetus in fetu a variation on conjoined twins in which one fetus is small and "parasitic" on or in the other relatively normal fetus.

FGF *See:* **fibroblast growth factor**.

FGF receptors tyrosine kinase receptors, which activate the MAP kinase pathway.

FGF2 a basic fibroblast growth factor, responsible for initiating vasculogenesis, required for the generation of hemangioblasts from splanchnic mesoderm. *See:* **angiopoietin-1 (Ang1)**, **vascular endothelial growth factor (VEGF)**.

fibroblast growth factor (FGF) a family of protein growth factors, first identified as mitogens for fibroblasts in tissue culture, stimulates proliferation of many cell types, inhibits differentiation of various types of stem cells, and acts as inductive signals in embryonic development, e.g., plays a key role in the induction of ventral mesoderm in *Xenopus* embryos; it is secreted by vegetal cells.

fibronectin an extracellular protein that plays a role in guiding cells, e.g., primordial germ cells in amphibians and mammals, in their migration through the body of the embryo.

field *See:* **morphogenetic fields**.

field of organization reference to the morphogenetic field by Hans Spemann.

field-specific selector gene   a specific class of selector genes that control the formation and patterning of morphogenetic fields such as the eye, leg, and wing.

filopodia   narrow extensions of the leading edge of a cell, by means of which the cell senses its environment, e.g., found on the growth cones of neurons extending axons and exhibited by the secondary mesenchyme cells of sea urchin gastrulae.

fimbria   literally, "a fringe"; the free end (near the surface of the ovary) of each fallopian tube is fimbriated (see Fig. 16).

fin fold   a median integumentary fold extending along the body of a fish embryo that gives rise to the dorsal, caudal, and anal fins.

first polar body   a tiny cell produced by the primary oocyte when it undergoes meiosis I.

FISH   *See:* **fluorescence in situ hybridization**.

fission   a type of asexual reproduction. The new individual is formed from a relatively large portion of the body of the maternal organism, and differentiated organs and tissues or their parts are passed on to the offspring; as in coelenterates and worms.

fistula   an abnormal connection between two internal structures or from an internal structure to the external surface of the body. *See:* **rectovaginal fistula**.

flagellum   also referred to as a tail; an organ of cellular locomotion, e.g., the organelle of the spermatozoon that provides it with the function of motility.

flexion   the bending of the longitudinal axis of the body around a transverse axis. When you try to touch your toes, you undergo flexion. Also, one of the seven cardinal movements of labor; a bending of the fetal head that places the chin against the chest. In the chick embryo, beginning at approximately 38 hours, flexion and torsion begin; flexion begins first in the head region, as the cranial flexure, as the forebrain is bent ventrally toward the yolk. Flexion proceeds caudally, affecting the cervical, dorsal, and caudal regions, with the spinal axis of the embryo gradually becoming C-shaped (from an originally straight longitudinal axis), by 4 days, with the head and tail lying close together. Flexion seems to be related to spatial limitations of the egg (chick) or the uterine cavity (mammals). Toward the end of the third week, human embryos show flexion in the cranial region; during the fourth week, unimpeded by a large yolk mass, human embryos show flexion already in the caudal region; although in certain details not quite as far advanced as a 50–55-hour chick embryo (see Fig. 40).

flexures   bends in the body axis of the embryo resulting from flexion, such as cervical flexure (see Fig. 8).

floor plate   the midventral wall of the embryonic neural tube; the ventral part of the embryonic neural tube. *See:* **roof plate**.

floral   pertaining to flower; the floral organs are sepals, petals, stamens, and carpels (see Plate 11 in the color insert).

floral meristem   those meristems that will actually produce flowers. *See:* **floral meristem identity genes**.

floral meristem identity genes   genes whose expression is necessary for the transition from an inflorescence meristem to a floral meristem; e.g., *LFY*, *AP1*, and *CAL* are floral meristem identity genes in *Arabidopsis*. Such genes initiate a cascade of gene expression that turns on cadastral genes.

floral organ identity genes   genes necessary to specify the four whorls of floral organs; e.g., *AP2*, *AGAMOUS*, *AP3*, and *PISTILLATA* in *Arabidopsis*; these genes code for transcription factors that initiate a cascade of events leading to the actual production of floral parts. These are homeotic genes.

flowering control gene *(FLC)*   a gene found in the wild mustard *Arabidopsis* (equivalent to the *VRN2* gene found in wheat; i.e., its repression allows flowering to occur).

fluorescence in situ hybridization   a means of hybridizing a fluorescent probe to whole chromosomes to determine the location of a gene or other DNA sequence within a chromosome.

foam   *See:* **spermicide**.

focus   the developmental organizer at the center of a butterfly eyespot field.

Fol, Hermann (1845–1892)   born and worked in Geneva as a cytologist; detailed the mechanism of spermatozoon entry into the egg; saw that it was only a single sperm that entered the egg; in 1879, he was the first researcher to see the fertilization of an animal ovum (Spiny Starfish, *Marthasterias glacialis*), an observation that Nathaniel Pringsheim earlier had made in plants. He and his compatriot Édouard Sarasin made experiments with the disapperance of light in water at the Riviera, and Fol himself used diving equipment during these studies. Fol later disappeared without any trace together with his crew and the research ship, which he had fitted out by his own means.

folic acid (vitamin B$_{12}$)   it has been estimated that 50% of human neural tube defects could be prevented by a pregnant woman's taking supplemental folic acid; the U.S. Public Health Service recommends that all women of childbearing age take 0.4 mg of folate daily to reduce the risk of neural tube defects during pregnancy.

follicle   a structure formed in ovaries, usually consisting of one germ cell and multiple follicle cells, in which germ cells are maintained and differentiate, culminating either in ovulation or death of the contained oocyte. In mammals, ovarian follicles are found in the cortical region of the compact type of mammalian ovary. In frogs, ovarian follicles are found in the periphery of the hollow type of amphibian ovary, between the outer mesovarium and vascular theca externa on the one hand and the inner theca interna, with smooth muscle fibers on the other hand. In insect ovarioles (egg tubes), a follicle or egg chamber is formed by distension of the ovariole resulting from the growth of an oocyte; each follicle is lined by a definite follicular epithelium (see Figs. 22 and 33).

follicle cells   somatic cells found in ovaries in association with germ cells as part of a structure referred to as a follicle; also called granulosa cells. Follicle cells in the ovary of *Drosophila* play a key role in patterning the oocyte; in *Drosophila* there are various types of follicle cells, expressing different genes and having different effects on the oocyte; follicle cells also secrete the vitelline envelope and the eggshell that surround the mature egg. The oocyte induces the follicle cells it contacts to adopt a posterior fate; the inductive signal is carried to the follicle cells by the gurken protein. The posterior follicle cells send a signal back to the oocyte that results in a reorganization of the oocyte's microtubule cytoskeleton (see Plate 8 in the color insert). *See:* **gurken protein**.

follicle stimulating hormone (FSH)   a glycoprotein hormone released by the anterior lobe of the pituitary gland that affects the gonads (gonadotrophin) in females and males. In females, FSH stimulates the development of a small group of ovarian follicles and stimulates follicle cells to produce estrogen. In males, FSH is specifically taken up by Sertoli cells within the seminiferous tubules; the Sertoli cells then synthesize an androgen-binding protein that helps maintain a high concentration of testosterone in the seminiferous tubule, which is necessary for spermatogenesis.

follicular stem cell   a multipotent adult stem cell that resides in the bulge niche of the hair follicle; it gives rise to the hair shaft, sheath, and sebaceous gland.

follistatin   an organizer-secreted protein; secreted by the organizer and prevents the bone morphogenetic protein (BMP) from binding to the ectoderm and mesoderm near the organizer. *See:* **chordin, noggin**.

fontanelles   the openings between the bony plates of the skull, most conspicuous at approximately the time of birth.

foot activator gradient   a proposed morphogenetic gradient in *Hydra*, of basal disc activator, which is highest at the basal disc; when basal disc tissue from one *Hydra* is transplanted into the middle of another *Hydra*, a new apical-basal axis is formed, with the basal disc facing outward. *See:* **foot inhibitor gradient, head activator gradient**.

foot inhibitor gradient   a proposed morphogenetic gradient in *Hydra*, of basal disc inhibitor, which is highest at the basal disc. *See:* **foot activator gradient, head inhibitor gradient**.

foramen cecum   the opening from the pharynx into the thyroglossal duct of the thyroid diverticulum.

foramen ovale   an oval opening that develops in the septum secundum. It becomes part of the important valvular mechanism between the atria in the prenatal heart. The foramen ovale allows the left atrium to receive blood so that it may have a load to pump and therefore get the exercise it needs for its normal development.

foramen primum   the first opening in septum primum, between the two atria of the developing heart, which subsequently closes.

foramen secundum   the second opening in septum primum, which persists during prenatal development, the edges of which provide the flaps for the valve at foramen ovale.

foramina of Monro   openings between the lateral ventricles of the cerebral hemispheres and the third ventricle of the brain through which cerebrospinal fluid circulates (see Fig. 7).

forebrain   the most rostral of the early brain vesicles of the three-vesicle brain (see Plate 5 in the color insert).

forebrain protuberance   the bulge found on the surface of the head of the embryo during the middle of the embryonic period, reflecting the dramatic development of the forebrain beneath the surface of the head.

foregut   the cephalic-most portion of the early embryonic gut, which gives rise to part of the mouth, pharynx, esophagus, stomach, liver, gall bladder, and pancreas. The foregut is the first region of the digestive tract established and the most advanced in differentiation. In the chick embryo, one can recognize a pharyngeal (with a larger lumen) and an esophageal portion; the pharyngeal region lies ventral to the myelencephalon and is encircled by aortic arches. At approximately 50 hours, there is an approximation of a ventral outpocketing near the anterior end of the pharynx to a depression in the adjacent ectoderm of the ventral surface of the head; the ectodermal depression is the stomodaeum; the thin layer of tissue resulting from the apposition of stomodaeal ectoderm to pharyngeal endoderm is the oral plate, which, later in development, breaks through. During the third day, the region where the oral plate was originally located in the embryo becomes, in the adult, the region of transition from the oral cavity to the pharynx. The oral opening is not established at the extreme cephalic end of the pharynx; part of the pharynx, extending rostral to the mouth opening, constitutes the pre-oral gut. The pre-oral gut eventually disappears, but it persists for a time as a small diverticulum called Seessel's pocket (see Plate 5 in the color insert as well as Figs. 6 and 19). *See:* **hindgut**, **midgut**.

founder animal   an animal that establishes a new population.

founder cell   a cell that gives rise to a particular group of cells; e.g., a cell clone consists of all surviving descendants of a single cell, which is called the founder cell of the clone.

fourth ventricle   the ventricle of the metencephalon and myelencephalon; ventricle IV connects the aqueduct of Sylvius with the neural (spinal) canal. *See:* **lateral ventricles**, **third ventricle**.

fraternal twins   twins derived from two different zygotes. Also called dizygotic twins.

free-central placentation   ovules attached to a free-standing column in the center of a unilocular ovary.

free nerve terminations   those nerve terminations that simply move in among the cells of epithelial or connective tissues and develop branches. *See:* **encapsulated nerve endings**.

freemartin   genetically female calf who is masculinized if it shares the same circulatory system *in utero* with a male co-twin.

French flag   a general mode of pattern formation illustrated by considering the patterning of the French flag, which has a simple pattern: one third blue, one third white, and one third red, along just one axis. Each cell in a line of cells has the potential to develop as blue, white, or red. The line of cells is exposed to a concentration gradient of some substance (morphogen), and each cell acquires a positional value defined by the concentration at that point; each cell then interprets the positional value it has acquired and differentiates into blue, white, or red, according to a predetermined genetic program, thus forming the French flag pattern.

frizzleds   the Wnt receptors.

frogs   animals belonging to the order Anura of the vertebrate class Amphibia. Extensively used for teaching and research in embryology and developmental biology because frogs (1) are an excellent transition between Amphioxus and the higher evolved vertebrates, (2) display all the fundamental vertebrate systems in a primitive condition, (3) have been thoroughly studied and are used in much current experimental work, and (4) are generally available and easy to handle. The genera *Rana* and *Xenopus* have especially been used extensively in teaching and research.

frontal plane   the plane that, when passed through the body, separates the dorsal part from the ventral part of the body.

frontonasal process   also called nasofrontal process; the anterior region of the embryonic head that later develops into the frontal, median nasal, and lateral nasal processes; the neural crest cells of the frontonasal process generate the bones of the face.

frontonasal prominence   the surface of the developing face of the embryo between and above the two nasomedial processes.

fructose   a simple sugar (monosaccharide) containing six carbons (hexose), preferentially used as an energy source by human spermatozoa, whereas other cells of the body preferentially use glucose.

fruit   develops primarily from the tissue of the ovary (or group of ovaries) of a flower; it plays an important part in the method of dispersal of the contained seeds.

Fucus   a genus of brown algae belonging to the phylum *Phaeophyta*; has been used by developmental biologists to study body axis determination by environmental factors, the importance of the cell wall as a potential source of developmental signals, and the developmental significance of the asymmetric cleavage of the zygote.

functional genomics   the science concerned with understanding the functions of genes; the study of the pattern of genome-wide gene expression at various times or under various conditions.

fundus   the upper (superior) part of the uterus, between the attachment sites of the two fallopian tubes. The height of the ascent of the fundus out of the pelvic cavity is used by obstetricians to judge the stage of pregnancy (gestation) (see Fig. 48).

Fungi   although once classified in the plant kingdom, and now classified as the Kingdom *Fungi*, molecular evidence suggests that fungi are organisms more closely related to animals than to plants; includes such familiar organisms as mushrooms, yeasts, and molds. Some fungi are of great economic (brewer's yeast) and medical (*Penicillium*) importance. Unlike most members of the plant kingdom, fungi are not photosynthetic but heterotrophic.

funiculus   the stalk connecting the ovule to the placenta; the stalk of a seed.

fusogen   an agent that causes cell fusion, e.g., polyethylene glycol, inactivated Sendai virus.

fusome   a large, spectrin protein-rich structure that spans the ring canals between the cystocytes; perhaps coordinates intracellular transport between the cystocytes, e.g., as in *Drosophila*.

G

G₀ a stage, not part of the cell cycle, occupied by cells that are not actively cycling, i.e., are not progressing toward cell division; such cells are referred to as being quiescent. The G_0 state is distinct from the state of proliferating cells in any phase of their cycle; e.g., the rate of protein synthesis is drastically reduced. It is the ability to enter G_0 that accounts for the enormous variability of the length of the cell cycle in multicellular organisms.

G₁ a stage of the cell cycle, after mitosis and before the initiation of DNA synthesis; constitutes part of the interphase. The greatest variation in the length of stages of the cell cycle, in most of the commonly studied cell types, occurs in the duration of G_1.

G₁ checkpoint the checkpoint normally controlled by growth factors; operated by cyclins A, D, and E and by Cdk2, 4 and 6; In the absence of growth factors, the cell will leave the cycle and enter a state called G_0, in which the Cdks and cyclins are absent. The G_1 checkpoint is also the point in animal cells at which cell size is assessed. *See:* **Rb**.

G₂ a stage of the cell cycle, after DNA synthesis and before mitosis; constitutes part of interphase.

G₂ checkpoint controls the transition from G_2 to M.

G proteins a cell regulatory important class of proteins; a protein that is activated by binding to GTP and inactivated by hydrolysis of the bound GTP to GDP by its inherent GTPase activity. The trimeric G proteins, consisting of an alpha, beta, and gamma subunit, are found on the inner surface of the plasma membrane and are activated, when some plasma membrane receptors bind their ligands, as an early step in an intracellular signal transduction pathway.

gain-of-function mutation a mechanism of dominance; a mutation that endows the protein product of a gene with a new function; e.g., the acquired ability of a cell-surface receptor to be activated even in the absence of its extracellular ligand; in thanatophoric dysplasia, a form of dwarfism, a mutation causes the FGF receptor to be constitutively active, causing development of the anomalous phenotype. *See:* **constitutive**, **dominant negative mutation**, **haploinsufficiency**, **loss-of-function mutation**.

gallbladder a sac-like organ, attached to the under surface of the liver, which stores and releases bile.

gametangium a container of gametes; which may be unicellular or multicellular, e.g., oogonium (unicellular egg container), antheridium (unicellular or multicellular sperm container), and archegonium (multicellular egg container) (see Plate 16 in the color insert).

gamete intrafallopian transfer (GIFT) artificial transfer of gametes (sperm and egg) into the normal site of fertilization in the ampulla of the fallopian tube.

gamete recognition proteins proteins that mediate species-specific interactions between sperm and egg, the fastest evolving proteins known in the animal kingdom. *See:* **bindin**, **bindin receptor**.

gametes a subset of germ cells directly involved in fertilization. Also called sex cells, sperm, and eggs (see Plate 16 in the color insert).

gametic meiosis occurs at the end of the life cycle, as in animals where it is involved in gametogenesis. *See:* **sporic meiosis**.

gametogenesis formation of the gametes. *See:* **oogenesis**, **spermatogenesis**.

gametophyte literally gamete plant; a multicellular, haploid plant that produces, by mitosis, gametes. Ranges in conspicuousness from the visible green moss plant, to the barely visible fern prothallium, to the microscopic embryos sacs and pollen grains of angiosperms (see Plates 15 and 17 in the color insert).

Dictionary of Developmental Biology and Embryology, Second Edition. Frank J. Dye.
© 2012 Wiley-Blackwell. Published 2012 by John Wiley & Sons, Inc.

gametophyte generation   in the typical plant life cycle, exhibiting alternation of generations, the gametophyte is the multicellular, haploid, gamete-producing generation.

ganglion   (pl. ganglia) an aggregate of nerve cell bodies outside of the central nervous system (see Fig. 31).

ganglion cells   of the retina, are neurons whose axons send electrical impulses to the brain; their axons meet at the base of the eye and travel down the optic stalk, which is then called the optic nerve.

ganglion mother cell   is formed by division of a neuroblast in *Drosophila* and gives rise to neurons.

gap genes   genes found in *Drosophila* that subdivide the embryo into regions along the anteroposterior axis; they all code for transcription factors expressed early in development and are zygotic, as opposed to maternal, genes. The mutant phenotype of a gap gene usually shows a gap in the anterioposterior pattern, in more or less the region in which the gene is normally expressed Their expression is initiated by the anteroposterior gradient of bicoid protein while the embryo is still a syncytial blastoderm. Bicoid protein activates anterior expression of the gap gene *hunchback* that, in turn, plays a role in switching on the expression of the other gap genes, including *giant*, *Krüppel*, and *knirps*, which are expressed in order along the anteroposterior axis. Interactions between the gap genes, all of which code for transcription factors, help to define their borders of expression. Gap genes regulate pair-rule gene expression in each parasegment of the embryo. Like pair-rule genes, the activity of gap genes is only temporary.

gap junctions   specialized contacts formed between cells that establish cytoplasmic continuity between the cells. Cells with these junctions rapidly communicate with each other. During compaction of early mammalian embryos, gap junctions form between the cells of the inner cell mass. *See:* **cytoplasmic bridges**, **tight junctions**.

Garstang, Walter (1868–1949)   professor of zoology at the University of Leeds. During his early work, he was interested in the connections among echinoderms, tunictes, and vertebrates; he was particularly interested in one question—why did planktonic larvae have such different shapes and lifestyles from the adults into which they metamorphose? Garstang's reasoning decided that it was from the different selective pressures that they must face. He decided that the larvae had to face two opposing pressures. One selective pressure was that of dispersal and that for greater dispersal the larval state would have to be maintained for a longer period of time. However, the amount of predation would increase as a result of this so that there would also be a pressure to shorten larval life in such a way that they must become sexually mature and reproduce as soon as they are able. He derived a new theory he called paedomorphosis, which is a process by which larval forms can become sexually active. This by itself is known as neotony and is observed in animals such as the axolotl and is brought about by the deficiency of nutrients. More accurately paedomorphosis occurs as a result of natural selection so that neotony is a process of evolutionary change. This theory meant that it would be possible for animals that abandoned their adult stage to take a new path of evolution. Garstang went on to suggest in 1926 that chordates were primitively free living and that the ancestor was not motile but sessile, transformed by paedomorphosis so that it would maintain its larval morphology, which he suggested was Amphioxus-Craniata. He then went on to describe the development of the jaws from branchial bars, paired limbs, and the evolution of the vertebrates. He first clearly conceived (in 1922) the principle of paleogenesis; i.e., the concept that descendent ontogenies tend to recapitulate ancestral ontogenies. He declared, according to S. F. Gilbert (2006), that ontogeny (individual development) does not recapitulate phylogeny (evolution); it creates phylogeny.

Gartner's ducts   the vestigial remnants of the mesonephric ducts in females; each is also called a ductus epoophori longitudinalis.

Gastraea Theory   of metazoan ancestry, attributed to Ernst Haeckel; the concept that the hypothetical ancestor of all Metazoa consisted of two layers (ectoderm and endoderm), similar to the gastrula stage in embryonic development; furthermore, that the endoderm originated as an invagination of the blastula composed of a single layer of cells. That is, the diploblastic stage of ontogeny was considered as the repetition of this ancestral form.

gastrocoel   the cavity of the gastrula, the primitive gut. *See:* **archenteron**.

gastrula   the stage of animal development after the blastula stage and before the neurula stage; during this stage of development, dramatic morphogenetic movements occur and the three primary germ layers are formed (see Plate 4 in the color insert).

gastrulate   to undergo gastrulation.

gastrulation   an early morphogenetic process, following cleavage, which results in the establishment of the three primary germ layers. Gastrulation is a process that involves integrated cell and tissue movements, during which dramatic rearrangements of the blastomeres occur; not only are the germ layers formed, but the stage is set for their interactions. In a memorable quote, Lewis Wolpert in 1986 said, "It is not birth, marriage, or death, but gastrulation, which is truly the most important time in your life." Although there are different patterns of gastrulation, relatively few cellular mechanisms, in various combinations, are involved. In the sea urchin, gastrulation consists of the following four stages: (1) ingression of primary mesenchyme cells (resulting from altered adhesion between cells and between cells and the extracellular matrix), (2) first stage of invagination (resulting from buckling of the bilaminar hyaline envelope), (3) second stage of invagination (resulting from convergent extension), and (4) third stage of invagination (resulting from contraction of the filopodia of secondary mesenchyme cells). Not all amphibians undergo gastrulation in the same way; the ways in which *Rana fusca* and *Xenopus laevis* gastrulate are different. Amphibian gastrulation has been intensively studied for the past century; fate maps using vital staining have provided insightful information about cell movements during amphibian gastrulation. Unlike in the sea urchin, where invagination begins at the vegetal plate, amphibian invagination begins in the marginal zone near the equator of the blastula, forms the dorsal lip of the blastopore in the region of the gray crescent, and is initiated by bottle cells. In insects, gastrulation takes place, during growth of the amniotic folds, as a ventral furrow-like ingrowth on the middle line of the germ band; it begins at the site of the future stomodaeum and gradually extends to the caudal end of the germ band. The cells of this furrow-like ingrowth become deployed as an inner or lower layer beneath the outer layer or ectoderm; this lower layer gives rise to mesoderm and endoderm (see Plate 4 in the color insert).

Gegenbaur, Karl (1826–1903)   German anatomist; one of the first to consider anatomy from an evolutionary standpoint; an authority on comparative anatomy of vertebrates. He showed (in 1861) that the eggs of all vertebrate animals, regardless of size and condition, are in reality single cells (Rudolph Albert von Kölliker (in the 1840s) described the formation of spermatozoa from testicular cells.) Gegenbauer and Ernst Haeckel both insisted on the centrality of evolutionary questions and on the value of using embryos in particular to construct phylogenetic trees. *See:* **His, Wilhelm**.

gel mobility shift assay   an assay based on electrophoresis, used to determine whether a transcription factor attaches to a particular DNA fragment; if two identical DNA fragments, one incubated with and one incubated without a transcription factor, are run on the same gel, the fragment that binds the transcription factor will run more slowly on the gel.

gemma   (pl. gemmae) a bud or bud-like structure, or cluster of cells that separate from the parent plant and propagate offspring plants.

gemmation   the process of reproduction by gemmae.

gemmule formation   a type of asexual reproduction found in freshwater sponges and bryozoans. The new individual is formed from a group of cells that become completely cut off from the maternal individual and are disseminated.

gemmules   bodies produced by freshwater sponges, during asexual reproduction, that can survive after the maternal organism dies off during an unfavorable season (e.g., winter or drought). When conditions become favorable, the gemmules develop into new individuals. It is a protective, overwintering structure that germinates the following spring. *See:* **statoblasts**.

gene   unit of hereditary material, made up of stretches of DNA (RNA in some viruses) nucleotide sequences.

gene alteration   generally genes are not altered during development; however, three cases of gene alteration are known: plasma cells, Trypanosoma brusei, and yeast mating types.

gene cloning   *See:* **molecular cloning**.

gene clusters   any group of two or more closely linked genes on a chromosome that are related functionally, e.g., the four Hox gene clusters in the mouse genome. Genes of a gene cluster usually code for various steps in a particular metabolic pathway.

gene complex   a group of adjacent genes related by gene duplication; e.g., in *Drosophila*, the homeotic selector genes that control segment identity are organized into two gene complexes, collectively known as HOM genes (which together are broadly homologous to a single Hox gene complex in vertebrates), because each codes for a transcription factor containing a homeobox. The two homeotic complexes in *Drosophila* are the bithoracic and Antennapedia gene complexes. The order of the genes from 3′ to 5′ in each complex reflects both the order of their spatial expression (anterior to posterior) and the timing of expression (3′ earliest).

gene divergence   the process through which two or more genes created by a duplication event(s) acquire distinct functions, e.g., the *Drosophila gooseberry/gooseberry-neuro* gene pair. Changes in regulatory and/or coding regions lead to differences in gene function.

gene duplication   the creation of additional genes from the template of one gene, e.g., the *Drosophila gooseberry/gooseberry-neuro* gene pair and deuterostome *engrailed* genes.

gene expression   in *Drosophila*, rare mitotic genes, such as *twist*, *snail*, and *decapentaplegic*, whose expression is regulated by the dorsal protein, contain binding sites for dorsal protein in their regulatory regions that activate or repress gene expression at particular concentrations of the dorsal protein; the threshold effect on gene expression is the result of the integrating function of these regulatory binding sites. It is likely that the threshold response involves cooperativity between the different binding sites; also, inhibitory interactions with other gene products are involved (e.g., snail protein represses expression of the *rhomboid* gene in ventral regions and confines its expression to the neuroectoderm). The presence of control regions that, when activated, can lead to gene expression in a specific position in the embryo is an important principle controlling gene action in development. *See:* **pair-rule genes**.

gene family   two or more genes with related sequences that are derived from a common ancestral gene, e.g., deuterostome *engrailed* genes. They are not necessarily linked, may be widely dispersed in the genome, and may be greatly diverged in function.

gene knockout   *See:* **knockout**.

gene targeting   *See:* **knockout**.

general transcription factors (GTFs)   in eukaryotic cells, RNA polymerases are not sufficient for regulating transcription; eukaryotic transcription also requires other proteins called general transcription factors. General transcription factors are involved in the formation of a preinitiation complex. The most common are abbreviated as TFIIA, TFIIB, TFIID, TFIIE, TFIIF, and TFIIH. *See:* **specific transcription factors**, **transcription factors**.

generative cell   one of two cells in a pollen grain; the generative cell undergoes mitosis to produce two spermatozoa. *See:* **tube cell**.

genetic accommodation   as phenotypic plasticity is beneficial, genes that produce plasticity are selected. *See:* **Baldwin effect**, **genetic assimilation**, **phenotypic accommodation**.

genetic assimilation   transfer of induction from an external inducer to an internal inducer; i.e., a trait that had been induced by the environment becomes part of the genome of the organism; e.g., the ostrich is born with calluses where it will touch the ground. *See:* **Baldwin effect**, **competence transfer**, **genetic accommodation**.

genetic code   the relationship between the codons of messenger RNA and the amino acids of the corresponding protein.

genetic diversity   the degree of variety of genes in a population.

genetic heterogeneity   the phenomenon whereby mutations in different genes produce similar phenotypes.

genetic mosaics   are organisms derived from a single genome, but in which there is a mixture of cells with rearranged or inactivated genes. In flies, genetic mosaics can be generated by inducing recombination events in the embryo or larva. *See:* **mitotic recombination**, *Minute* **technique**.

genetic redundancy   two genes are redundant when the absence of the function of one gene causes little or no effect because a second gene can operate in its place.

genetic screen   a procedure or test to identify and select individuals who possess a phenotype of interest. A genetic screen for new genes is often referred to as forward genetics as opposed to reverse genetics, the term for identifying mutant alleles in genes that are already known. A basic screen involves looking for a phenotype of interest in the mutated population; e.g., one might screen for obvious phenotypes such as fruit flies with no wings or an *Arabidopsis* flower with no petals.

genetic sex   the normal pattern of sex; in humans, two X chromosomes result in a female and one X and one Y chromosome result in a male. *See:* **phenotypic sex**.

genetic toolbox   those genes that determine the overall body plan and the number, identity, and pattern of body parts.

geneticist   a person engaged in the study of heredity (genetics).

genetics   the study of heredity and variation.

genistein   an estrogenic compound found in soy, and in tofu; a possible endocrine disruptor.

genital folds   in males, give rise to the floor of the penile urethra and, in females, to the labia minora (see Fig. 49).

genital herpes   a viral, highly contagious, sexually transmitted disease.

genital papilla   a papilla in the cloaca of the male chicken that carries the opening of the sperm duct.

genital pores   the openings on the aboral surface of the sea urchin through which the gametes are shed.

genital ridge   a medial ridge or fold on the ventromedial surface of the mesonephros in the embryo, produced by growth of the peritoneum; the primordium of the gonads and their ligaments.

genital swellings   in males, give rise to the scrotum and, in females, to the labia majora (see Fig. 49).

genital tubercle   the embryonic precursor of the penis in the male or the clitoris in the female (see Fig. 49).

genitalia   reproductive organs.

genome   the genetic makeup of an individual, found in a haploid set of chromosomes; the complete DNA sequence of an organism; one complete set of genetic information from a genetic system; e.g., the singular, circular chromosome of a bacterium is its genome. *See:* **proteome**, **transcriptome**.

genome duplication   a form of polyploidy in which an organism has duplicate copies of its entire genome.

genomic equivalence   the concept that every cell of an organism has the same set of genes.

genomic imprinting   the nonidentity of maternal and paternal chromosomes received by a zygote, because of differential chemical modification (methylation) of the DNA of the chromosomes by the two parents. Because of genomic imprinting, both maternal (necessary for embryonic development) and paternal (necessary for extraembryonic development) genomes of mammals are necessary.

genomics   the study of the structure and function of whole genomes; comparative analysis of the complete genomic sequences from different species; used to assess evolutionary relations between species and to predict the number and general types of proteins produced by an organism.

genotype   the fundamental hereditary constitution of an individual; the set of genes that an individual carries.

Geoffroy Saint-Hilaire, Etienne (1772–1844)   French naturalist; propounded a theory of organic unity that held that a single plan of structure prevails throughout the animal kingdom (strongly opposed by Georges Cuvier, who held that there were four types of structure); he believed that many animal varieties, including humans, were formed following embryonic transformations during geological time. He thought embryos were transformed because they breathed and were thus sensitive to environmental variations; i.e., organic modifications must have resulted from environmental variations acting directly on embryos. Experimentally, he exposed chicken eggs to various "environmental modifications" (hypothermia, vertical egg incubation, etc.) and would not have been surprised if a bird egg gave rise to a reptile or vice versa; the malformations he believed he had produced (exencephaly) were normally produced by the crested chicks he used as his experimental organism.

germ band   a limited ventral region of the insect egg where the cells of the blastoderm become higher to form a cylindrical epithelium (i.e., the blastoderm thickens); which later gives rise to the embryo. The rest of the egg consists chiefly of yolk enclosed by the extraembryonic blastoderm. The germ band forms the ventral part of the developing insect, and to complete the embryonic body, the margins begin to grow upward; the final result is the completion of the embryo on the dorsal side. The upward growth involves not only the ectoderm or body wall but also the epineural sinus and the mesoderm, while the developing mid-intestine ultimately encloses the yolk. The embryonic membranes later rupture and, becoming contracted, are fully resorbed. In *Drosophila*, the ventral blastoderm is the initial germ band; the band of germ layers that forms on the ventral side of the embryo during gastrulation in insects.

germ band stage   a stage in insect embryogenesis. *See:* **germ band**, **phylotypic stage**.

germ cell determinants   cytoplasmically localized determinants responsible for the determination of germ cells; although cytoplasmically localized determinants are found throughout the animal kingdom, those responsible for the determination of germ cells are most frequently observed.

germ cell-less (gcl)   a gene found in *Drosophila*; the mRNA of this gene is part of the pole plasm of *Drosophila*; transcribed by ovarian nurse cells, the mRNA is transported into the egg and resides in what will become the pole plasm; the *gcl*-encoded protein is essential for pole cell production.

germ cells   ova or spermatozoa or their antecedent cells; those cells, as opposed to somatic cells, which have the responsibility for continuity of the species; that is, they provide the cellular basis for reproduction.

germ layers   cellular layers resulting from gastrulation, which make up the developing body of the early embryo; i.e., ectoderm, mesoderm, and endoderm. *See:* **neural crest**.

germ line   a lineage of cells ancestral to the gametes that, during the development of an organism, are set aside as cells destined to give rise to gametes.

germ-line mutation   a mutation in germ cells that may be carried to the next generation. *See:* **somatic mutation**.

germ plasm   a visible specialization of the cytoplasm, with which a cytoplasmic determinant is associated, that programs cells that inherit it to become germ cells, e.g., polar granules in *Caenorhabditis elegans* and pole plasm in *Drosophila*.

germ plasm theory   only in those cells of the embryo destined to become gametes were all types of inherited determinants retained; the nuclei of all other cells would have only a subset of the original determinant types (Weismann, 1883).

germ ring   in fish embryos, the outer, thickened edge of the blastodisc or blastoderm. The posterior prominence of the germ ring is the embryonic shield. *See:* **embryonic shield**.

germ wall   in chick embryos, is the marginal area, where the expanding germ layers merge with the underlying yolk. The advancing peripheral boundary of the chick blastoderm.

germarium   that part of an insect ovariole that is terminal and contains the primordial germ cells or oogonia; the oogonia eventually differentiate into oocytes and trophocytes.

germinal crescent   in chick development, when cells destined to become foregut displace the hypoblast cells in the anterior portion of the embryo and confine these cells to a region in the anterior portion of the area pellucida, this region is known as the germinal crescent. The germinal crescent does not form any embryonic structures, but it does contain the precursors of the germ cells.

germinal disc   a disc-shaped region of active cytoplasm found on the nucleus of Pander of the chick oocyte; also called blastodisc and, when cleavage is underway, blastoderm. *See:* **blastodisc**.

germinal epithelium   historically called such because it was believed that the germinal epithelium was where germ cells originated; however, they do not originate here; the epithelial covering of the ovary. Primordial germ cells, in human development, are first observable on the yolk sac of the 21-day-old embryo; i.e., germ cells do not originate in the germinal epithelium but in an extra-gonadal site.

germinal neuroepithelium   the original, one-cell-layer-thick wall of the neural tube.

germinal vesicle   the enlarged nucleus of the primary oocyte before reduction divisions are complete.

germination   the beginning or resumption of development, usually after a period of dormancy. *See:* **pollen germination**, **seed germination**, **spore germination**.

germplasm   germ cells, collectively.

gerontology   the study of aging.

gestate   to carry the young in the uterus from conception to delivery.

gestation   pregnancy.

gestation period   the period in mammals from fertilization to birth.

gestational mother   a woman who carries a pregnancy, regardless of the origin of the conceptus. *See:* **biological mother**, **surrogate mother**.

gibberellic acid (GA)   a plant hormone; formed in the plant embryo, it stimulates the production of some cereal grains. Gibberellins also promote stem elongation in plants by stimulating both cell elongation and cell division. Gibberellins are also important in breaking seed dormancy.

gill arches   the thickened masses of tissue that make up the walls of the pharynx in the early developing embryo. *See:* **branchial arches**, **pharyngeal arches**.

gill (branchial) chamber   in the frog, between hatching and metamorphosis, ventral fusion of the opercula results in the formation of a gill chamber.

gill clefts   *See:* **hyomandibular furrows**.

gill plates   bilateral, thickened regions on the surface of the frog neurula, toward the anterior end of the embryo, just posterior to the sense plate. A pair of indentations, the hyoman-dibular furrows, appear between the gill plates and the sense plate and a pair of indenta-tions, the fourth branchial furrows, appear at the posterior margin of each gill plate. Subsequently, the first, second, and third branchial furrows appear on each gill plate; the solid areas between the furrows are called the branchial arches.

gill slits   openings, through the lateral walls of the pharynx, from the pharynx to the outside environment; in the frog there are formed altogether four pairs of actual gill slits, from the second through the fifth pairs of visceral pouches (see Plate 1 in the color insert).

gills   thin, highly vascular tissue, across which exchanges of respiratory gases occurs in aqueous environments; external gills occur on the branchial arches of frog tadpoles before hatching, between hatching and metamorphosis internal gills will replace the exter-nal gills.

glands of Littré   multiple male auxiliary sex glands found along and emptying into the penile urethra (see Fig. 4).

glaucoma   this disease is also generally associated with aging, but congenital glaucoma results from a defect of the eye that impairs the flow of aqueous humor, leading to increased intraocular pressure and eye tissue damage. Glaucoma may result from rubella virus infec-tion during the embryonic period.

glial cells   one of three kinds of cells derived from the neuroepithelium of the neural tube. Originally thought to be a sort of connective tissue in the central nervous system, glial cells have been found to play a more dynamic role in the physiology of the nervous system. Also, mesenchyme cells give rise to glial cells (microglial cells) outside of the central nervous system.

glial growth factor (GGF)   a neural derived mitosis-stimulating factor; released by neurons and increases the proliferation of cells of salamander limb regeneration blastemas.

glial guidance   a mechanism thought to be important for positioning young neurons within the developing mammalian brain; throughout the cortex, neurons are observed to ride glial processes to their respective destinations.

glioblasts   cells derived from the neuroepithelium of the neural tube, which give rise to the glial cells of the central nervous system (see Fig. 18).

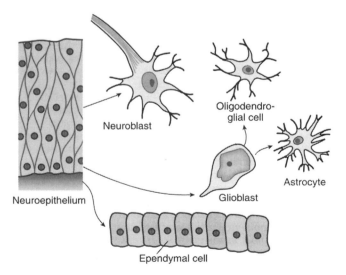

Figure 18. Cell types derived from the neuroepithelium: ependymal cells, neuroblasts, and glioblasts. Ependymal cells line the central canal of the brain and spinal cord, neuroblasts give rise to various kinds of neurons of the central nervous system, and glioblasts give rise to glial cells of the CNS. Reprinted from Frank J. Dye, *Human Life Before Birth*, Harwood Academic Publishers, 2000, fig. 13-1, p. 106.

globular stage embryo   an early stage in the embryogenesis of angiosperms; resulting from transverse and longitudinal cell divisions of the early embryo. *See:* **heart stage embryo, torpedo stage embryo**.

glomerular filtrate   the liquid that passes from the glomerulus into Bowman's capsule of the uriniferous tubule.

glomerulus   the tuft of capillaries that projects into Bowman's capsule; found in the kidneys.

-glossia   refers to the tongue, e.g., ankyloglossia.

glucose   an example of a simple type of sugar called a monosaccharide; glucose is the sugar most commonly used by the cells of the body.

Gluecksohn-Schoenheimer, Salome   (also known by Salome Gluecksohn-Waelsch) (1907–2007) a founder of developmental genetics; along with Conrad Hal Waddington, considered to be one of the two embryologists that brought about the first successful reintegration of genetics and embryology (in the late 1930s)—they attempted to find mutations that affected early development and to discover the processes that these genes affected; she showed that mutations in the *Brachyury* genes of the mouse caused the aberrant development of the posterior portion of the embryo, and she traced the effects of these mutant genes to the notochord, which would normally have helped induce the dorsal axis.

glutathione   a tripeptide (glutamate-cysteine-glycine) that binds heavy metals, such as mercury and cadmium, which are toxic to cells and embryos; glutathione also provides defenses against oxidant damage. *See:* **metallothioneins, phytochelatins**.

glycoprotein   a molecule that is partially carbohydrate and partially protein; extremely important molecules within cells, on the surface of cells and in the extracellular matrix.

glycosaminoglycans (GAGs)   the polysaccharide uints of proteoglycans; long, linear, highly charged polymers of a repeating disaccharide in which many residues are often sulfated; as units of proteoglycans, GAGs are major components of the extracellular matrix.

gnathostomes   vertebrates with jaws. *See:* **agnatha**.

Goldschmidt, Richard   (1878–1958)   German American biologist; author of *The Material Basis of Evolution* (1940); challanged the modern synthesis of evolutionary biology and Mendelian genetics; maintained that macroevolutionary change requires another evolutionary method other than that of sheer accumulation of micromutations. Goldschmidt saw homeotic mutants as "macromutations" that could change one structure into another and possibly create new structures or new combinations of structures; these mutations would be in regulatory genes rather than in structural genes. Goldschmidt was largely ignored because, among other things, he did not believe in Thomas Hunt Morgan's notion of the gene as a particulate entity.

Golgi apparatus   a cellular organelle, concerned with glycosylation of proteins and macromolecular traffic in the cell.

gonadal ridges   paired mesodermal ridges from which gonads develop in vertebrates.

gonadotrophin   or gonadotropin.

gonadotropin   a hormone that acts to stimulate the gonads.

gonadotropin-releasing hormone (GnRH)   peptide hormone released from the hypothalamus that stimulates the pituitary to release the gonadotropins, follicle-stimulating hormone, and luteinzing hormone, which are required for mammalian gametogenesis and steroidogenesis.

gonads   the organs responsible for the production of gametes; ovaries producing eggs in females and testes producing sperm in males.

gonial syncytia   during spermatogenesis, the products of meiotic divisions are connected by cytoplasmic connections (bridges).

gonidia   large, asexual reproductive cells formed in the life cycle of *Volvox*.

gonocytes   mammalian primordial germ cells (PGCs) that have arrived at the genital ridge of a male embryo and have become incorporated into the sex cords.

gonorrhea   a bacterial, highly contagious, sexually transmitted disease.

gonotome   the part of an embryonic somite involved in gonad formation.

goosecoid   a *Xenopus* gene, turned on by higher concentrations of the morphogen activin, responsible for instructing cells to become dorsal mesoderm, e.g., notochord. *See: Brachyury*.

Graaf, Regnier de   (1641–1673)   Dutch, after whom the Graafian follicle is named; observed sections of mammalian ovaries and observed fluid-filled spaces that he erroneously considered to be eggs. He also reported (in 1672) that he had succeeded in tracing the mammalian egg down the Fallopian tubes to the uterus, and he was able to describe ovulation in general terms; thus, the generative process was associated with changes occurring in the ovary.

Graafian follicle   a mature ovarian follicle, consisting of a single oocyte and numerous follicle cells (and their products). During the 18th century, Graafian follicles were believed to be eggs, and efforts to trace them led to frustration and imagination.

gradient   the variation in the concentration of a substance, with increasing distance from the source corresponding to decreasing concentration of the substance. To move from a region of lower to higher concentration is said to move up (against) the concentration gradient; movement in the opposite direction is said to move down (with) the concentration gradient; e.g., *Dictyostelium* myxamoebae move up a gradient of extracellular cyclic adenosine monophosphate (cAMP).

gradualism   the view that most species evolved gradually and continuously in direct resonse to selective pressures from the environment. *See:* **punctuated equilibrium theory**.

granulocytes   white blood cells distinguished by granules in their cytoplasm; basophils, eosinophils, and neutrophils.

granulosa cells   also called follicle cells. *See:* **follicle cells**.

gray crescent   on frog eggs, a slightly lighter pigmented area, formed opposite the spermatozoon entrance point; the gray crescent will become the dorsal side of the future animal. An indicator of the significance of the gray crescent is that if the first cleavage division is such that one blastomere receives all of the gray crescent and the other none, and these blastomeres are experimentally separated, the blastomere with the crescent will form a complete embryo, while the other blastomere forms only a few abortive parts. On fertilized *Styela* eggs, a gray crescent also occurs naturally; by following the fate of this gray material, it has been determined that the gray crescent of the fertilized egg has two developmental fates: (1) the upper (toward the animal pole) portion consists of presumptive neural ectoderm and (2) the lower (toward the vegetal pole) portion consists of presumptive notochord . *See:* **cortical rotation**, **yellow crescent**.

gray matter   those portions of the central nervous system that have a large concentration of nerve cell bodies and therefore appear gray in the fresh state (see Fig. 43).

great fault of embryology   the tendency to explain any and every operation of development as merely the result of inheritance (Locy, 1908); surrounding conditions have much to do with individual development, and the course of events may depend on external stimuli and not only on an inherited tendency.

green fluorescent protein   a fluorescent reporter protein that is usually made in jellyfish.

grex   the pseudoplasmodium of *Dictyostelium discoideum*. *See:* **slime molds**, **slug**.

Grobstein, Clifford (1916–1998)   American developmental biologist; defined the basic rules of the tissue interactions that support development—cell differentiation and morphogenesis—in embryos of mammals (and. as we know today, all vertebrates). The roles of extracellular materials and matrix during such development and the ability of different cell populations (epithelial and mesenchymal) to interact at a distance are landmark findings that have stimulated and guided experimentation worldwide as deeper understanding of development in embryos and developmental phenomena in adults has been gained.

ground meristem   the primary meristem that gives rise to the ground tissue of plants; begins to become visible in the heart-shaped embryo of angiosperms.

ground tissue   one of three tissue systems that emerge at the globular stage of angiosperm embryogenesis; the ground tissue forms the cortex and pith. *See:* **dermal tissue**, **procambium**.

growth   the process of growing, an increase in size by the addition of material; in insects, growth may be by (1) cell multiplication (tissues that grow by this method are not, generally, subjected to dissolution and are carried over into the adult insect) or (2) increase in cell size (tissues growing by this method are usually subjected to histolysis during the pupal stage).

growth cone   the leading edge of the forming axon of a neuron; the locomotory organelle of the growing axon.

growth factor   a secreted protein that binds to a receptor and initiates signals to promote or retard cell division and growth (Table 1).

growth hormone   a protein normally secreted into the bloodstream by the pituitary gland, which promotes an increase in body size. *See:* **adipocytes**, **chondrocytes**.

Table 1. Examples* of Growth Factors and Their Functions

Growth Factor	Function
Bone morphogenetic proteins	Morphogenetic signals
Epidermal growth factor	Cell proliferation
Erythropoietin	Red blood cell production
Fibroblast growth factor	Mitogens
Hepatocyte growth factor	Angiogenesis
Insulin-like growth factor	Neurotrophic factor
Nerve growth factor	Neuron survival
Platelet-derived growth factor	Angiogenesis/Mitogen
Vascular endothelial growth factor	Vasculogenesis/Angiogenesis
Placental growth factor	Vasculogenesis/Angiogenesis

*This is not a comprehensive list of growth factors or their functions.

growth plate closure   cessation of bone growth, when epiphyseal growth plates no longer form cartilage.

guard mother cell   the direct precursor of the two guard cells that flank a stoma; in both dicots and monocots, an asymmetric division precedes the differentiation of guard cells; the smaller cell is the guard mother cell; in most plants, the guard mother cell influences adjacent cells to divide asymmetrically to form subsidiary cells. Thus, the final stomatal complex consists of a group of clonally related as well as of recruited adjacent cells.

Gurdon, Sir John Bertrand (1933–)   English geneticist; professor of cell biology at Cambridge. In 1968, he transplanted a nucleus derived from frog gut into an enucleated, fertilized egg and produced a normal tadpole; this demonstrated that fully differentiated animal cells retain the genetic information to become any cell type under the correct environmental stimuli.

gurken gene   the setting up of the _Drosophila_ egg's dorsoventral axis involves a set of oocyte-follicle cells interactions later than those for the anteroposterior axis. The oocyte nucleus moves (along microtubules) to a site on the anterior margin of the oocyte; here the _gurken_ gene is expressed again; the locally secreted gurken protein acts as a signal to adjacent follicle cells on one side of the oocyte, specifying them as dorsal follicle cells; the side away from the nucleus becomes the ventral region by default; the ventral follicle cells produce proteins that are only deposited in the ventral vitelline envelope.

gurken protein   the protein synthesized and secreted by the posterior end of the _Drosophila_ oocyte; the inductive signal, directing the surrounding follicle cells to adopt a posterior fate, is transmitted by the gurken protein, which binds to a receptor in the follicle cell plasma membrane encoded by the torpedo gene; the torpedo protein is a transmembrane receptor tyrosine kinase.

gut   a term commonly used to refer to the digestive tube.

gut-associated lymphoid tissue (GALT)   mediates mucosal immunity and oral immune tolerance; malfunction has been associated with allergies and with Crohn's disease in humans. In rabbits, a set of bacteria (_Bacteroides fragilis_ and _Bacillus subtilis_) appears to be required together to _induce formation_ of GALT.

GVBD   germinal vesicle breakdown; follows progesterone stimulation and precedes ovulation in the frog.

gymnosperms   literally naked seeds; these plants do not have their seeds enclosed in an ovary, as in the angiosperms (vessel seeds) but borne "naked" on the surfaces of cone scales.

-_gyn_-   a woman, e.g., as in polygyny.

gynandromorphs   animals in which certain parts of the body are male and other regions are female.

gyne   the reproductive queen of ants, which has functional ovaries.

gynoecium   collectively, all of the carpels or pistils of a flower.

gynogenesis   development of a fertilized egg through the action of the the egg nucleus, without participation of the sperm nucleus.

gynomerogony   development of a fragment of a fertilized egg containing the haploid egg nucleus.

gyri   the ridges found on the surface of the cerebrum, separated from each other by the sulci (furrows).

H

Haeckel, Ernst (1834–1919) German biologist, professor of zoology and comparative anatomy at Jena; established the phylum Chordata (in 1874), incorporating the subphyla Urochordata, Cephalochordata, and Vertebrata. His students included Wilhelm Roux and Hans Driesch. In his comparative studies, Karl Ernst von Baer observed corresponding stages in the development of different animals, but he did not relate these observations to evolution; Haeckel made the connection and popularized the phrase "ontogeny recapitulates phylogeny." Expanding the idea of his mentor, Johannes Müller, Haeckel argued that the embryological stages of an animal were a recapitulation of its evolutionary history, and indeed that there had once been complete animals resembling the embryonic stages of animal forms living today. More recently, developmental biology has treated Haeckel's recapitulation theory with caution. He proposed the Gastraea Theory. *See:* **Gastraea Theory**.

hair germ a hair follicle primordium.

hairy **gene** the expression pattern of this gene correlates with the positioning of the place where a somite will separate from the unsegmented paraxial mesoderm; this gene encodes a transcription factor.

Haldane, J. S. (1860–1936) declared that there is "no *spatial* limit to the life of an organism."

Haller, Albrecht von (1708–1777) Swiss scientist and writer; he had already won distinction as botanist and poet when he was appointed (in 1736) professor of anatomy, medicine, and botany at the University of Göttingen, where he carried on the research in experimental physiology for which he is especially known. He considered parthenogenesis (as in aphids) to be one of the most powerful arguments in favor of ovism.

halteres the hindwings of two-winged insects (Diptera), e.g., flies and mosquitoes, which functions as a small balancing organ. Have been studied in the developmental regulation of and evolutionary diversification of insect wing morphology.

Hamburger, Viktor (1900–2001) German experimental embryologist, student of Hans Spemann; made fundamental contributions to our understanding of the nervous system and, with Rita Levi-Montalcini, discovered the nerve growth factor. Published, with Harold Hamilton (in 1951), staged series of morphological landmarks of normal development of the chick, which provided the standard by which chick development continues to be described.

Hammerling, Joachim (1901–1980) Danish biologist; carried out classical experiments with *Acetabularia*, providing evidence that the nucleus provides the information determining phenotype.

haplodiploidy a reproductive system in which males are haploid and develop from unfertilized eggs, while females are diploid and develop from fertilized eggs.

haplodiplontic life cycle life cycles exhibited by plants, in which mitotic cell divisions are involved in both the haploid and diploid generations. *See:* **diplontic life cycle**, **haplont**.

haploid half the diploid number of chromosomes. For humans, the normal haploid number is 23.

haploinsufficiency a mechanism of dominance, in which one copy of the gene (haploid condition) is not sufficient to produce the required amount of product for normal development, e.g., Waardenburg syndrome type 2, with roughly half the wild-type amount of MITF. *See:* **dominant negative mutation**, **gain-of-function mutation**.

haplont a plant with only haploid somatic cells; the zygote is diploid, e.g., many algae and fungi.

Dictionary of Developmental Biology and Embryology, Second Edition. Frank J. Dye.
© 2012 Wiley-Blackwell. Published 2012 by John Wiley & Sons, Inc.

haptotaxis   migration on preferred substrates; the ability of a cell, or the growth cone of a neuron, to move up a gradient of adhesivity; may play a role in guidance of migratory cells.

hard palate   the anterior part of the palate, containing bone.

harelip   *See:* **cleft lip**.

Harris, J. Arthur (1880–1930)   "natural selection may explain the survival of the fittest, but it cannot explain the arrival of the fittest." *See:* **"arrival of the fittest."**

Harrison, Ross Granville (1879–1959)   American biologist and anatomist; went to Yale as professor of comparative anatomy in 1907 and held various honorary positions there until his death. He is known for his work on nerve development in the embryo and on nerve regeneration as well as for his discovery of a method of tissue culture that permits study of isolated living cells in a controlled environment; perfected the transplantation techniques (as did Hans Spemann) used to discover when the body and limb axes are determined; author of *Organization and Development of the Embryo* (1969).

Hartsoeker, Nicolaas (1656–1725)   Dutch scientist, an animalculist; even went as far as to suggest that each sperm contained a tiny preformed man, a homunculus; stated (in 1694) that each spermatic "worm" of a bird enclosed a male or female bird of the same species and, in copulation, a single "worm" entered an egg where it was nourished and grew. *See:* **homunculus**.

Harvey, William (1578–1657)   English physician and anatomist, born in Kent, England; is believed to have coined the word "epigenesis" to refer to the process that drives the egg. His book, *The Development of Animals*, published in 1651, did not lead immediately to any significant advance in embryology. The best remembered part of Harvey's book was the dictum, *Ex ovo omnia* ("All creatures come from an egg."). According to E. J. Gardner (1965), Harvey had not observed a mammalian egg and he did not know with certainty what an egg was. Acording to William A. Locy (1908), although there were observers in the field of embryology before Harvey, little of substantial value had been produced. According to Cleveland P. Hickman (1966), Harvey was opposed to the preformation theory that was held by many biologists of his and later times. *See:* **epigenesist**.

hatching   escape of an embryo from the confines of the original egg coats; e.g., frog tadpoles hatch out of the jelly coats of the original egg; fish embryos hatch out of the chorion; mammalian embryos hatch out of the zona pellucida; bird eggs hatch out of the egg shell; and sea urchin embryos hatch out of the fertilization membrane (see Plate 7 in the color insert).

hCG   human chorionic gonadotropin.

head activator gradient   a proposed morphogenetic gradient in *Hydra*, of head activator, which is highest at the hypostome; when hypostome tissue from one *Hydra* is transplanted into the middle of another *Hydra* a new apical-basal axis is formed, with the hypostome facing outward. *See:* **foot activator gradient**, **head inhibitor gradient**.

head fold   one of four body folds found in chick development; the head fold sculpts the head of the embryo out of the cephalic region of the embryonic disc (see Plate 2a in the color insert and Fig. 6).

head inhibitor gradient   a proposed morphogenetic gradient in *Hydra*, of head inhibitor, which is highest at the hypostome; extra heads do not form in *Hydra* because the presence of the hypostome prevents the formation of any other hypostome. *See:* **foot inhibitor gradient**, **head activator gradient**.

head kidney   the pronephros.

head mesoderm   mesoderm of the head in the chick, derived from mesenchyme.

head process   historically, the name for the young notochord of chick embryos.

head process stage   historically, the young notochord was called the head process; in the 18-hour chick embryo, the notochord is a conspicuous structure so the 18-hour chick embryo has been referred to as the head process stage.

heart development   in the frog, the heart begins to pulsate at the 5-mm stage; in the chick embryo, the paired primordia of the heart originates from thickened splanchnic mesoderm at the lateral margins of the anterior intestinal portal. In the early stage of heart development (29 hours), the heart is a straight tubular structure, at the level of the rhombencephalon, in the midline, ventral to the foregut, cephalically continuous with the ventral aorta and caudally continuous with the paired omphalomesenteric veins. By 33 hours, rupture of the dorsal mesocardium has occurred at the midregion of the heart and the midregion of the cardiac tube is dilated and bent to the right. The fundamental regions of the heart are, from blood intake to discharging end: sinus venosus, atrium, ventricle, and truncus arteriosus. At 36–38 hours: the sinus venosus is still paired, the atrium is held (by the dorsal

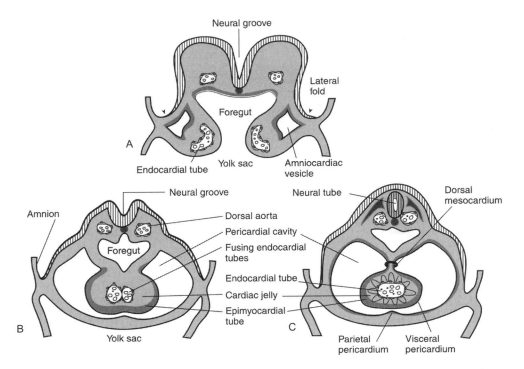

Figure 19. Formation of the heart (endocardial) tube is illustrated in diagrams of transverse sections through the heart-forming regions of the early embryo. (A) 20 days, (B) 21 days, (C) 22 days. Reprinted from Frank J. Dye, *Human Life Before Birth*, Harwood Academic Publishers, 2000, fig. 15-1, p. 121.

mesocardium) close beneath the caudal part of the foregut, the ventricle makes a U-shaped bend to the right, and the truncus arteriosus is in the midline. The formation of the tubular heart is a sequential process: first the truncoventricular part, then the atrium, and finally the sinus venosus. Between 30 and 55 hours, the heart grows more rapidly than the body of the embryo where it lies; as its cephalic and caudal ends are fixed, the unattached midregion first becomes U-shaped and then twisted to form a loop. In this twisting of the heart, the atrial region undergoes little change in position and the ventricular region is carried over the right side of the atrium and comes to lie caudal to it. Beginning at 50–55 hours, the myocardium shows irregular projections extending into the cardiac jelly; this is the beginning of the formation of the trabeculae carneae. As the trabeculae grow, the endothelial lining of the heart tends to extend between them and follow closely the configuration of the muscular strands, separated from them by a thin layer of cardiac jelly. The ventricular wall becomes honeycombed by tortuous intertrabecular spaces that bring blood to growing cardiac muscle before coronary circulation has begun (see Plates 2b, 5, and 13 in the color insert as well as Figs. 19 and 20).

heart murmur   a roaring sound coming from the heart; detectable with a stethoscope.

heart stage embryo   an intermediate stage in the embryogenesis of angiosperms. At this stage, the earliest development of the cotyledons (dicots have two and monocots have one cotyledon) can be observed; this gives the dicot embryo a heart-shaped appearance; the two lobes at the top of the embryo go on to become the cotyledons in the mature embryo. The axial body plan of the embryo is evident by this stage of embryogenesis. *See:* **globular stage embryo, torpedo stage embryo**.

heat shock   a technique for inducing phenocopies and ecotypes, used as early as the 1890s to disrupt the pattern of butterfly wing pigmentation. *See:* **ecotype, phenocopy**.

heat shock proteins (Hsp)   cellular proteins that bind to unstable, partially denatured, proteins and stabilize them, e.g., Hsp90. *See:* **canalization**.

hedgehog   a segment polarity gene in *Drosophila*; a mutation of it results in an altered denticle pattern, similar to that observed in a mutation of *wingless*. *Hedgehog* encodes a

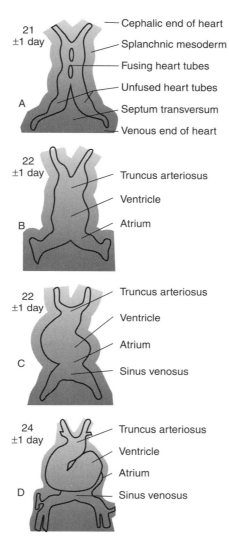

Figure 20. Regions of the early heart are shown in diagrams of the ventral side of the heart during the fourth week. (A) 21 Days, (B) 22 days, (C) 23 days, (D) 24 days. Reprinted from Frank J. Dye, *Human Life Before Birth*, Harwood Academic Publishers, 2000, fig. 15-2, p. 122.

secreted signaling protein and is related to the *Sonic hedgehog* gene in vertebrates. Hedgehog and wingless proteins are highly conserved developmental signaling molecules.

hedgehog receptor also called patched; a member of the G-protein-linked class of receptors.

hedgehog signal transduction pathway an important pathway in limb and neural differentiation in vertebrates.

Heilbrunn, Lewis Victor (1892–1959) "calcium release Heilbrunn"; noted a requirement for calcium in order for embryos to survive mechanical damage (in 1943); he suggested that calcium ions were required for "clotting" the cytoplasm, noting the similarity to the then recently discovered requirement of calcium ions for blood clotting.

Hellin's Law a mathematical prediction of the occurrence of multiple births; for example, twins once in 80^1 pregnancies, triplets once in 80^2 pregnancies, and so on.

hemangioblasts the common precursor cells shared by blood vessels and blood cells; true blood cell stem cells, of the chick embryo, derived from the splanchnopleure and formed within nodes of mesoderm that line the mesentery and the major blood vessels.

hematopoiesis   formation of blood cells.

hematopoietic   blood-forming.

hematopoietic inductive environments (HIMs)   regions where blood-forming cells undergo determination; the developmental pathway taken by the descendants of a pluripotential stem cell depends on which growth factors it is exposed to, which, in turn, is determined by the stromal cells to which it is exposed. Stem cells in the spleen become predominantly erythroid, while those in the bone marrow become predominantly granulocytic. Stromal cells of the bone marrow create HIMs by their ability to bind hematopoietic growth factors; without these growth factors, the stem cells die.

hematopoietic stem cells (HSCs)   those cells that generate the different types of blood cells.

Hemichordata   a phylum of animals that has a combination of invertebrate and chordate characteristics.; includes the classes Enteropneusta (acorn or tongue worms) and Pterobranchia (pterobranchs).

hemichordate   an animal belonging to the phylum Hemichordata.

Hemichordate Theory   of the origin of vertebrates. *See:* **Echinoderm Theory**.

Hemimetabola   comprise all the lower orders of insects or those with external wing-growth (Exopterygota); they undergo simple metamorphosis. The insect orders making up the Hemimetabola are the Orthoptera, Isoptera, Plecoptera, Embioptera, Dermaptera, Ephemeroptera, Odonata, Psocoptera, Anoplura, Thysanoptera, and Hemiptera.

hemimetabolous   descriptive of the life cycles of some species of insects that show the egg, nymphal, and imago (adult) stages of development; these insects do not undergo a dramatic metamorphosis, e.g., dragonflies and stoneflies; they are said to undergo simple metamorphosis, described as being direct or incomplete. The organs of the nymph become transformed into those of the imago with little change other than an increase in size and limited structural elaboration. *See:* **holometabolous**.

hemisected   cut into two halves.

hemizygous   genes that are found on the male Y chromosome and not on the female X chromosome, its pairing partner; genes usually come in pairs because they are found on paired chromosomes; because such genes are only present in a single dose (i.e., on the Y chromosome), they are said to be hemizygous (half in the zygote).

hemochorial placenta   a type of placenta having the maternal blood in direct contact with the chorionic trophoblast; exhibited by primates (including humans) and rodents, for example.

hemodichorial placenta   a placenta with a double trophoblastic layer.

hemoendothelial placenta   a placenta having the endothelium of vessels of chorionic villi in direct contact with the maternal blood, e.g., exhibited by rabbits.

hemogenic endothelial cells   primary endothelial cells of the dorsal aorta that are derived from angioblasts that migrated from the sclerotome; they give rise to the hematopoietic stem cells that migrate to the liver and bone marrow and become the adult hematopoietic stem cells.

hemoglobin   the oxygen- and carbon dioxide-carrying protein found in blood; packaged in red blood cells, this protein is a cell differentiation product of red blood cells. *See:* **adult hemoglobin**, **fetal hemoglobin**.

hemolytic disease of the newborn   also called erythroblastosis fetalis, a hemolytic anemia observed in the newborn, caused by transfer of maternal antibodies across the placenta in response to Rh incompatibility.

hemorrhage   abnormal loss of blood.

hemotrichorial placenta   a placenta with a triple trophoblastic layer.

Hensen's node   thickening formed by a group of cells at the anterior end of the primitive streak in higher vertebrate gastrulae; the mound of cells at the cephalic end of the primitive streak in the chick embryo. Cells passing through the primitive node during gastrulation give rise to the notochord. Also called the primitive node; Hensen's node in the chick embryo is considered to be the homologue of the dorsal lip of the amphibian blastopore.

hepatectomy   removal of the liver. *See:* **partial hepatectomy**.

hepatic diverticulum   an evagination of the embryonic foregut that gives rise to the liver, gallbladder, and their ducts.

hepatic duct   duct that transmits bile to the common bile duct.

hepatocyte growth factor (HGF)　an important protein for returning liver cells to the cell cycle; induces many embryonic proteins.

Herbst, Curt (1866–1946)　the first practitioner of chemical experimental embryology; made the discovery (in 1893) that sea urchin embryos would exogastrulate if lithium salts were added to the sea water in which they were kept, thereby introducing the phenomenon of exogastrulation to experimental embryology.

He found (in 1900) that when cleaving echinoderm eggs were raised in calcium-free water, the cleavage cells separated completely and cleanly from one another; according to Jane M. Oppenheimer (1994), artificial separation of blastomeres was an important embryological practice at the time; laboratory procedures were considerably facilitated and improved by the introduction of Herbst's new method of making it possible for blastomeres to separate themselves more gently than embryologists had previously been able to do it for them.

Ernst Haeckel was one of his teachers, and Hans Driesch was one of his fellow students. Herbst began his experments on sea urchin eggs by raising them in calcium-free sea water (the skeleton of the echinoderm is calcareous in constitution) and found that if their calcareous skeletal rods were defective or absent, the arms of the plutei (larvae) were absent; his *interest* in such experiments had been aroused by Julius Sachs' view that the morphogenesis of an organism is affected by its chemical constitution, and Herbst *began* his experimental investigations on the influence of an altered chemical constitution of the surrounding medium on the development of animals stimulated by three short communications by Georges Pouchet and Laurent Chabry, that dealt with the influence on sea urchin development of a greater or lesser degree of precipitation of the calcium from the sea water by potassium or sodium oxalate. Herbst felt that his own interpretation went beyond that of Pouchet and Chabry in that he thought of the effect of the skeletal spicules not in a mechanical analogy, but in terms of morphogenetic stimuli and *Auslösung* (the word that most embryologists of the early 1900s used when they encountered evidence for interactions between embryonic parts).

In 1901, Herbst published his book *Formative Reize in der tierischen Ontogenese* (Formative Stimuli in Animal Ontogenesis), in which Herbst, as Hans Spemann, postulated that the formation of the vertebrate lens is brought about as the result of contact between the ectoderm that forms it and the subjacent optic vesicle; unlike Spemann, Herbst did not do his own experiment; he interpreted the naturally occurring, developmental, vertebrate anomaly of cyclopia.

In 1895, Herbst made it very clear that he then felt that the strongest influence on him emanated from the work of Rudolf Virchow. It was Virchow, claimed Herbst, who introduced the idea of formative stimuli into science. According to Oppenheimer (1994), Virchow the pathologist was thinking of the formation of tumors; Herbst applied the idea to embryonic development. Karl Ernst von Baer wrote in no uncertain terms that the developing eye evokes the formation of other structures by the skin. According to Oppenheimer (1994), "To try to pin down the ultimate origins of the concept of induction would be a silly exercise; in logic, the idea reaches back into antiquity." (p. 83)

Herbst did not interpret induction in terms of cells; "The reference of developmental processes to the cell," wrote Ross Granville Harrison in 1937, "was the most important step ever taken in embryology." (p. 372) Herbst stopped short of taking that step, according to Oppenheimer (1994). Herbst's contribution to induction theory was fleshing it out into an elaborately conceived hypothesis; he was not, according to Oppenheimer, its only begetter. According to Margaret Saha (1994), Herbst argued that it eventually might be possible to resolve the entire process of embryogenesis into a series of inductions exerted by one region of the embryo on another.

According to Oppenheimer (1994), Herbst's studies, published in 1892 and 1893, on the effects of an altered constitution of the ambient media on embryonic development were among the most important early contributions toward the initiation of chemical embryology; according to Viktor Hamburger (1988), he "deserves credit as the first practitioner of chemical experimental embryology." (p. 14) *See:* **Pfeffer, Wilhelm Friedrich Philipp.**

hermaphrodite　an individual in which the germ cells will produce both sperm and eggs.

hermaphroditic　(1) generally, describes a species in which male and female reproductive function occur in the same individual; (2) a plant; with pistils and stamens in the same flower; bisexual; perfect.

hermaphroditism   in Greek mythology, Hermaphroditus', a son of the Greek gods Hermes and Aphrodite, body coalesces with that of a nymph who is in love with him. In each afflicted individual, there is the presence of recognizable ovarian and testicular tissue and sexual structures, and secondary sex characteristics may show any combination of maleness and femaleness. Approximately 50% of these individuals have an XX sex chromosome makeup, approximately 25% have an XY, and the remainder are XX/XY mosaics.

hernia   abnormal protrusion of an organ or a part through the containing wall of its cavity is called a hernia or rupture; e.g., umbilical hernia. *See:* **-cele**.

herpes simplex virus (HSV)   a virus that may be passed from mother to baby during birth as the baby passes through the birth canal, with the resulting infection and possible death of the baby.

Hertwig, Wilhelm August Oscar (1849–1922)   German zoologist, professor at Jena University; he (in 1890) and Theodor Boveri separately discovered the real nature of reduction division in the egg; (in 1876) demonstrated the penetration of the egg by the spermatozoon and the union of their nuclei, in the sea urchin. Hertwig was the first to observe actual cellular fertilization in animals. He performed essentially Roux's famous experiment, using the sea urchin instead of frogs, but he succeeded in separating the two blastomeres with a fine needle; both lived, and each formed a whole embryo, indicating that the two cells were not predetermined.

Hertwig's Law   the mitotic spindle always tends to lie so that its longitudinal axis coincides with that of the yolk-free cytoplasm of the cell. The consequence of this law for frog eggs is that the plane of cleavage (for the first two cleavage divisions) will coincide with the egg's polar axis and so passes through its animal and vegetal poles.

heteroblastic   developing from different tissues or germ layers, in referring to similar organs in different species.

heterochromatin   condensed and, generally, inactive chromatin; found, for example, in Barr bodies. *See:* **constitutive heterochromatin, facultative heterochromatin**.

heterochronic mutation   a mutation that changes the timing of developmental events.

heterochrony   a shift in the relative *timing* of two developmental processes from one generation to the next; a phenomenon exhibited by animals where there is a change in the relative time of appearance and rate of development of characters already present in their ancestors; in marsupials, the forelimbs develop at a faster rate relative to the embryonic forelimbs of placental mammals. The mutations affect the regulatory regions of the genes. *See:* **heterometry, heterotopy, heterotypy**.

heterocyberny   the concept that environmentally induced changes to a phenotype, when adaptive over long periods of time, can become the genetic norm for a species: a contribution of eco-devo.

heterogamy   See **anisogamy**.

heterogeneous nuclear RNA (hnRNA)   a class of large, heterogeneous-sized RNAs found in the nucleus, including unspliced mRNA precursors. *See:* **nuclear RNA**.

heterogony   See **allometry**.

heterokaryon   literally different nucleus; the condition wherein two nuclei occupy the same cell but do not fuse to form a single nucleus; characteristic of some fungi.

heterometry   the change in the *amount* of a gene product; e.g., the correlation between the beak shape of finches and the amount of *Bmp4* expression; the mutations affect the regulatory regions of the genes. *See:* **heterochrony, heterotopy, heterotypy**.

heteromorphosis   the condition in regeneration wherein one type of appendage is amputated and another kind regenerates in its place. *See:* **Loeb, Jacques**.

heteroplastic transplantation   transplantation between individuals of different species or genera.

heterosporous   literally different spores; heterosporous plants produce more than one kind of spore; e.g., a microspore that gives rise to the male gametophyte and a megaspore that gives rise to the the female gametophyte. Mosses, some ferns, and angiosperms are heterosporous. *See:* **homosporous**.

heterotaxis   the state in which internal organs of normal and inverted symmetry are present in the same animal.

heterothallus   literally different thalli; heterothallic plants produce separate male and female thalli. *See:* **homothallus**.

heterotopic graft   a graft to a different position in the host embryo; represents a test for determination. *See:* **orthotopic graft**.

heterotopy changing the *location* of a gene's expression; i.e., altering in which cells it is expressed; e.g., the webbing in embryonic duck feet remains intact as a result of heterotopy. The mutations affect the regulatory regions of the genes. *See:* **heterochrony**, **heterometry**, **heterotypy**.

heterotypy mutations that affect the protein that binds to the regulatory regions of the gene; the changes of heterotypy affect the actual coding region of the gene and change the functional properties of the protein being synthesized. *See:* **heterochrony**, **heterometry**, **heterotopy**.

heterozygous descriptive of an organism in which its two alleles for a given characteristic are not identical (different in the zygote).

Heuser's membrane on day 8 of human development, a thin layer of hypoblast cells, known as Heuser's membrane, migrates out over the inner, cellular layer of the trophoblast (cytotrophoblast), thus creating the primitive yolk sac that occupies the bulk of the early blastocyst cavity. It is later sloughed off and replaced by a definitive yolk sac, which will become the extraembryonic endoderm.

hibernal flowering in the winter.

hierarchy the organization of developmental regulatory circuits into two or more tiers; gene products in one tier control the expression of genes in lower tiers. *See:* **developmental genes**, **regulatory circuit**.

high mobility group (HMG)-box factors transcription factors; do not have a specific activation or repression domain; instead they work by bending the DNA to bring other regulatory sites into contact with the transcription complex, e.g., the testis determining factor SRY.

higher vertebrates amniotic vertebrates; reptiles, birds, and mammals.

hindbrain the most caudal of the early brain vesicles of the three-vesicle brain (see Plate 5 in the color insert).

hindgut the caudal-most portion of the early embryonic gut, which gives rise to the rectum and part of the cloaca. In the frog, the rectum originates, with slight growth, from the region of the archenteron remaining between the yolk mass and the posterior body wall. Before hatching, a perforation at the proctodaeum gives rise to the anus (this is the point at which the rectum may be judged to end; i.e., to open into the cloaca). In the chick embryo, the hindgut appears at approximately 50 hours (see Plate 5 in the color insert and Fig. 6). *See:* **foregut**, **midgut**.

hinge regions regions where the neural tube contacts surrounding tissues and which play a role in bending of the neural plate of vertebrates. In birds and mammals, they consist of the medial hinge point (MHP) (cells of which become anchored to the underlying notochord) and dorsolateral hinge points (DLHPs) (cells of which are anchored to the surface ectoderm of the neural folds).

His, Wilhelm (1831–1904) German anatomist and embryologist; argued that the causes of germ layer movements must be sought in the mechanics of cell growth and the interacting pressures of expanding germ layers, rather than in the biogenetic law. His improvement of the microtome allowed for successful serial sectioning, so important to embryology. By his polemical attacks on, what he believed were, the phylogenetic excesses of Ernst Haeckel, he stimulated the rise of experimentation in embryology. His insisted (Maienschein) that individual embryonic development must be explained in mechanical terms directly affecting the individual itself. He called for a physiological approach to development, which would base embryology strictly on a study of developmental processes taking place within the individual and not on evolutionary patterns of structural change, in distinction to Karl Gegenbauer and Haeckel. By claiming that embryology could provide causal explanations of development, His gave embryology a new purpose and suggested a program of research. Evidence gathered with techniques including His's microtome improvements confirmed His's views that different organisms differ in their *earliest* stages; according to Haeckel, all organisms are *initially* essentially alike. *See:* **Gegenbauer, Karl**; **Haeckel, Ernst**.

histioblasts small and active cells derived from divisions of the archeocytes of sponge gemmules; as the archeocytes divide, they give rise to smaller and smaller cells (histioblasts), much like the way that blastomeres diminish in size as cleavage progresses.

histoblasts undifferentiated cells, which retain the potentialities of embryonic tissue; the primary elements in the histogenesis that takes place during insect metamorphosis. Until metamorphosis sets in, the histoblasts remain dormant; when they become functional, they undergo cell division and form the rudiments of the future (adult) organs.

histodifferentiation differentiation of cell groups into tissues.

histogenesis the developmental process by which the definitive cells and tissues that make up the body of an organism develop from embryonic cells; the generation of new tissues, as in the replacement of the larval organs and tissues by the organs of the adult insect during metamorphosis.

histolysis a variable degree of dissolution of the larval organs and tissues that takes place during the pupal stage of the holometabolic insect life cycle.

histone acetylation the addition of acetyl groups (negatively charged) to histones that neutralizes the basic charge of lysine, loosens the histones, and activates transcription.

histone acetyltransferase an enzyme that transfers acetyl groups, from acetyl CoA, to histones in nucleosomes, which destabilizes the nucleosomes and permits gene expression.

histone deacetylases enzymes that remove acetyl groups, stabilize the nucleosomes, and prevent transcription.

histone methylation the addition of methyl groups to histones; can either activate or further repress transcription.

histone methyltransferases enzymes that add methyl groups to histones.

histones small, basic, proteins that combine with DNA to form chromatin; chemical modification of histones affects differential gene expression by modifying nucleosome packing. *See:* **acetylation**.

Historical Periods of Embryology "five periods, each marked by an advance in general knowledge: (1) the period of William Harvey and Marcello Malpighi, (2) the period of Friedrich Wolff, (3) the period of Karl Ernst von Baer, (4) the period from von Baer to Francis Balfour, and (5) the period of Balfour. Among all the leaders, von Baer stands as a monumental figure at the parting of the ways between the new and the old—the sane thinker, the great observer" (page 196). From William A. Locy (1908).

History of Evolutionary Biology can be observed as nested subsets of syntheses, each one explaining more phenomena: (1) Charles Darwin's synthesis of biogeography, breeding, paleontology, and taxonomy was superceded by the (2) Fisher-Wright version of the Modern Synthesis, which integrated Darwinian evolution with population genetics and showed the mechanisms through which Darwin's natural selection can take place through changes in alleles within a population; as population genetics matured, the (3) Mayr-Dobzhansky version of the Modern Synthesis explained more about how speciation events could develop in natural populations; this version remains the core of evolutionary biology onto which other evolutionary theories have been appended . . . more than 50 years old . . . . due for a major revision . . . . . developmental biology will be the core of such a new synthesis . . . . . . ecological developmental biology will be a major bridge linking development and evolution. That time has come. From S. F. Gilbert and D. Epel (2009).

histotypic aggregation is the reconstruction of complex tissues from single cells, as when, for example, a suspension of enzymatically separated cells of embryonic mouse skin are maintained in rotary culture; initially the epidermal cells migrate to the periphery and the dermal cells migrate inward; subsequently, the skin is reconstructed.

HIV human immunodeficiency virus; the causative agent of AIDS (acquired immunodeficiency disease).

HMG protein a nuclear protein with a high electrophoretic mobility (high-mobility group); some HMG proteins have been implicated in control of transcription.

holandric genes genes found only on the Y chromosome.

holoblastic a type of cleavage pattern wherein the entire egg undergoes cleavage; human eggs undergo holoblastic cleavage. The holoblastic cleavage pattern comes in two general varieties, equal and unequal; in equal holoblastic cleavage, the resulting blastomeres are of the same size, as in the sea cucumber, *Synapta*; in unequal holoblastic cleavage, the resulting blastomeres are of different sizes, as in amphibians. The holoblastic cleavage pattern, of microlecithal eggs, comes in four varieties, radial, spiral, bilateral, and rotational, indicating that there are inherited patterns of cell division superimposed on the restraints as a result of the amount of yolk (see Fig. 12). *See:* **meroblastic**.

Holometabola comprise all the higher orders of insects (Endopterygota) or those in which the developing buds of the genitalia, wings, and other appendages lie concealed beneath the body wall; they undergo complex metamorphosis. The insect orders making up the Holometabola are the Neuroptera, Mecoptera, Trichoptera, Lepidoptera, Coleoptera, Strepsiptera, Hymenoptera, Diptera, and Aphaniptera.

holometabolous   descriptive of the life cycles of some species of insects that show the egg, larval, pupal, and imago (adult) stages of development; these insects undergo a dramatic metamorphosis during the pupal stage, e.g., fruit flies and butterflies. They are said to undergo complex metamorphosis, described as being indirect or complete. A variable degree of dissolution of the larval organs and tissues takes place during the pupal stage, the process being known as histolysis; simultaneously, they become gradually replaced by the organs of the adult insect, this generation of new tissues being termed histogenesis. Histolysis and histogenesis intergrade and involve a gradual, successive transformation without destroying the continuity of the parts concerned. Among Hymenoptera and Diptera, histolysis and histogenesis are processes of considerable complexity. *See:* **hemimetabolous**.

Holtfreter, Johannes (1901–1992)   born in Richtenberg, Germany; professor of biology at the University of Rochester; according to S. F. Gilbert (2009), it was Holtfreter's work that "devitalized" the concept of the organizer by devitalizing the organizing tissue itself; his also was the "solution" (Holtfreter solution) that enabled experiments to be performed on amphibian embryos without enormous mortality.

Holtfreter's doctoral thesis, concerning the early development of the liver in the frog embryo, was sponsored by Hans Spemann, whom Holtfreter considered to be one of the founders of experimental embryology. He felt that Spemann was a remote person and did not think of him (Holtfreter) as a prospective scientist. Holtfreter, believing that his future was with the amphibian egg, went to work with Otto Mangold, who later married Hilde (Pröschold) Mangold. Holtfreter felt his "Holtfreter solution" (an ideal culture medium for experimental amphibian embryos), discovered empirically, was, perhaps more than anything else, how he became known among embryologsts; together with his sterilization techniques, these methodological improvements opened up new avenues of experimentation.

Holtfreter also devised the sandwich method by wrapping explants in a mantle of pure ectoderm and cultivating the explants *in vitro*. Using this method, and others, he provided clear evidence that heating, boiling, freezing, or desiccating chorda-mesoderm did not abolish, or diminish, its capacity to elicit in the adjacent ectoderm massive neural formations. These results (in the 1930s) constituted a heavy blow to Spemann's vitalistic concept of "the organizer," for they indicated that the chorda-mesoderm tissues, alias the organizer, are merely the source of certain chemical substance(s) that initiate neural differentiation in the responding ectoderm. Holtfreter later went as far as to declare the term "organizer" to be a misnomer; however, he did maintain that there was sufficient evidence in support of the notion that the chorda-mesoderm area has the properties of a self-regulatory "morphogenetic field."

A joint communication (in 1932), by Hermann Bautzmann, Holtfreter, Spemann, and Otto Mangold, ushered in a new era of research—the search for the chemical constitution of the inducing agent(s). Holtfreter maintained that embryonic induction is the most important device by which differentiation comes about in amphibian development.

Holtfreter thought that gradient theories have a place in sea urchin development, but felt there was no foundation for assuming that they are present in amphibian development; he thought, regarding the amphibian egg or embryo, that none of the proposed gradients existed in reality. He thought that perhaps the most fundamental problem of embryology is how cytoplasmic patterns of potentialities and of actual differentiation come about and what their relation is to the genetic information. He was the first to obtain total exogastrulation in amphibians by including removal of the vitelline membrane as part of his technique.

Hom-C   *See:* **homeotic complex**.

homeobox   that portion of a homeotic gene (DNA) that is common to homeotic genes and that serves to define them as such. (In *Drosophila*, the homeobox consists of 180 base pairs and is responsible for specifying the 60 amino acids of the homeodomain portion of the homeoprotein.) The homeobox encodes a particular class of DNA-binding domains; approximately 20 families of homeobox-containing genes exist. The homeobox has been used as a probe to find other homeobox-containing genes in insects and vertebrates. It has been shown that the anterior-posterior axis of developing mammals is specified by the same homeotic genes that specify the *Drosophila* body axis. Because the homeobox enables one to transfer genetic knowledge of *Drosophila* embryos into the less known realm of mammalian development, the homeobox has been called the Rosetta stone of developmental biology. *See:* **TATA box**.

homeobox genes genes containing a homeobox domain; these genes are the most striking example of a widespread conservation of developmental genes in animals.

homeodomain that portion of a homeoprotein specified by its gene's homeobox.

homeodomain family one of four major families of transcription factors; including the Hox subfamily, which play an important role in axis formation, e.g., Hoxa-1, Hoxb-2, etc.

homeodomain protein a protein possessing a homeodomain.

homeoprotein a protein specified by a homeotic gene. *See:* **homeodomain protein**.

homeorhesis how the organism stabilizes its different cell lineages while it is still constructing itself.

homeosis the transformation of one body part into another, homologous, body part, inappropriate for its location in the body; e.g., the transformation of antennae on the head of a fly into legs. These bizarre transformations develop out of the homeotic selector genes' key role as positional identity specifiers; they control the activity of other genes in the segments.

homeothermic able to maintain a constant body temperature independent of the environment; homoiothermic vertebrates include birds and mammals. Homeothermy is regarded as an essential factor in the evolutionary radiations of birds and mammals. *See:* **poikilothermic**.

homeotic complex (Hom-C) the chromosome region, of chromosome 3 of *Drosophila*, containing both the Antennapedia complex and the bithorax complex.

homeotic gene a gene that when mutated results in homeosis, the transformation of a whole segment or structure into another related one; a master regulatory gene, responsible for the control of the formation of a part of the body; homeotic genes specify the identity and sequence of body segments. Homeotic genes are master control genes that are important in development and evolution (by duplication and mutation, they lead to diversification of the respective body segments); the products of these genes control batteries of genes. The *eyeless* (*ey*) gene is a paradigm for a master control gene; this single gene apparently can switch on the eye developmental program, which involves more than 2500 other genes needed for eye morphogenesis. As the neural tube is laid down, it becomes specified as to the type of neural tube it will be, forebrain, midbrain, hindbrain, or spinal cord; the mesoderm and endoderm are similarly patterned. Recent studies suggest that this specification is accomplished by the same homeobox-containing genes that specify the anterior-posterior axis in *Drosophila*. While the gap and pair-rule proteins control the pattern of HOM gene expression, these proteins disappear after approximately 4 hours; the continued correct expression of the homeotic genes involves two groups of genes, the *polycomb* and the *trithorax* groups; i.e., these genes maintain expression of the HOM complex. *See:* **homeotic mutation**.

homeotic mutation causes a homeotic transformation; a mutation that transforms one body part into another body part, inappropriate for its location in the body; e.g., *Antennapedia* mutation causes legs rather than antennae to grow out of the head; *Ultrabithorax* mutation results in a fly with four wings rather than two wings.

homeotic selector genes genes that regulate the identity of body regions; mutations in these genes cause the transformation of one body region or part into the likeness of another. In *Drosophila* development, the same segment polarity genes are turned on in each segment; therefore, what makes the segments different from each other? Specification of segment identity is carried out by a class of master regulatory genes, the homeotic selector genes; each of which controls the activity of other genes and is required throughout development to maintain this pattern of gene expression.

homoiogenetic of a determined part of an embryo, capable of inducing formation of a similar part when grafted into an undetermined field.

homolecithal eggs a type of egg that has its yolk uniformly distributed throughout its volume; human eggs are homolecithal (see Fig. 12). *See:* **telolecithal eggs**.

homologous this refers to structures that have the same evolutionary and developmental origin but serve different functions, e.g., the wing of a bird and the flipper of a seal.

homologous chromosomes the members of a pair of chromosomes that undergo synapsis (pairing) during prophase I of meiosis.

homologous developmental pathways homologous signal transduction pathways, composed of homologous proteins arranged in a homologous manner. Homologous pathways form the basic infrastructure of development; however, the targets of these pathways may differ among organisms; e.g., the Dorsal-Cactus pathway used in *Drosophila* for specifying dorsal-ventral polarity is also used by the mammalian immune system to activate inflammatory proteins. *See:* **deep homology, intracellular signal transduction**.

homologous genes   genes derived from a common ancestral gene and sharing similarity in their nucleotide sequences; e.g., the *HOM-C* genes of *Drosophila*, the *Hox* genes of the mouse, and the *HOX* genes of a human are homologous to each other.

homologous organs   body organs/parts of different species that have the same evolutionary and developmental origins but have come to serve different purposes because the animals possessing them have undergone adaptive radiation (e.g., mammalian: hand, hoof, flipper, or wing).

homologous recombination   the enzymatic incorporation, by enzymes involved in DNA repair and replication, of a mutant gene in the place of a normal copy.

homologous regulatory genes   genes that share similarities in their sequence because they evolved from a common ancestral gene; genes responsible for particular developmental functions have been conserved over hundreds of millions of years; e.g., the *Pax6* gene orchestrates eye development in species as distant as flies and humans. The eyes of insects and vertebrates did not come from separate origins, but they developed in the distant past from a common ancestor. The *Pax6* gene is not the only developmental regulator that seems to be homologous in insects and mammals, e.g., *Csx/tinman*.

homologs   genes that have evolved from a common ancestral gene; includes orthologs and paralogs. *See:* **homologous genes**.

homology   similarity between species that results from inheritance of traits from a common ancestor.

homonomous   a body plan or body part composed of repeating, similar parts.

homophilic binding   when cells are bound together by the same type of cadherin on the cells; e.g., cells with E-cadherin bind to other cells with E-cadherin.

homoplasy   similarity in the characters found in different species that is attributed to convergent evolution, parallelism, or reversal—not common descent.

homosporous   literally same spores; homosporous plants produce one kind of spore; a spore that gives rise to a gametophyte that produces both male and female gametangia; most ferns are homosporous. *See:* **heterosporous**.

homothallus   literally same thalli; homothallic plants produce a single thallus that produces both male and female gametangia. *See:* **herterothallus**.

homozygous   descriptive of an organism in which its two alleles for a given characteristic are identical (same in the zygote).

homunculus   literally "little man"; the preformed human that supposedly was visible inside the head of the spermatozoon. *See:* **Hartsoeker, Nicolaas**.

hopeful monster   according to Richard Goldschmidt, a new species would start out as a "hopeful monster." He argued that although macromutations generally result in deformed organisms, some of these monstrositites have potential if they happen to develop in the right environment. *See:* **Goldschmidt, Richard**; **Metchnikoff, Elie**.

hopeless monsters   evolutionary biologist Ernst Mayr's term of derision for Richard Goldschmidt's idea of hopeful monsters.

horizontal division   *See:* **latitudinal division**.

horizontal transmission   the metazoan host is born free of symbionts but subsequently becomes infected, either by its environment or by other members of the species, e.g., the *Euprymna scolopes* (a squid)–*Vibrio fischeri* (bioluminescent bacterium) symbiosis. *See:* **vertical transmission**.

hormone response elements   enhancers that respond to cytoplasmic or nuclear receptors bound to their ligands.

hormones   a chemically diverse group of *messenger* molecules secreted by endocrine cells and designed to act on cells at some distance from the hormone-secreting endocrine cells. *See:* **second messenger**.

hornworts   a class of nonvascular plants or bryophytes, e.g., *Anthoceros*.

horseshoe kidney   a greater or lesser degree of congenital fusion of the two kidneys, usually at the lower poles.

Horstadius, Sven (1898–1996)   Swedish biologist who carried out a series of experiments with sea urchin embryos from 1928 to 1935, showing that Hans Driesch was not 100% correct (regarding regulative development in sea urchins). Horstadius showed that even in sea urchin embryos there appears to be some degree of mosaicism, at least along the animal-vegetal axis. He separated various layers of early sea urchin embryos with fine glass needles. Using the 8-cell stage, he found that when the embryo was divided meridionally, through the animal and vegetal poles, each half produced a pluteus larva, but when the embryo was divided equatorially, separating animal and vegetal hemispheres, neither part developed into a complete larvae. This experiment was also carried out using sea urchin *eggs* (making merogones), with similar results. Thus, even sea urchin embryos demonstrate some degree of mosaic development. The blastomeres will stay regulative so long as they are not completely formed from animal or vegetal cytoplasm; even in an embryo that undergoes regulative development, there comes a time when the potencies of its cells are resticted. Horstadius could relate this restriction in potency to the plane of cleavage. Horstadius also separated and recombined the six tiers of blastomeres of the 64-cell stage of sea urchin embryos; e.g., by combining the animal hemisphere with the micromeres (leaving out much of the vegetal hemisphere), he was able to obtain a recognizable larva, thus demonstrating regulation in sea urchin development. Horstadius carried out a series of experiments suggesting the ability to suppress "animalization" was localized as a gradient; there appeared to be a gradient of "animalization" proceeding from an1 (blastomeres closest to animal pole) to the micromeres (blastomeres closest to vegetal pole). *See:* **Driesch, Hans**; **Roux, Wilhelm**; **Runnström, John**.

housekeeping genes   genes that encode proteins required for basic functions (e.g., protein synthesis and energy metabolism) required in all cells.

***Hox* genes**   homeobox-containing genes, apparently found in linked clusters in all bilaterians, concerned with patterning the anteroposterior body axis. The *Hox* gene complexes of vertebrates, whose ancestors diverged from those of arthropods hundreds of millions of years ago, show the same correspondence between gene order and order of expression as is found in the HOM genes of *Drosophila*; assumedly, this is related to the mechanisms that control the expression of these genes. Once gastrulation begins, anterior-posterior polarity in all vertebrates becomes specified by the expression of *Hox* genes. *Hox* genes provide one of the most remarkable pieces of evidence for deep homologies among all the animals of the world. So far, no metazoans have been found to develop without expression of HOX gene clusters.

Hox proteins   a subset of homeodomain proteins that have a special role in the control of anteroposterior pattern in animals.

hub   a regulatory microenvironment in *Drosophila* testes where the stem cells for sperm reside.

Hubel, David (1926–)   Nobel laureate (for physiology or medicine, 1981, shared with Roger Sperry and Torsten Wiesel); demonstrated that there is competition between the retinal neurons of each eye for targets in the cortex and that their connections must be strengthened by experience.

human accelerated regions (HARs)   human homologues, of conserved sequences, that diverge from the common mammalian norm; 202 of these have been identified; two of these, *HAR1* on chromosome 20 and *HAR2* on chromosome 2, are dramatically distinct in humans; the *HAR1* region appears to be adjacent to a region of genes encoding transcription factors, and it encodes a regulatory RNA that is expressed in the cortex of the mammalian brain.

Human Genome Project   an international, multiyear project that discovered the sequence of all 3 billion base pairs that make up the human genome.

humoral immunity   that portion of immunity that involves antibodies produced by cells (B-lymphocytes); as opposed to cell-mediated immunity that directly involves cells (T-lymphocytes).

hunchback   a gap gene; a key posterior group maternal gene in early *Drosophila* development. *Hunchback* is also expressed zygotically in the early embryo. In the early embryo, zygotic *hunchback* is activated at the anterior end of the embryo by high concentrations of bicoid protein (a transcription factor), resulting in an anteroposterior gradient of the hunchback protein. The bicoid protein is a member of the homeodomain family of transcriptional activators and activates the *hunchback* gene by binding to regulatory sites within the promoter region.

hunchback protein   acts, in the *Drosophila* embryo, as a morphogen for the next stage of patterning. To establish a clear anteroposterior gradient of zygotic hunchback protein, nanos protein inhibits translation of posterior maternal *hunchback* mRNA. The hunchback protein, itself a transcription factor, acts as a morphogen to which other gap genes respond. The bands of gap gene expression are delimited by mechanisms that depend on the gene control regions being sensitive to different concentrations of hunchback protein, and to other proteins, including the bicoid protein.

Hunter, John (1728–1793)   English comparative anatomist; traced the descent of the testes in the male fetus, observed the nature of placental circulation, and prepared the best series of drawings on the embryology of the chick before the 19th century.

Hutchinson–Gilford progeria syndrome   caused by a dominant mutant gene; causes children to age rapidly and to die as early as 12 years. *See: klotho* **gene**.

Huxley, Thomas Henry (1825–1895)   took a great step toward unifying the idea of germ layers throughout the animal kingdom when he maintained that the two cell layers in animals like the *Hydra* correspond to the ectoderm and endoderm of higher animals; wrote in 1893 that "[e]volution is not a speculation but a fact; and it takes place by epigenesis." *See:* **Klinenberg, James R**.

hyalin   a protein released by the cortical reaction of the eggs of some marine invertebrates, which makes up the hyaline membrane.

hyaline membrane   a membrane formed from the released contents of cortical granules during the cortical reaction of fertilization in some marine invertebrates, including sea urchins; this membrane holds blastomeres together during early development and plays a role in sea urchin gastrulation.

hyaline membrane disease   also called respiratory distress of the newborn, is associated with the initial ventilation of the lungs that lack adequate surfactant stores. Lack of surfactant leads to uneven distribution of ventilation, inadequate gas exchange, and respiratory distress. Hyaline membrane refers to a layer of fibrin and cellular debris found adhering to the walls of alveoli, alveolar ducts, and respiratory bronchioles upon postmortem examination.

hyaluronic acid   an extracellular polysaccharide that seems to play an important role in the migration of embryonic cells; it seems to be important in keeping individually migrating mesenchymal cells separated from one another.

hyaluronidase   an enzyme, released by the sperm of some species, which hydrolyzes hyaluronic acid, a major component of extracellular matrix; believed to play a role in the fertilization of some species by allowing sperm to breach egg cell coats.

hybrid merogony   the fertilization of cytoplasmic fragments of the egg of one species by the sperm of a related species.

hydatidiform mole   a birth defect where the fetus is absent and only a very abnormal placenta is present; a human tumor that resembles placental tissue and develops when a haploid sperm fertilizes an egg missing the female pronucleus and the entire genome is derived from the sperm; evidence of genomic imprinting.

Hydra   a genus of freshwater cnidarians; used extensively in studies of regeneration. *See:* **morphallactic regeneration**.

hydro-   a word part referring to water; in the context of development referring to an abnormal accumulation or deficit of water.

hydrocephalus   a condition involving distension of the ventricular system (cavities) of the brain, which may be caused by overproduction or defective absorption of cerebrospinal fluid. In infants, it leads to an increased head size and thinning of skull bones with suture separation.

hydrocephaly   a birth defect originating from an abnormal accumulation of cerebrospinal fluid in the ventricles of the developing brain.

hydrogen bonds   weak chemical bonds resulting from proton (hydrogen ion) sharing by oxygen and/or nitrogen atoms.

hydrolytic   to split (break a chemical bond) by means of water; digestion is hydrolytic.

hydronephrosis   a birth defect originating from an abnormal accumulation of urine in the developing kidneys; a condition of gross dilation of the kidney pelvis and calices with urine, caused by an obstruction of the urinary tract.

20-hydroxyecdysone   a steroid; one of two insect hormones that control insect molting and metamorphosis. 20-hydroxyecdysone initiates and coordinates each molt and regulates the changes in gene expression that occur during metamorphosis. *See:* **juvenile hormone**.

Hymenoptera   insects with membranous wings, the hind pair smaller and connected with the fore pair by hooklets; saw-flies, ants, bees, wasps, ichneumon flies, and their allies.

hyoid arches   the second pair of branchial arches. *See:* **branchial arches**.

hyomandibular furrows   the pair of furrows located between the first (mandibular) and second (hyoid) pairs of branchial arches.

hyomandibular pouches   portions of the endodermal lining of the pharyngeal cavity that separate the paired hyoid and mandibular arches in vertebrate embryos.

hypaxial muscles   formed from those myoblasts, of the dermamyotome, farthest from the neural tube; muscles of the body wall, limbs, and tongue.

hyperactivation   sperm suddenly swimming at higher velocities and generating greater force; e.g., certain mammalian sperm become hyperactivated when they pass from the uterus into the oviducts.

hypermetamorphism   type of embryological development in certain insects in which one or more stages have been interpolated between the full-grown larva and the adult.

hypermorphosis   the principle that developmental patterns tend to change more by terminal or subterminal addition than by substitution, omission, or preterminal addition; changes that do occur with evolution in ontogeny tend primarily to be added onto previous ontogenies, rather than occurring in earlier developmental stages.

hyperplasia   an increase in the size of a tissue or organ resulting from an increase in the number of cells.

hypertrichosis   excessive growth of hair.

hypha   a filament; as produced by many kinds of fungi, including bread mold and mushrooms.

hypoblast   also called primitive endoderm; the early human embryo is composed of two layers; the lower layer is the hypoblast. In human development, the hypoblast gives rise to extraembryonic endoderm and extraembryonic mesoderm. In chick development, the hypoblast induces the formation of the primitive streak. In mammalian and avian development, the hypoblast does not give rise to any part of the newborn organism (see Fig. 44). *See:* **epiblast**.

hypobranchial eminence   a small bulge, posterior to the foramen cecum, which gives rise to the posterior third (pharyngeal part) of the tongue.

hypocotyl   that portion of the angiosperm embryonic axis, below the attachments of the cotyledons.

hypomere   lateral plate mesoderm.

hypophysis   In angiosperm embryogenesis, the cell found at the interface between the suspensor and the embryo proper (gives rise to some of the root cells in many species). In *Arabidopsis*, the hypophysis (uppermost cell of the suspensor) gives rise to the root meristem. *See:* **pituitary gland**.

hypoplasia   underdevelopment of a tissue or organ, usually associated with a decreased number of cells.

hypostome   the conical region of the "head" of *Hydra* (which contains the mouth, surrounded by a ring of tentacles); has been used in transplant studies of regeneration. *See:* **basal disc**.

I

ichthyosis congenita a severe form of a skin condition characterized by dry, harsh skin with adherent scales; the fetus may be stillborn or may live a few days.

ICM *See:* **inner cell mass**.

ICSI *See:* **intracytoplasmic sperm injection**.

identical twins twins derived from the same zygote. Also called monozygotic twins.

IGF *See:* **insulin-like growth factor**.

IL *See:* **interleukin**.

***Illyanassa obsoleta* (*Nassarius obsoletus*)** the mud snail; used in studies of the role of polar lobes in cell differentiation during early embryonic development (see Plate 18 in the color insert).

imaginal descriptive of the adult (imago) stage of insect development.

imaginal buds *See:* **imaginal discs**.

imaginal discs the rudiments of adult organs formed before or during insect metamorphosis; small sets of epithelial cells, rudiments of future imaginal (adult) organs, formed from histoblasts; in the higher Diptera, including *Drosophila*, the imaginal discs are already laid down in the embryo.

imaginal molt the molt undergone by the pupa of holometabolous insects to become an imago (adult); the imago sheds the pupal coat and emerges. *See:* **eclosion**.

imago the adult stage, the final instar, in the insect life cycle.

imbibition the uptake of water; developmentally important, for example, in seed germination.

immature ovum an oocyte.

immediate adaptive responses alteration of development in response to a particular crisis that is already occurring in the embryo, e.g., oxygen deprivation causing changes in blood flow to spare brain development. *See:* **predictive adaptive responses**.

immortality of the germplasm the concept, proposed by August Weismann in the 1880s, that unlike somatic cells, which die during each generation, the cells of the germplasm (germ cells) are immortal and are passed from generation to generation. *See:* **Weismann, August**.

immunoglobulin superfamily CAMs cell adhesion molecules whose cell-binding domains resemble those of antibody molecules, e.g., exhibited by muscle, nerve, and kidney cells; Ng-CAM, exhibited by glial and nerve cells; and Cell-CAM, exhibited by hepatocytes.

imperfect flower a flower not possessing both male and female parts; staminate flowers lack carpels (pistils), and carpellate (or pistillate) flowers lack stamens.

imperforate anus is an example of an abnormal closure (atresia) of a normal opening; i.e., the anal opening.

imperforate hymen is an example of an abnormal closure (atresia) of a normal opening; i.e., the vaginal opening into the vestibule.

implantation the attachment of the conceptus to, and its incorporation into, the lining of the uterus (endometrium). The trophoblast cells of the blastocyst are noticeably sticky, and the blastocyst tends to adhere to the uterine lining, by the cells directly overlying the inner mass cells. Almost as soon as adhesion has occurred, the trophoblast cells begin to proliferate rapidly and to erode the underlying uterine mucosa. Cells in contact with the uterine mucosa rapidly change their character, from small and thin to large, darkly staining cells; as these cells continue to enlarge, they lose their cell boundaries and constitute the trophoblastic syncytium. This syncytial layer has the ability to eat its way into the maternal tissues, and this activity is responsible for the way the embryo burrows into the mucosa. Once the implantation process has commenced, it progresses rapidly; by approximately 8

Dictionary of Developmental Biology and Embryology, Second Edition. Frank J. Dye.
© 2012 Wiley-Blackwell. Published 2012 by John Wiley & Sons, Inc.

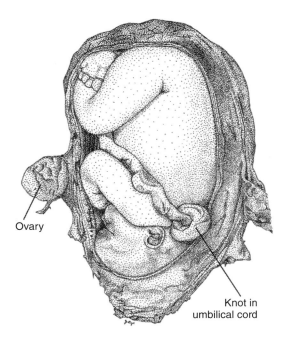

Ovary

Knot in
umbilical cord

Figure 21. Week 28, third trimester. Although this fetus appears to be sleeping, the eyes by now are open again. Reprinted from Frank J. Dye, *Human Life Before Birth*, Harwood Academic Publishers, 2000, fig. 9-8, p. 79. Original artwork by John C. Dye.

days, the embryo has become almost completely buried in the mucosa; by the ninth day, the uterine epithelium is beginning to heal over the area through which the embryo eroded its way into the mucosa. As primary villi grow out of the chorion, they destroy adjacent maternal tissue, producing a liquefied material, embryotroph, which may nourish the embryo. The trophoblast eventually comes into contact with small, maternal, blood vessels and breaks down their walls; there is a continuing oozing of blood from the invaded vessels. The time of establishment of human embryonic circulation is toward the close of the third or the beginning of the fourth week after fertilization; i.e., approximately 2 weeks after implantation.

imprinting *See:* **genomic imprinting**.

IMZ *See:* **involuting marginal zone (IMZ) cells**.

in cis a condition in which two genes are located on the same chromosome.

in situ literally, "in place"; refers to a stucture in its normal location, as the heart in the thoracic cavity as opposed to the heart removed from the thoracic cavity.

in situ **hybridization** the hybridization of a DNA probe to a DNA nucleotide sequence in its *in situ* location on a chromosome; a technique that allows one, for example, to locate genes on a specific chromosome. Also, the hybridization of a DNA probe with mRNA in the embryo or organ itself; a technique for the study of the time and location of gene expression. The DNA probe, itself a DNA nucleotide sequence, is tagged, e.g., with a radio-isotope or is an antigen for an immunofluorescent antibody, so the location of the probe hybridized to the target DNA nucleotide sequence is revealed.

in trans a condition in which two genes are located on separate chromosomes.

in utero in the uterus (see Fig. 21).

in utero **surgery** surgery on the fetus while it is in the uterus; both hydrocephalus and hydronephrosis, which involve an abnormal accumulation of fluid, have been successfully treated with *in utero* surgery.

in vitro literally, "in glass"; as, in the case of a biochemist, carrying out a biochemical reaction in a glass vessel such as a beaker, or, in the case of a cell biologist, maintaining cells in a glass cell culture vessel; however, in recent years, cells are more often cultured in plastic rather than in glass cell culture vessels.

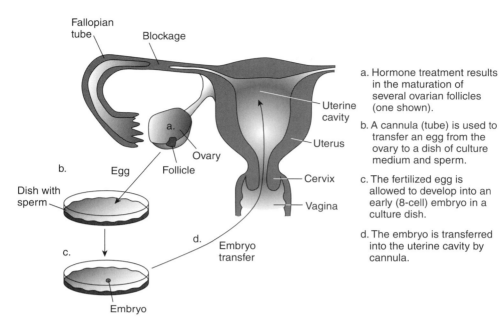

a. Hormone treatment results in the maturation of several ovarian follicles (one shown).

b. A cannula (tube) is used to transfer an egg from the ovary to a dish of culture medium and sperm.

c. The fertilized egg is allowed to develop into an early (8-cell) embryo in a culture dish.

d. The embryo is transferred into the uterine cavity by cannula.

Figure 22. In vitro fertilization. This simplified diagram shows the bringing together of the egg and sperm in a laboratory dish so that in vitro fertilization may occur. Also shown is transfer of the resulting embryo back into the woman's uterus. Reprinted from Frank J. Dye, *Human Life Before Birth*, Harwood Academic Publishers, 2000, fig. 21-1, p. 188.

in vitro **fertilization (IVF)**   human fertilization occurring under artificial conditions outside of the body (see Fig. 22).

in vivo   literally, "in life"; that which occurs in a cell or organism.

inborn errors of metabolism   congenital defects in metabolism attributable to variation in genetic constitution.

incomplete flower   a flower not containing all four kinds of floral parts: sepals, petals, stamens and pistils.

incubation   depending on whether a chicken egg reaches the uterus of the oviduct before or after noon on a given day, the egg will be deposited by the hen or retained until the next morning; therefore, eggs that have been incubated the same length of time will exhibit striking differences in the development of embryos.

indeciduate placenta   a placenta having the maternal and fetal elements associated but not fused.

independent development   also called autonomous specification.

indeterminate cleavage   cleavage in which all the early cells have the same potencies with respect to development of the entire zygotye. *See:* **determinate cleavage**.

India ink   an ink traditionally used to demonstrate the jelly coat around the sea urchin egg, by means of a type of negative staining; the index of refraction of the jelly coat and of sea water are nearly identical so, in sea water, the jelly coat is invisible; however, the presence of the jelly coat excludes the India ink particles, thereby revealing the presence of the jelly coat.

indifferent stage   also called the bipotential stage; in mammals, the gonadal rudiment has two normal "options," to develop into either an ovary or a testis; before this "decision" is made, the mammalian gonad develops through the indifferent stage during which it has neither female nor male characteristics.

indirect development   embryonic development that includes a larval stage with characteristics very different from those of the adult organism.

individualization   formation of distinct organs or structures through the interaction of adjacent tissues.

individuation   individualization.

indoleacetic acid (IAA)   a plant hormone, of the auxin family, found in the embryo of seeds, meristems of apical buds, and young leaves; its functions include stimulation of stem elongation; root growth, differentiation, and branching; development of fruit; apical dominance; phototropism; and gravitropism.

induced labor   the initiation of labor by artificial means, such as by injection of the hormone, oxytocin.

induced pluripotent stem cells (iPS cells)   adult cells that have been converted to cells with the pluripotency of embryonic stem cells, generally accomplished by the activation of specific transcription factors; are a type of pluripotent stem cell artificially derived from a non-pluripotent cell, typically an adult somatic cell, by inducing a "forced" expression of certain genes.

inducer   tissue that produces a signal (or signals) that induces a cellular behavior in some other tissue.

inducing factors   mainly proteins (although a few are small molecules such as retinoic acid), released from a signaling center (inducer) and exerts an effect on a responding cell group (responder). *See:* **induction**.

induction   restriction in the potency of cells; any system of embryonic induction has at least two components: (1) an inducer and (2) a responder; in both primary and secondary inductions, there is restriction in the potency of ectodermal cells. Here we find determination to be dependent on interactions between groups of cells; however, it is not surprising to find organisms in which determination is not progressive but immediate.

indusium   a thin epidermal outgrowth from a fern leaf, which covers over a group of sporangia called a sorus.

inferior   below; as the chin is inferior to the forehead.

inferior vena cava   a major venous (vein) blood vessel returning blood from the body (inferior to the heart) to the right atrium of the heart.

inflorescence   the flowering part of a plant; a flower cluster; the arrangement of the flowers on the flowering axis.

inflorescence meristem   a meristem that initiates axillary meristems that can produce floral organs but does not directly produce floral parts itself; derived from a vegetative shoot meristem. Apparently, the inflorescence meristem develops through the action of a gene that suppresses terminal flower formation, e.g., the *CEN* gene in snapdragons. *See:* **floral meristem**.

infundibulum   funnel-shaped; the funnel-shaped opening at the ovarian end of the fallopian tube; also, an evagination of the floor of the diencephalon that gives rise to the neurohypophysis (see Fig. 16).

ingression   a basic morphogenetic movement; the movement of individual cells into a preformed cavity, as in the formation of primary mesenchyme cells during sea urchin gastrulation.

inguinal ring   the passageway between the scrotum and the abdominal cavity, through which the spermatic cord passes into the abdominal cavity.

inheritance of acquired characteristics   the hypothesis that phenotypic changes in the parental generation can be passed on, intact, to the next generation; may have a mechanism if the inherited morphological alteration can be mediated by epigenetic changes in the DNA methylation of germ cells.

initials   in plants; are cells that divide in such a way that one of the sister cells remains in the meristem as an initial while the other becomes a new body cell or derivative; derivatives usually divide near the root or shoot tip before undergoing differentiation. Initials along with their sister cells constitute the apical meristems.

inner cell mass (ICM)   that portion of the blastocyst that gives rise to the embryo. Also called embryoblast (see Plate 7 in the color insert).

innervate   to supply with and control with nerves.

inositol triphosphate (IP$_3$)   *See:* **phosphoinositol pathway**.

insectivores   insect eating.

instar   the form assumed by a developing insect during a particular stadium.

instars   the successive stages of larval development punctuated by molts that allow growth in the larva.

instructive interaction   occurs when a signal from the inducing cell is necessary for initiating new gene expression in the responding cell; without the inducing cell. the responding cell would not be capable of differentiating in that particular way; when the optic vesicle is experimentally placed under a new region of the head ectoderm and causes that region of the ectoderm to form a lens. *See:* **permissive interaction**.

insulator a DNA element that shields a gene from the positive effect of an enhancer or the negative effect of a silencer.

insulator sequences also called boundary sequences, DNA sequences that bind proteins that prevent enhancers from interacting with promoters on the other side of them; thus, they could establish boundaries. It has been proposed that regulatory signals in one domain do not cross the boundary into the next.

insulin-like growth factor I (IGF-I) a protein growth factor; promotes cell survival, stimulates cell metabolism, and collaborates with other growth factors to stimulate cell proliferation.

integrins cell receptors for extracellular matrix molecules; these proteins span the plasma membrane and bind the fibronectin of the extracellular matrix and provide anchorage sites for the actin microfilaments of the cytoskeleton; i.e., they integrate the extracellular and intracellular scaffolds.

integuments the covering of the ovule that will become the seed coat.

intercalary meristems meristems, found in monocot stems, inserted between mature tissues (monocot stems do not have lateral meristems).

intercalated discs the specialized cell junctions between the contractile cells of the heart.

intercalation a basic morphogenetic movement; occurs in multilayered tissues when nearby cells intermix and spread out, thereby producing a thinner sheet.

intercourse the insertion of the male penis into the female vagina and subsequent ejaculation of semen into the vagina.

interdigital necrotic zone region of cell death involved in sculpting the fingers/toes out of the hand/foot plate.

interior necrotic zone region of cell death that separates the ulna and radius from each other.

interkinesis a stage between the first and second meiotic divisions, which superficially resembles interphase; however, unlike during interphase, during interkinesis there is *no* synthesis of DNA.

interlaminar jelly in the chick embryo, the noncellular, gelatinous material found between the germ layers that serves as a substratum for the movement of mesenchymal cells.

Interleukin-2 (IL-2) a protein growth factor; stimulates the proliferation of activated T lymphocytes.

Interleukin-3 (IL-3) a protein growth factor; stimulates proliferation and survival of various types of blood cell precursors.

intermediate filaments one of the three, fibrous, proteinaceous, components of the cytoskeleton; made up of fibrous proteins such as keratin.

intermediate inheritance descriptive of an organism in which two alleles for a given characteristic are not identical and both are expressed in the phenotype.

intermediate layer (1) skin: a layer of skin produced during the early fetal period between the outer periderm and the inner basal layer. (2) spinal cord: the portion of the wall of the spinal cord made up of the cell bodies of nerve cells between the inner ventricular layer and the outer marginal layer, that is, gray matter.

intermediate mesoderm that portion of the mesoderm found bilaterally between the somitic mesoderm and the lateral (or lateral plate) mesoderm (see Figs. 28 and 41).

intermediate spermatogonia spermatogonia committed to becoming spermatozoa; they divide mitotically once to form type B spermatogonia.

internal rotation one of the seven cardinal movements of labor; the rotation of the fetal head necessary for it to fit through the pelvic outlet (see Fig. 40).

internodes lengths of a stem found between successive nodes.

interphase that portion of the cell cycle when the cell is not dividing; i.e., G_1, S, and G_2 collectively (see Fig. 10).

intersex a condition in which male and female traits are observed in the same individual.

interspecies epigenesis development coordinated between two or more species; e.g., bacteria help form our guts and immune systems. *See:* **transgenerational continuity**.

interspecific incompatibility the failure of pollen from one species to germinate and/or grow on the stigma of another species.

interstitial cells of Leydig testosterone-secreting cells found between the seminiferous tubules of the testes.

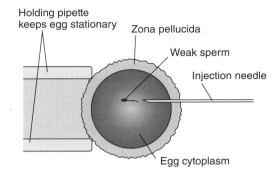

Holding pipette
keeps egg stationary
Zona pellucida
Weak sperm
Injection needle
Egg cytoplasm

Figure 23. Intracytoplasmic sperm injection (ICSI). By this procedure, a spermatozoon that is incapable of reaching the egg or penetrating the egg coats or the plasma membrane of the egg may still deliver its genetic "payload" to the egg. Reprinted from Frank J. Dye, *Human Life Before Birth*, Harwood Academic Publishers, 2000, fig. 21-3, p. 191.

interstitial growth   the growth of an object by the addition of material throughout its volume, as in the growth of the liver.
interstitial tissue   tissue found between the seminiferous tubules of the testes (see Fig. 45).
intervening sequence   *See:* **intron**.
interventricular septum   the partition between the two ventricles of the developing heart.
interventricular sulcus   the groove on the surface of the ventricular region of the developing heart opposite the internal, forming interventricular septum.
intervertebral discs   discs of soft tissue found between successive vertebrae in the vertebral column, which originate from the somites and notochord.
intervillous spaces   placental spaces, filled with maternal blood, found between the chorionic villi of the placenta. The intervillous spaces are derived mainly from the lacunae of the early invasive syncytiotrophoblast (see Fig. 35).
intracellular signal transduction   developmentally important cellular activities include cell division and differential gene expression; in a multicellular organism, these cellular activities are under the control of the cell's environment, including signals from other cells. Although some of these signals, e.g., steroid hormones, are nonpolar and are able to pass through the plasma membrane of the cell, other signals are polar, e.g., the hormone epinephrine, and attach to receptors in the plasma membrane. To convey information carried by the signal from the cell surface to the interior of the cell (e.g., the nucleus) where the information is acted upon, the cell makes use of cascades of chemical reactions that make up intracellular signal transduction pathways.
intracytoplasmic sperm injection (ICSI)   the injection of a sperm, incapable of fertilizing an egg in the normal way, directly into the cytoplasm of an egg (see Fig. 23).
intraembryonic   within the embryo.
intramembranous ossification   ossification occurring within condensed mesenchyme, usually involved in the formation of flat bones, such as in the calvaria.
intraspecific incompatibility   the failure of pollen from one member of a species to germinate and/or grow on the stigma of another member of the same species. *See:* **self-incompatibility**.
intrauterine device   *See:* **IUD (intrauterine device)**.
intron   intervening sequence; the portion of a gene that is not expressed, as part of a protein molecule, but is spliced out of the primary transcript to form the definitive mRNA transcribed from the gene. Introns are characteristic of eukaryotic genes. *See:* **exon**.
invagination   a basic morphogenetic movement; an inpocketing, the movement of a layer of cells into a preformed cavity; e.g., the optic cups develop from invaginations of the optic vesicles.
inversion   to turn inside out.
invertebrates   animals without backbones (vertebral columns); the ability of the mesoderm to form a notochord and its overlying ectoderm to become a neural tube separated the chordates from the remaining invertebrates.

involuting marginal zone (IMZ) cells   those cells of the marginal zone of the amphibian embryo that undergo involution during gastrulation.

involution   a basic morphogenetic movement; the inward movement of cells over a rim, as in the inward movement of cells over the dorsal lip of the blastopore during the gastrulation of some frogs.

ionophores   literally to bear or carry ions; also called ionophorous antibiotics; these are small molecules that allow inorganic ions, such as calcium ions, to pass through cell membranes that would otherwise be impermeable to them, e.g., A23187. Used, for example, in studying the role of calcium ions in fertilization.

iris   the pigmented membrane located between the cornea and the lens of the eye, surrounding the pupil, and capable of contraction.

irregular cleavage   division of a zygote into random masses of cells, as in certain coelenterates.

islets of Langerhans   scattered accumulations of insulin-secreting cells found in the pancreas.

isogamy   sexual reproduction by the union of gametes that are similar in size; found in some algae and fungi. *See:* **anisogamy, oogamy**.

isolation experiment   involves removing a part of an embryo and observing the development of the isolated part and the defective embryo.

isolecithal   (see Figs. 12 and 14). *See:* **isolecithal eggs**.

isomerism   when the parts on the two sides of the body of a bilaterally symmetrical animal are wholly equivalent. *See:* **situs solitus**.

isometric growth   growth where all components grow at the same rate so that shape is preserved.

isthmus   the narrow segment of the mammalian oviduct adjacent to the uterus.

iteroparous   describes a species or population in which individuals experience more than one bout of reproduction over the course of a typical lifetime, e.g., humans.

IUD (intrauterine device)   a birth control device inserted into the uterus that prevents the embryo from implanting into the endometrium. *See:* **contragestion**.

IVF   *See: in vitro* **fertilization (IVF)**.

J

Jacob, Francois (1920–)   French biologist, co-recipient of the 1965 Nobel Prize in Physiology or Medicine for work on the genetic control of enzyme and virus synthesis. In 1960, Jacob and Jacques Monod introduced the term "operon" for a closely linked group of genes, each of which controls a different step in a given biochemical pathway. In 1961, Jacob and Monod postulated the existence of a molecule, messenger RNA, that carries the genetic information necessary for protein synthesis from the operon to ribosomes, where proteins are made. In 1977, Jacob proposed that evolution rarely creates a new gene; rather, evolution combines existing parts in new ways rather than creating new parts; he predicted that such "tinkering" would be most likely to occur in those genes that construct the embryo, not in the genes that function in adults.

JAK-STAT signal transduction pathway   an intracellular signal transduction pathway important in development. This pathway is extremely important in the differentiation of blood cells and in the activation of the casein gene during milk production. The STAT pathway is very important in the regulation of human fetal bone growth.

jaundice of newborn   jaundice in infants during the first few days after birth, resulting from various causes.

jelly coat   a type of noncellular, jelly, coat found around the eggs of some species, e.g., echinoderms and amphibians.

Jenkinson, J. W. (1871–1915)   by 1909, experimental embryology had achieved full status with its own textbook, *Experimental Embryology*, by the British embryologist J. W. Jenkinson. The first page of this book states that the field really dates from Wilhelm Roux's production of a half-embryo from a half-blastomere and the consequent formulation of the "Mosaic-Theorie" of self-differentiation. J. Maienschein (1994) expresses the view that statements like Jenkinson's produced the *false* impression that Roux alone, or at least primarily, had provided the framework.

jervine   a teratogen, found in the plant *Veratrum californicum*, known to cause cyclopia in vertebrates; blocks the synthesis of cholesterol. *See:* **cyclopamine**.

jugular veins   *See:* **cardinal veins**.

Just, Ernest J. (1883–1941)   made critical discoveries on fertilization; spurned genetics and emphasized the role of the cell membrane in determining the fates of cells.

juvenile   physiologically immature or undeveloped; a young individual resembling an adult of its kind except in size or reproductive activity (see Plate 12 in the color insert).

juvenile hormone   a lipid-like insect hormone produced by the corpora allata, a pair of glands attached to and controlled by the brain. The nature of a molt is determined by the concentration of juvenile hormone: a high concentration results in a larval molt, a reduced concentration results in a pupal molt, and the absence of juvenile hormone results in an adult molt. Juvenile hormone (JH) prevents the ecdysone-induced changes in gene expression that are necessary for metamorphosis. *See:* **20-hydroxyecdysone**.

juvenile phase   a period gone through by some plants, especially woody perennials, during which the plant cannot produce reproductive structures even if all the appropriate environmental signals are present.

juxtacrine factors   molecules that function in juxtacrine signaling, e.g., Bride of sevenless protein.

juxtacrine signaling   a mode of cell-cell communication in which signaling molecules are retained on the surface of the signaling cell and interact with receptor proteins on adjacent cell surfaces, e.g., interaction between Bride of sevenless protein and its receptor Sevenless. *See:* **autocrine signaling**, **endocrine signaling**, **paracrine signaling**.

Dictionary of Developmental Biology and Embryology, Second Edition. Frank J. Dye.
© 2012 Wiley-Blackwell. Published 2012 by John Wiley & Sons, Inc.

K

kairomones   chemicals that are released by a predator and can induce defenses in the prey.

Kallman syndrome   some infertile men have no sense of smell; the Kallman syndrome gene produces a protein that is necessary for the proper migration of both olfactory axons and hormone-secreting nerve cells.

Kartagener's syndrome   also Kartagener triad syndrome; these individuals lack dynein on all their ciliated and flagellated cells making them immotile; males with this disease are sterile (immotile sperm), are susceptible to bronchial infections (immotile respiratory cilia), and have a 50% chance of having the heart on the right side of the body.

karyogamy   the conjugation (fusion) of nuclei; fusion of gametic nuclei, as in fertilization.

karyokinesis   division of the nucleus; mitosis. Karyokinesis can occur in the absence of cell division; either experimentally by using the fungal metabolite, cytochalasin B, or naturally, as occurs during normal insect cleavage.

karyoplasm   *See:* **nucleoplasm**.

karyotype   the arrangement of the chromosomes of a given species according to internationally agreed to specifications; made by isolating individual chromosome images from an image of a metaphase chromosome spread. In a human karyotype, the 22 pairs of autosomes are arranged in order of decreasing size; the sex chromosomes are placed after the 22 pair of autosomes; all chromosomes are placed with their short (p) arm up and their long (q) arm down; human chromosomes are also arranged in seven groups, A through G. One use of karyotypes is to look for chromosome anomalies that may have developmental consequences (see Figs. 24, 25, and 46). *See:* **metaphase chromosome spread**.

keel   in fish embryos, the thickened, elongated, dorsal midline, antero-posteriorly extended, ridge formed from the neural plate. The keel gradually sinks toward the underlying notochordal tissue, develops a lumen, and gradually becomes transformed into an elongated neural tube. In this fashion, the thickened-keel method, neurulation, is accomplished in fish.

Keller explant   an explant of dorsal tissue, extending from the dorsal lip of the amphibian blastopore to the animal pole, which displays elongation driven by cell intercalation without the other features of gastrulation.

keratin   a fibrous protein produced by cells of the epidermis. A component of intermediate filaments of the cytoskeleton of epithelial cells.

keratinocyte growth factor (KGF)   also known as fibroblast growth factor 7; a growth factor needed for epidermal production; a paracrine factor produced by fibroblasts of the underlying dermis. *See:* **transforming growth factor α (TGF-α)**.

keratinocytes   epidermal skin cells.

KGF   *See:* **keratinocyte growth factor**.

kinases   enzymes that phosphorylate substrates using adenosine triphosphate (ATP) as the phosphate donor; e.g., protein kinases use ATP to phosphorylate protein substrates.

kinesin   *See:* **motor proteins**.

kinetochore   that part of a chromosome to which the microtubules of spindle fibers attach; necessary for chromosome movement during mitosis and meiosis; made visible with the electron microscope.

Klinefelter's syndrome   a syndrome, including a male phenotype, tall stature, and sterility; resulting from an XXY sex chromosome constitution (see Fig. 25).

***klotho* gene**   mutations of this gene in mice cause a syndrome similar to Hutchinson–Gilford progeria syndrome in humans.

knockin   a genetic engineering method that involves the insertion of a protein coding cDNA sequence at a particular locus in an organism's chromosome; the difference between

Dictionary of Developmental Biology and Embryology, Second Edition. Frank J. Dye.
© 2012 Wiley-Blackwell. Published 2012 by John Wiley & Sons, Inc.

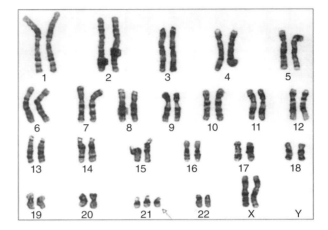

Figure 24. Karyotype of a patient with Down's syndrome; note the extra number 21 chromosome (arrow). (Courtesy of Jacqueline Burns, PhD, the Danbury Hospital.) Reprinted from Frank J. Dye, *Human Life Before Birth*, Harwood Academic Publishers, 2000, fig. 19-3A, p. 164.

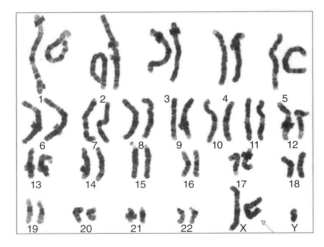

Figure 25. Karyotype of a patient with Klinefelter's syndrome; note the three sex chromosomes (XXY). (Courtesy of Jacqueline Burns, PhD, the Danbury Hospital.) Reprinted from Frank J. Dye, *Human Life Before Birth*, Harwood Academic Publishers, 2000, fig. 19-4B, p. 164.

knock-in technology and transgenic technology is that a knock-in involves a gene inserted into a specific locus and is a "targeted" insertion; a common use of knock-in technology is for the creation of disease models.

knockout   or targeted gene knockout; the targeted disruption of function in a cloned gene; the gene is usually introduced back into the organism in such a way as to replace a normal gene copy, by homologous recombination, in order to analyze the effect of the mutation; the purose of knockout experiments is to find out precisely what molecule or molecules are playing a role in a given situation in development. Interestingly, in many cases, knocking out a normal gene has had no effect at all, even when the gene was thought to be essential. *See:* **redundancy**.

Koller's sickle   the posterior margin of the blastoderm (embryo) in birds and mammals; the organizer originates at Koller's sickle in birds and mammals.

Kölliker, Albrecht von (1817–1905)   traced the origin of tissues from the segmenting ovum through the developing embryo. He encompassed the whole field of embryology and published, in 1861 and again in 1876, a general treatise on vertebrate embryology of high merit.

He is said to have done more to establish the cell theory than anyone else; he was one of the first to recognize the ovum as a cell (in 1840 in his doctoral thesis); the first to point out that spermatozoa were cells (In 1840, Kölliker realized that sperm from the male and ova from the female were cells.); earned his doctorate with work that traced the development of spermatozoa from stem cells to their release from the testes (in 1841). His *Entwicklungsgeschichte* was the first book on comparative embryology; it was he who first applied the cell theory to embryology.

Kolreuter, Joseph (1733–1806)   found that hybrids between plant varieties might resemble one or the other parent or appear intermediate between them.

Kovalevsky, Aleksandr O. (1840–1901)   Russian zoologist; best known for his studies in embryology and structure of ascidians and of Amphioxus, and in embryology and postembryological development of insects. Worked with Amphioxus (in 1867); is given credit for setting forth the idea that a primary single-layered condition changes gradually into a double-layered condition; this soon became regarded as a fundamental embryological sequence of development. In 1866, he showed the practical identity, in the earliest stages of development, between amphioxus and a tunicate; the effect of Kowalevsky's (note different spellings of the name) observations was to break down the sharply limited line ("the great divide") supposed to exist between the invertebrates and the vertebrates. Kowalevsky also founded the generalization that all animals in development pass through a gastrula stage—a doctrine associated, since 1874, with the name of Ernst Haeckel under the title of Gastraea Theory.

Kowalevsky, Aleksandr O. (1840–1901)   *See:* **Kovalevsky, Aleksandr O.**

Kupffer's cells   cells found in the liver; immobile macrophages.

Kupffer's vesicle   in fish embryos, a slight cavity extending up from the archenteron toward the region where the lips of the blastopore close. Believed by some to be homologous to the neurenteric canal; however, the fish incipient spinal cord does not yet possesss a lumen.

L

labeled pathways hypothesis suggested method of axonal pathfinding in insects wherein a given neuron can specifically recognize the surface of another neuron that has grown out before it.

labiogingival lamina a fold of tissue between the lips and the gums.

labioscrotal folds the pair of bulges that make up part of the external genitalia during the indifferent stage, and that give rise to the labia majora in the female and the scrotum in the male (see Fig. 49).

labor the process during which contractions of the smooth muscle of the uterine walls dilate the uterine cervix and force the fetus out of the uterine cavity through the birth canal.

labyrinth a system of intercommunicating canals and cavities that makes up the inner ear; it has membranous and bony components.

labyrinthodonts extinct amphibians, considered to be the amphibian ancestors of the reptiles.

lactation the production of milk.

lactiferous ducts the ducts that carry milk from the mammary glands to the nipples.

lacunae fluid-filled spaces found within the syncytiotrophoblast during implantation, which become the intervillous spaces of the placenta (see Fig. 5).

***lacZ* gene** the *Escherichia coli* gene for ß-galactosidase; commonly used as a reporter gene.

lamellipodia broad flat extensions of the leading edge of a cell; the locomotaory organelle of cells in culture. Also called ruffled membranes.

lamina basalis the basal plate of the neural tube.

lamina terminalis the, originally, anterior-most part of the wall of the neural tube; the median strip of the anterior wall of the prosencephalon, which persists in a relatively unchanged state as the cerebral hemispheres grow out forward beyond the original anterior end of the prosencephalon.

laminin a major component of the type of extracellular matrix called the basal lamina; plays a role in: assembling the extracellular matrix, promoting cell adhesion and growth, changing cell shape, and permitting cell migration.

lampbrush chromosomes diplotene, meiotic chromosomes of oocytes; especially well studied are those of amphibian oocytes. Their appearance is attributed to laterally extended loops of DNA along the length of the chromosomes; lampbrush chromosomes are transcriptionally active.

lancelets *See:* **cephalochordates**.

Langerhans cells cells of the immune system found in the skin.

Lankester, Ray (1847–1929) advanced views (in 1877) supporting an early segregation from the fertilized egg of "already formed and "individualized" substances, provided incentive for considerable research in "cell lineage" investigations.

lanugo fine, downy-type hair found on the skin of the fetus. *See:* **vellus**.

laparoscopy a technique for examining the interior of the peritoneal (abdominal) cavity by means of a peritoneoscope.

laparotomy the making of an incision through the abdominal wall.

large offspring syndrome increased placental and birth weights exhibited by cloned mammals.

larva an immature (pre-adult) stage in the development of many animal species. Some examples of larval forms and their phyla are amphiblastula (Porifera), planula (Coelenterata), miracidium (Platyhelminthes), trochophore (Annelida), veliger (Mollusca),

Dictionary of Developmental Biology and Embryology, Second Edition. Frank J. Dye.
© 2012 Wiley-Blackwell. Published 2012 by John Wiley & Sons, Inc.

nauplius (Crustacea), nymphs, caterpillars, grubs, maggots (Insecta), and pluteus, bipinnaria (Echinodermata); ammocoete is the larva of the Cyclostomata. An insect during the immature, larval, stage of development; a stage in the life cycle of insects exhibiting the holometabolous type of metamorphosis, e.g., Hymenoptera and Diptera. As a rule the embryo is large in relation to the amount of yolk present in the egg, which results in the insect hatching in a relatively early stage of morphological growth known as a larva. In general, a larva differs fundamentally in form, structure, and behavior from the adult insect. In Holometabola, the larval instars are preparatory to the development of the pupa. In the development of most of the invertebrate phyla (excluding long germ-band insects), cleavage patterns are invariant, transcription is activated early in cleavage, and signaling that results in specification of various lineages occurs before gastrulation; in most instances, the embryos form a ciliated swimming larva, which often bears little or no resemblance to the adult and the juvenile develops from a small population of precursor cells embedded within the larva; i.e., the larva is a "life support system" for the developing juvenile, which becomes independent only as a result of metamorphosis. Most terrestrial vertebrates show direct development, while, generally, marine invertebrates use indirect development and set-aside cells. *See:* **tadpole**.

Larvacea   a class of the subphylum Urochordata, consisting of minute planktonic animals in which the tail, with dorsal nerve cord and notochord, persists throughout life.

larval molt   the molts between the instar (larval) stages of holometabolous insects.

laryngotracheal groove   an evagination of the floor of the embryonic pharynx that gives rise to the lining of the respiratory system.

latebra   formed in chicken oocytes by the growth of the yolk nucleus and having a light yellowish appearance. *See:* **yolk nucleus**.

laterad   to move in a lateral direction.

lateral   side.

lateral body folds   two of four body folds; the lateral folds sculpt the sides of the embryonic body out of the blastoderm or embryonic disc (see Figs. 6 and 19).

lateral geniculate nucleus   the region of the brain in mammals where the axons from the retina terminate.

lateral lingual swellings   two eminences, found anterior to the foramen cecum, which give rise to the anterior two thirds (oral part) of the tongue.

lateral meristem   a meristematic part of a plant, the activity of which leads to growth in girth of a plant part, such as the growth in girth of a tree trunk. *See:* **meristem**.

lateral mesoderm   that part of the mesoderm, found bilaterally, lateral to the intermediate mesoderm (see Figs. 28 and 41).

lateral palatine processes   bilateral processes that grow out from the inner, medial borders of the maxillary processes and fuse with each other and the median palatine process to give rise to the hard palate.

lateral plate mesoderm   bilaterally, that mesoderm that is beyond (lateral to) the mesomeric mesoderm or nephrotome; gives rise to, e.g., blood, kidneys, and heart.

lateral somitic frontier   the boundary between the somite-derived and the lateral plate-derived dermis.

lateral ventricles   the ventricles of the cerebral hemispheres; ventricle I is the cavity of the left cerebral hemisphere, and ventricle II is the cavity of the right cerebral hemisphere.

latitudinal division   a cleavage division where the plane of cleavage is perpendicular to the animal-vegetal axis of the embryo; sometimes referred to as a horizontal division.

Law of Independent Assortment   one of Gregor Mendel's profound generalizations about heredity; the way in which members of one pair of alleles segregate (separate) in the formation of the gametes has no influence on the way in which members of another pair of alleles segregate; this is not always true, e.g., if pairs of alleles for two different genes are on the same chromosome.

Law of Segregation   one of Gregor Mendel's profound generalizations about heredity; members of pairs of alleles segregate (separate) in the formation of the gametes; consequently, a gamete has a single allele for each gene.

LD$_{50}$ the dose of a chemical (e.g., toxin) that is lethal for 50% of a population of organisms.

leaf axils   the junction between the leaf and the main stem.

leaf development   leaves originate as lateral outgrowths of the growing point of a stem; leaf primordia undergo cell division, cell elongation, and cell differentiation, but compared with roots and shoots, (1) the duration of cell division is shorter and (2) all leaf cells eventually differentiate; i.e., unlike roots and shoots, leaves do not have a persistent meristem.

leaf primordia clusters of cells, initiated at the periphery of shoot meristems, that will form leaves.

lectin a carbohdrate-binding protein.

lectin interaction the concept that bindin, a protein on spermatozoa, binds with glycoprotein receptors on egg coats and plays a role in species specific fertilization.

Leeuwenhoek, Antony van (1632–1723) Dutch, was unquestionably (according to Gardner, 1965) a philosophical preformationist and an animulculist; at least he professed belief in an intangible preformationism in the sperm. He observed (in 1677) human sperm, the association of sperm with eggs in frogs and fish, and maintained (in 1683) that eggs were impregnated by seminal animalculae. The assertion that fertilization is accomplished when one animalcule gets into one egg was originally made by Leeuwenhoek in 1683.

leghemoglobin a molecule made cooperatively by a legume (provides the globin protein) and rhizobia bacteria (provides the heme group); allows for the symbiotic relationship and for the functioning of nitrogenase that is necessary for biological nitrogen fixation; i.e., leghemoglobin sequesters oxygen away from the oxygen-sensitive nitrogenase complex. *See:* **mycorrhiza**.

lens placodes a pair of bilateral thickenings of the ectoderm of the head of the early embryo; induced by the optic vesicles, they give rise to the lenses of the eyes.

lens vesicles bilaterally symmetrical, paired invaginations of the surface ectoderm of the head of vertebrate embryos, induced to form by the pair of optic vesicles, that give rise to the lenses of the eyes.

leptotene the first stage of prophase I of meiosis, during which the chromosomes become visible as elongated threads; sister chromatids are not usually individually visible.

level-interactive modular array a hierarchically nested system; e.g., cells are parts of tissues, which are parts of organs, which are parts of systems, and so forth.

Levi-Montalcini, Rita (1909–) Italian neurologist, discovered nerve-growth factor (first in the salivary glands of developing mouse embryos), a substance that controls how many cells make up the adult nervous system. She was the co-recipient of the Nobel Prize in 1986. *See:* **Hamburger, Viktor**.

levonorgestrel one form of a variety of progestogen hormones found in birth control pills; has activity similar to the natural hormone progesterone.

levotropic cleavage spiral cleavage with the cells displaced counterclockwise. *See:* **dextrotropic cleavage**.

Lewis, Edward B. (1918–2004) co-recipient, with Eric F. Wieschaus (United States) and Christiane Nüsslein-Volhard (Germany), of the 1995 Nobel Prize for Physiology or Medicine for discovery of genes that control the early stages of the body's development.

Lewis, Warren (1870–1964) American embryologist; According to Margaret Saha (1994), Hans Spemann notwithstanding, the textbook paradigm for general embryonic induction has remained Lewis's model of the optic vesicle inducing the lens in a single step.

Leydig cells *See:* **interstitial cells of Leydig**.

lie one of four descriptions of the fetus' alignment in the uterus; lie of the fetus refers to the relationship of the long axis of the fetus to that of the mother (see Fig. 3). *See:* **attitude, position, presentation**.

life expectancy the amount of time a member of a species can expect to live; not a characteristic of species but of populations; the age at which half the population still survives. *See:* **maximum life span**.

ligand a substance that attaches to the binding site of a protein; examples of which are the substrates of enzymes, the antigens of antibodies, the ligands of membrane receptors, and the allosteric effectors of enzymes. *See:* **cooperativity**.

ligation, tubal *See:* **tubal ligation**.

Lillie, F. R. (1870–1947) a professor at the University of Chicago, was one of the first to recognize that secondary sexual characteristics of the male are dependent on a circulating factor, based on his studies of the masculinized freemartin. He spent much of his career directing the Marine Biological Laboratory at Woods Hole. In 1898, at Woods Hole, he presented a lecture that argued that "modern" evolutionary studies would do better to concentrate on changes in embryonic development that allowed for survival in particular environments than to focus on ancestral homologies that united animals into lines of descent. He gathered evidence that egg capsules (coats) did not just protect eggs but also played a functional role in fertilization (in 1912). *See:* **Wilson, Edmund Beecher**.

***LIM* genes**   genes that encode transcription factors (structurally related to those encoded by the *Hox* genes) expressed by younger motor neurons as they migrate through the region of older motor neurons in the intermediate zone of the spinal cord, as a result of a retinoid (e.g., retinoic acid) signal secreted by the early-born motor neurons. As a result of their differing birthdays and migration patterns, motor neurons form three major groupings (those of the column of Terni, lateral motor column, and medial motor column); each group of neurons is characterized by a particular constellation of LIM transcription factors.

LIM-homeodomain proteins   transcription factors; the LIM domain is responsible for protein-protein interactions; these proteins possess two LIM domains together with the DNA-binding homeodomain, e.g., Lim-1 in the organizer and Islet-1 in motor neurons.

limb buds   the first indications of the developing limbs; found bilaterally on the flanks of the early 4- to 5-week human embryo. In the chick embryo, the early limb bud consists of an apical ectodermal ridge and a core of mesenchyme.

limb field   an area of the embryo containing all of the cells capable of forming a limb.

limbs   arms and legs.

linea nigra   black line; a line of pigmentation, running from the pubic region to the umbilicus (navel), in pregnant women.

lineage   a descent in a line from a common progenitor. *See:* **progenitor**.

lineage-based information   information passed on from a mother cell to her daughters. *See:* **positional information**.

lineage diagram   a linear version of a fate map, showing the fates of each cell of the embryo.

lineage labels   markers applied to cells to tag them for lineage studies; traditional examples include vital staining with neutral red and nile blue; more recent examples include the injectable horseradish peroxidase and fluorescein-dextran-amine (FDA) or the surface marker DiI.

lineage-restricted stem cells   stem cells that can produce only one type of cell in addition to renewing itself, e.g., the BFU-E (burst-forming unit, erythroid).

lineage studies   tracing the fate of a cell or group of cells during a period of development; the cells in question may be naturally marked (by, for example, pigmentation) or artificailly marked (by, for example, the injection of a fluorescent dye).

lipid   a class of organic molecules that carries out various functions in the cell; an especially important function is to serve as the basic structure of cell membranes—for both the plasma membrane and the membranes of various organelles.

lipovitellin   a lipoprotein yolk protein found in frog oocytes, derived from the splitting of vitellogenin and packaged into yolk platelets.

liquor folliculi   the fluid filling the atrum of the mature graafian follicle (see Fig. 33).

litter   a collective term for the multiple progeny, resulting from a single pregnancy, of some animals, such as cats and dogs.

Littré's glands   *See:* **glands of Littré**.

liver bulge   the bulge of the abdominal region found on the surface of the human embryo during the middle of the embryonic period.

liver primordium   the ventrally directed outgrowth (diverticulum) of the foregut, immediately anterior to the midgut; the anlage for the liver, bile duct, and gall bladder in the frog embryo. In the frog embryo, the anterior wall of the diverticulum becomes folded and thickened to give rise to the liver proper, while the posterior part is partially constricted away as the gall bladder; the original connection with the foregut remains as the bile duct.

liverworts   a class of nonvascular plants or bryophytes, e.g., *Marchantia*.

lobster-claw deformity   also called bidactyly, is congenital absence of all fingers or toes except the first and the fifth.

locule   the chamber or cavity of an ovary containing the ovules.

locus   (pl. loci) the site (location) along the length of a chromosome where an allele of a specific gene is located.

locus control regions (LCRs)   regions of DNA that function as "super enhancers"; these LCRs establish an "open" chromatin configuration, inhibiting the normal repression of transcription over an area spanning several genes.

Loeb, Jacques (1859–1924)   born in Germany and settled in America (in 1891); held professorships at Bryn Mawr and universities of Chicago and California; and became a member of the Rockefeller Institute for Medical Research (in 1910). In 1891, introduced the word "heteromorphosis" to describe the condition in regeneration wherein one type of append-age is amputated and another kind regenerates in its place. He carried out important work on artificial parthenogenesis, showing that unfertilized frogs' eggs could be induced to divide by altering their environment.

long-day plants   plants that flower only after being exposed to light periods longer than a certain critical length, as in summer; e.g., spinach, lettuce, and some varieties of wheat are long-day plants.

long-germ development   a type of insect development in which the blastoderm corresponds to the whole of the future embryo; all of the segments form at approximately the same time. *Drosophila* is a long-germ insect. *See:* **short-germ development**.

Lophotrochozoa   a great clade of protostomes including the annelids, molluscs, flatworms, and other phyla that produce a particular form of larva (trochophore) or feeding structure (lophophore).

loss-of-function mutation   a mutation that incapacitates a gene so that no functional product is produced; also called a forward, knockout, or null mutation. The protein product of the mutant gene is less active than the wild type. *See:* **gain-of-function mutation**, **null mutation**.

lower vertebrates   anamniotic vertebrates; fishes and amphibians.

lumbar rib   a common type of accessory rib formed by abnormal development of lumbar vertebrae.

lumen   the cavity of an organ, e.g., uterine cavity.

luminal surface   the lining of the lumen of an organ, e.g., endometrium of the uterus or ependyma of the spinal cord.

lung bud   a primary outgrowth of the embryonic trachea; the anlage of a primary bronchus and all its branches.

lungs   organs of respiration found in adult amphibians and the higher vertebrates. In the frog, lungs appear as a pair of solid, posteriorly directed proliferations from the ventral side of the pharynx, just posterior to the heart rudiment.

luteinizing hormone (LH)   a glycoprotein hormone released by the anterior lobe of the pituitary gland that affects the gonads (gonadotrophin) in females and males. In females, LH stimulates follicle cells and corpus luteum cells to produce progesterone. In males, LH stimulates Leydig cells to secrete testosterone.

luteotropin   a hormone secreted by the trophoblast that causes the corpus luteum to remain active and serum progesterone levels to remain high.

lymphoid precursor cell   the pluripotent stem cell, which gives rise to pre-T and pre-B cells; these cells, in turn, give rise to T cells and B cells, respectively; the terminally differentiated cells of this lineage are activated T cells and plasma cells, respectively.

Lyon hypothesis   the assumption that mammalian females are X-chromosome mosaics as a result of the inactivation of one X chromosome in some embryonic cells and the other in the rest.

Lysenko, Trofim (1898–1976)   born into a peasant family in the Ukraine, achieved acclaim working as a plant biologist in the late 1920s; in 1935, he started a campaign against Mendelian genetics; in 1940, he became Director of the Institute of Genetics within the USSR Academy of Sciences, where he used his political influence to dismiss Mendelian genetics as "bourgeois science" and "pseudoscience;" his theories were largely rejected outside the USSR.

lysins   lytic enzymes of spermatozoa, which allow spermatozoa to pass through egg coats. *See:* **acrosin, corona penetrating enzyme (CPE), hyaluronidase**.

lysosomal storage diseases   diseases resulting from the lack of specific enzymes in the lyso-somes of a cell; as a consequence the lysosome is not able to digest (hydrolyze) specific kinds of molecules, resulting in an accumulation of these molecule with often fatal conse-quences for the cell and, ultimately, the organism.

lysosomes   cellular organelles, containing a wide variety of hydrolytic (digestive) enzymes, which help the cell remove waste materials.

lysozyme   an enzyme that hydrolyzes glycoproteins on surfaces of bacteria, killing them, e.g., found in the cortical granules of carp eggs and transferred to the fertilization envelope after fertilization. Lysozyme is abundant in several secretions, such as tears, saliva, human milk, and mucus; it is also present in cytoplasmic granules of the polymorphonuclear neu-trophils (PMN); large amounts of lysozyme can be found in egg white. *See:* **embryo defenses**.

M

M checkpoint   a checkpoint in the cell cycle that must be passed for the cell to enter into mitosis; to pass the M checkpoint, a complex of cyclin B and Cdk1, called M-phase promoting factor, has to be activated. Active MPF phosphorylates and thereby activates the various components required for mitosis (nuclear breakdown, spindle formation, and chromosome condensation); it also initiates its own destruction by activating the breakdown of cyclin B, so that by the end of the M phase, the active MPF has disappeared.

M-phase promoting factor (MPF)   *See:* **maturation-promoting factor (MPF)**.

macrocephaly   also called megalocephaly, is an abnormal largeness of the head.

macroevolution   the large morphological changes observed among species, classes, and phyla; large evolutionary change, usually in morphology; typically refers to the evolution of differences among populations that would warrant their placement in different genera or higher level taxa. *See:* **microevolution**.

macrogamete   the larger, usually female, gamete produced by a heterogamous organism.

macroglossia   literally, "a large tongue" to the extent that it is abnormally large.

macrolecithal eggs   eggs with relatively large amounts of yolk. In general, reptiles, birds, and fishes have macrolecithal eggs (see Figs. 12 and 14).

macromeres   in those species that have blastulas made up of various-sized blastomeres, the largest blastomeres are called macromeres.

macromolecules   large molecules; three classes are found in cells and organisms: polysaccharides (some carbohydrates), polypeptides (proteins or subunits of proteins), and polynucleotides (nucleic acids).

macrophyll   also called megaphyll.

macrostomia   an abnormally large mouth.

macula   a spot; the macula acustica is the acoustic nerve termination in both the sacculus and the utriculus.

maintenance methylases   DNA methyl transferase enzymes that methylate the other CG (cytosine, guanine) of sites bearing a methyl group on only one strand. *See: **de novo methylases**.

malformations   congenital abnormalities caused by genetic events (mutations, aneuploidies, and translocations), e.g., aniridia (absence of the iris of the eye) and Down's syndrome. *See:* **disruptions**.

Malpighi, Marcello (1628–1694)   Italian anatomist; his two papers on chick embryology were titled, *On the Formation of the Chick in the Egg* (1673) and *Observations on the Incubated Egg* (1689). He was a preformationist. He supplied an illustrated account of the actual stages in the development of the chick from the end of the first day to hatching; his pictures represent what he was able to see. Considered by some to be the founder of embryology.

Malpighian layer   the layer of the epidermis composed of two sublayers: the inner basal layer (stratum germinativum) and the outer spinous layer.

Malpighian tubules   in insects, develop as ectodermal outgrowths of the proctodaeum, close to its union with the mesenteron; excretory tubes; they lie in the hemocoel and extract from the blood the excretory products that pass into the hindgut and exit with the feces.

Mammalia   the class of vertebrates composed of mammals; includes two subclasses, Prototheria and Theria.

mammalian embryos   human as an example: external features (4–6 mm); on the cephalic region, nasal pits are present, bulgings caused by growing optic cups, and auditory vesicles are observable (in cleared and stained preparations). The oral region and the region caudal

Dictionary of Developmental Biology and Embryology, Second Edition. Frank J. Dye.
© 2012 Wiley-Blackwell. Published 2012 by John Wiley & Sons, Inc.

to it are compressed against the thorax, including the mandibular arches and maxillary processes, hyoid and unnamed third and fourth postoral arches; between the arches are the furrows, which do not ordinarily break through into the pharynx. The entire region around the third and fourth clefts is deeply depressed as the cervical sinus. The neck is not yet elongated, and the thorax shows a cardiac prominence. Caudal to the cardiac prominence is a shallow groove, beneath which is the septum transversum, the primordium of the diaphragm. The belly stalk becomes delineated a little later. The somites (which begin to appear at the end of the third week and attain nearly their full number during the fourth week) begin to be recognizable caudal to the auditory vesicles. The arm buds lead the leg buds slightly in their developmental progress. At this stage, the human embryo has every bit as well developed a tail as a pig embryo.

mammary glands   glands found in the breasts of female mammals that secrete milk for the nourishment of young offspring.

mammary ridges   approximately parallel ridges of ectodermal tissue extending from the armpits to the groins of the embryo, along which mammary glands may develop at any point; normally a single pair of mammary glands form on the chest of a human female.

mandible   the lower jaw.

mandibular arches   the first pair of branchial (pharyngeal) arches; these give rise to the upper and lower jaws. *See:* **branchial arches**.

Mangold, Hilde (1898–1924)   German embryologist; was a graduate student at the University of Freiburg in Germany. In 1924, Hans Spemann and Mangold demonstrated the self-differentiation of the dorsal lip of the blastopore; namely, if this tissue was transplanted, during early gastrula stage, to a presumptive epidermal region of a second early gastrula, the recipient embryo would undergo two invaginations: the expected (primary) invagination and an extra (secondary) invagination caused by the transplant; furthermore, two embryonic axes would develop and two fused salamander embryos would result. This self-differentiating tissue of the dorsal lip of the blastopore was referred to as the organizer, and this experiment is considered by many to be the most important experiment carried out in embryology (developmental biology). Tragically, Hilde Mangold died at an early age, the victim of an accidental explosion; many believe that had she lived she would have shared the 1935 Nobel Prize with Hans Spemann. *See:* **Mangold, Otto**.

Mangold, Otto (1891–1962)   German embryologist; demonstrated (in 1933) regional specificity of neural induction. Was Hilde Mangold's husband.

mantle layer   (see Fig. 43). *See:* **intermediate layer**.

MAP   mitogen-activated protein.

MAP kinase   *See:* **ERK**.

MAPK   *See:* **mitogen-activated protein kinase**.

MAPs   *See:* **microtubule-associated proteins**.

marginal cells   cells found in the region near the equator (marginal zone) of the amphibian late blastula and gastrula; are of two general types: (1) superficial marginal cells, found in the surface of the marginal zone, and (2) deep marginal cells, found beneath the surface of the marginal zone.

marginal layer   the outer layer of the wall of the spinal cord, made up of the processes of the nerve cells; that is, white matter; the outer, cell-poor, axon-rich, region of the neural tube/spinal cord (see Fig. 43).

marginal sinus   *See:* **sinus terminalis**.

marginal zone   the region near the equator of the amphibian blastula, i.e., the margin between the animal and vegetal hemispheres; a thin layer of cells between the area pellucida and the area opaca of the developing chick embryo.

marine invertebrates   in an embryological context, those invertebrates, such as echinoderms, which have been used extensively during the twentieth century to advance our understanding of descriptive, comparative, and experimental embryology. The eggs and embryos of marine invertebrates are particularly suited for such studies because they are (1) small, (2) transparent, and (3) develop synchronously.

marrow   *See:* **bone marrow**.

marsupial   *See:* **Metatheria**.

marsupium   any pouch-like structure in which the young of an animal complete their development, e.g., the abdominal pouch of marsupials.

Mash-1   a transcription factor; specifies the sympathetic and parasympathetic neurons from the neural crest.

MASH1 **gene**   the mammalian homologue of the *Drosophila* genes that encode the achaete-scute transcription factors; expressed in subsets of neurons and may influence neuronal differentiation in olfactory receptor cells.

master regulatory genes   a phrase that refers to any set of genes that may have many "targets," or other genes that they turn on or off; they sit at the top of a large hierarchy of genes that are really controlling the morphology and development of an organism. There seems to be a relatively small handful of genes that we give this title of master regulatory gene; e.g., *Pax6* genes are important for eye development throughout the animal kingdom; *tinman* genes are responsible for heart development in flies, worms, and mammals; and *Hox* genes are critical for specifying the axes in all animals studied so far.

maternal effect genes   genes in the mother, expressed during oogenesis, which produce maternal proteins and maternal RNAs that are deposited in the egg and function in the embryo. In *Drosophila*, approximately 50 maternal genes are involved in setting up the anteroposterior and dorsoventral axes and a framework of positional information, which is then interpreted by the embryo's own genetic program. Maternal gene products establish the axes and set up regional differences along each axis in the form of spatial distributions of RNA and proteins, in turn, these proteins activate zygotic genes. Of the 50 or so maternal genes, the products of four, *bicoid, hunchback, nanos,* and *caudal,* become distributed along the anteroposterior axis and are crucial in establishing it; intranuclear dorsal protein is distributed along the dorsoventral axis. These protein distributions, graded along the embryonic axes, constitute a maternally derived framework of positional information, which is interpreted and elaborated on by zygotic genes, resulting in giving each region of the embryo an identity. Sporophytic and gametophytic maternal effect genes have been identified in *Arabidopsis*; how significant maternal effect genes are in establishing the sporophyte body plan is still unknown. *See:* **zygotic genes.**

maternal effects stage   the cleavage stage embryo, because the phenotype of the cleavage stage embryo depends entirely on the genotype of the mother and not on that of the embryo itself.

maternal-fetal co-development   maternal-fetal interactions. The fetus and the trophoblast induce changes in the structure and physiology of the uterus (the decidua reaction); the uterus induces proteins on the trophoblast that allows it to adhere to and ingess into the uterus.

maternal inheritance   the inheritance of genetic information from one's mother through the cytoplasm of the egg (instead of through the typical route, the nuclear genes); e.g., mitochondrial genes are inherited only from one's mother not from one's father.

maternal message   an mRNA produced in an oocyte before fertilization; many maternal messages are kept in an untranslated state until after fertilization.

maternal mRNA   *See:* **maternal message.**

maternal proteins   in *Drosophila*, soon after fertilization of the egg, several gradients of maternal proteins have been established along the anteroposterior axis; two gradients, bicoid and hunchback proteins, run in an anterior to posterior direction, while caudal protein is graded posterior to anterior.

mathematical modeling   as applied to developmental biology, seeks to describe developmental phenomena in terms of equations.

matrotrophy   pertaining to a form of nutrition provided by the maternal gametophyte as, for example, in the case of a moss gametophyte providing nutrients to the zygote and developing sporophyte.

maturation divisions   meiotic divisions making up part of gametogenesis in animals; during oogenesis in vertebrates, generally, the first maturation division occurs at the time of ovulation and the second maturation division occurs with fertilization of the egg (actually a secondary oocyte).

maturation-promoting factor (MPF)   also called mitosis-promoting factor; a regulator of the cell cycle, composed of two subunits: Cdc2 (a catalytic subunit) and cyclin B (a regulatory subunit). Activation of the Cdc2/cyclin B complex (by Cdc 25) brings about the transition from G2 to M of the cell cycle.

maxilla   the upper jaw.

maxillary processes   a pair of processes originating from the mandibular arches that give rise to the maxilla (upper jaw).

maximum life span   a characteristic of species; the maximum number of years a member of that species has been known to survive. *See:* **life expectancy.**

mazopathy disease of the placenta.

mechanoreceptors those sensory receptors responding to the sensations of sound and pressure.

Meckel, Johann Friedrich (1781–1833) German anatomist, especially known for work on comparative anatomy; in 1821 (and later Karl Ernst von Baer) indicated the close similarity between embryonic stages of widely different animals.

Meckel's cartilage the lower jaw of lower vertebrates, and, in higher vertebrates, the axis around which membrane bones of the lower jaw are arranged and formed.

Meckel's diverticulum the persistent blind end of the yolk stalk forming a tube connected with the lower ileum.

meconium the waste material that accumulates in the gut of the fetus and that, normally, is not excreted until after birth; also the waste products of a pupa or other embryonic form.

mediad toward the median plane of the body.

medial palatine process grows out from the innermost part of the premaxilla and fuses with the lateral palatine processes to give rise to the anterior, medial, part of the hard palate.

medial plane an important meridional plane; an imaginary plane that separates the right and left sides of the body from each other.

mediastinum testis a septum in the posterior portion of the testis, formed by an inward projection of the tunica albuginea.

mediolateral axis the left-right axis. *See:* **axes**.

medulla (1) general: the inner portion of something, as in the ovarian medulla or medulla of the adrenal gland; (2) brain: the caudal-most portion of the brain, derived from the myelencephalon of the early five-vesicle brain.

medullary cord(s) (1) the primary invaginations of the germinal epithelium of the embryonic gonad that differentiate into rete testis and seminiferous tubules or into rete ovarii; (2) forms by condensation of mesenchymal cells and then mesenchymal-to-epithelial transition in the caudal region of the embryo during the process of secondary neurulation, it will then cavitate to form the caudal section of the neural tube. *See:* **rete cord**.

medullary folds neural folds. *See:* **neural folds**.

medullary plate the neural plate. *See:* **neural plate**.

megagametogenesis the formation of the embryo sac by a haploid megasopre undergoing three successive mitotic divisions.

megagametophyte the embryo sac of angiosperms.

megalecithal eggs *See:* **macrolecithal eggs**.

megalocephaly *See:* **macrocephaly**.

megaphyll literally, "large leaf." A leaf with a branched vascular system in the blade; typical of many higher plants. *See:* **microphyll**.

megasporangium the ovules of the ovaries of flowering plants.

megaspore literally, "large spore," such as those produced by angiosperms that give rise, by mitosis, to the female gametophyte, the embryo sac.

megaspore mother cell a cell in the sporophytes of angiosperms that, by undergoing meiosis, gives rise to four megaspores.

megasporocyte also called megaspore mother cell; the embryo-sac mother-cell, diploid cell in the ovary that undergoes meiosis, producing four haploid megaspores.

megasporogenesis the formation of megaspores in the nucellus.

megasporophyll a leaf or leaflike structure bearing a megasporangium or megasporangia; a spore-bearing leaf developing megasporangia; carpel.

meiolecithal eggs eggs with a more or less equal proportion of yolk and cytoplasm; occurs in the protochordates.

meiosis one of two general types of cell divisions undergone by eukaryotic cells; an integral part of sexual reproduction. It was J. B. Farmer, who in a paper with J. E. Moore in 1905, proposed the name meiosis based on a greek verb, to lessen; their form of the term was spelt "maiosis," which was altered by later writers to "meiosis." The following activities set meiosis apart from mitosis: (1) pairing of homologous chromosomes, (2) crossing over between homologues, (3) reduction of chromosome number, (4) the slow pace of meiotic prophase, (5) the requirement of two cell divisions to complete the process, and (6) the lack of an S period between the two divisions. Regarding the chromosome separation that takes place during meiosis: (1) reduction separation, homologous segments of non-sister chromatids separate (paternal chromosome moving away from maternal chromosome), and (2) equational separation, homologous segments of sister chromatids separate

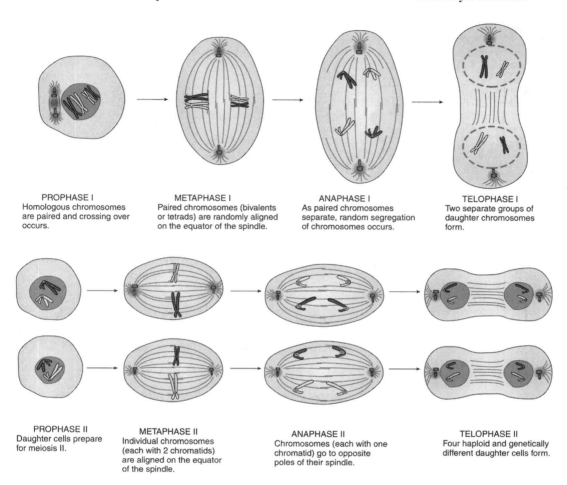

PROPHASE I
Homologous chromosomes are paired and crossing over occurs.

METAPHASE I
Paired chromosomes (bivalents or tetrads) are randomly aligned on the equator of the spindle.

ANAPHASE I
As paired chromosomes separate, random segregation of chromosomes occurs.

TELOPHASE I
Two separate groups of daughter chromosomes form.

PROPHASE II
Daughter cells prepare for meiosis II.

METAPHASE II
Individual chromosomes (each with 2 chromatids) are aligned on the equator of the spindle.

ANAPHASE II
Chromosomes (each with one chromatid) go to opposite poles of their spindle.

TELOPHASE II
Four haploid and genetically different daughter cells form.

Figure 26. Phases of meiotic cell division. Reprinted from Frank J. Dye, *Human Life Before Birth*, Harwood Academic Publishers, 2000, fig. 3-9, p. 18.

(maternal chromatid moves away from maternal chromatid). The terms "reductional" for the first and "equational" for the second division are misnomers and would be correct if crossing over did not occur. The most commonly used nomenclature for the two divisions are the terms meiosis I and meiosis II (see Fig. 26).

meiosis and the life cycle   meiosis may occur at three different points in the life cycle: (1) at the end (terminal or gametic), as in animals; (2) at the beginning (zygotic), as in *Chlamydomonas* (a green alga); or (3) intermediate (sporic), as in most plants.

meiotic chromosome terminology   during a regular meiotic cycle: (1) monad chromosome, at the end of premeiotic telophase (= unit chromatid); (2) dyad chromosome, after the S of premeiotic interphase (2 sister chromatids); (3) tetrad chromosome (bivalent), after synapsis of prophase (4 chromatids or 2 homologous chromosomes); (4) dyad chromosome, after anaphase I (2 chromatids), and (5) monad chromosome, after anaphase II (unit chromatid).

MEK   mitogen-activated, ERK-activating kinase; a kinase in the MAP kinase signal transduction pathway. *See:* **MAP**.

melanin   a dark pigment produced by melanocytes of the skin.

melanocyte stem cell   adult stem cell derived from trunk neural crest cells that resides in the bulge niche of the hair or feather follicle that gives rise to the pigment of the skin, hair, and feathers.

melanocytes pigment cells derived from the neural crest, which produce the pigment melanin (see Plate 13 in the color insert).

melanoma a malignant tumor of melanocytes.

-melia a part of a word referring to the limbs, e.g., phocomelia.

meltrins a set of metalloproteinases that mediate fusion of myoblasts into myotubes. *See:* **fertilin proteins**.

membrana granulosa several layers of granulosa (follicle) cells, that rest on the basement membrane of the graafian follicle, and, also, border the fluid-filled antrum of the follicle.

membranous labyrinth *See:* **labyrinth**.

memory cells a subset of cells that differentiate from B cells (B lymphocytes), upon B cell stimulation by antigen binding; populate lymph nodes and can respond rapidly when exposed to the same antigen at a later time.

menarche the beginning of menstrual cycles with the onset of puberty in a female child.

Mendel, Gregor (1822–1884) the nineteenth century, Austrian monk considered to be the father of genetics; he proposed that the hereditary material has a particulate nature (which we call genes), and he proposed two laws for their distribution (Law of Segregation and Law of Independent Assortment).

mendelian refers to Gregor Mendel.

meningocele a protrusion of the cerebral or spinal meninges (membranes) through a defect in the skull or vertebral column, forming a cyst filled with cerebrospinal fluid.

meningocephalocele a protrusion of part of the cerebellum and meninges through a defect in the skull.

meningomyelocele a protrusion of a portion of the spinal cord and membranes through a defect in the vertebral column.

menopause the end of menstrual cycles in a woman at the end of her child-bearing years.

menses the menstrual flow or discharge that occurs during the menstruation portion of the menstrual or uterine cycle.

menstrual age the age of an embryo or fetus calculated from the first day of the mother's last normal menstruation preceding pregnancy.

menstrual cycle the cyclic changes, especially in the lining of the uterus, undergone by a woman during her potential child-bearing years. Also called uterine cycle (see Fig. 27).

mericlinal chimeras results when a genetically marked cell gives rise to a sector of an organ or of a whole plant.

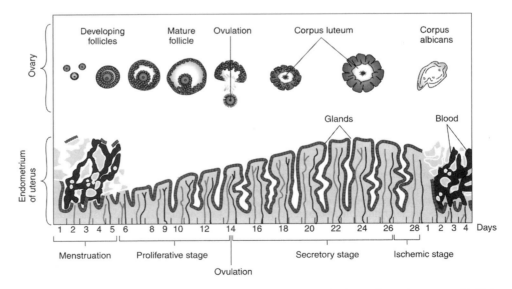

Figure 27. The ovarian and menstrual (uterine) cycles. The top of the figure shows the events of follicle development in the ovary. The bottom of the figure shows the corresponding events of the menstrual cycle. Reprinted from Frank J. Dye, *Human Life Before Birth*, Harwood Academic Publishers, 2000, fig. 5-10, p. 41.

meridional division   a cleavage division where the plane of cleavage is parallel to the animal-vegetal axis of the embryo; sometimes referred to as a vertical division.

meridional plane   an imaginary plane parallel to the animal-vegetal axis of the embryo; sometimes referred to as a vertical plane.

meristem   a part of a plant characterized by dividing cells; these regions of plants contain stem cell populations that produce cells, some of which go on to differentiate into plant tissues and some of which constantly renew the stem cell population, Generally, two varieties are found: (1) apical meristems, found at the tips of plant parts, e.g., at the growing ends of roots and shoots and (2) lateral meristems, found in the sides of plants; e.g., vascular cambium and cork cambium are lateral meristems found under the bark of tree trunks. The primary root of a flowering plant has three meristematic regions: the procambium, the ground meristem, and the protoderm. *See:* **central zone**, **inflorescence meristem**, **intercalary meristem**, **peripheral zone**, **primary tissues**.

meristic variation   differences in the number of repeating structures.

Merkel cells   cells derived from the neural crest, which give rise to mechanoreceptors in the skin.

Merkel's tactile disc   a type of free nerve termination; each composed of a Merkel cell and the leaf-like expansion of a nerve terminal; Merkel cells are thought to play a role in sensation—as skin mechanoreceptors, sensitizing the skin to touch.

meroanencephaly   partial absense of the brain.

meroblastic   a type of cleavage pattern wherein only part of an egg undergoes cleavage; chicken eggs undergo meroblastic cleavage. The meroblastic cleavage pattern comes in two general varieties; meroblastic discoidal and meroblastic superficial. In meroblastic discoidal, characteristic of macrolecithal eggs, cleavage is initially confined to a relatively yolk-free, disc-shaped region close to the animal pole; in meroblastic superficial cleavage, characteristic of insects, cleavage is initially confined to a superficial layer of relatively yolk-free cytoplasm, which surrounds the central yolk (see Fig. 12). *See:* **holoblastic**.

merogone   an artificially created *part* of an egg.

merogony   the normal or abnormal development of a part of an egg following cutting, shaking, or centrifugation of the egg before or after fertilization.

meroistic oogenesis   a type of insect oogenesis in which cytoplasmic connections remain between the cells produced by the oogonium, e.g., exhibited by *Drosophila*.

meroistic ovariole   ovariole containing nutritive or nurse cells.

meromelia   refers to deficits of limb development so that parts of limbs are missing; a type of birth defect caused by the drug thalidomide.

MESA   *See:* **microsurgical epididymal sperm aspiration**.

mesectoderm   the portion of the mesenchyme originating from the ectoderm.

mesencephalon   the midbrain; the middle portion of the embryonic vertebrate brain; gives rise to fiber tracts between the anterior and posterior brain, optic lobes, and the tectum (see Plate 5 in the color insert and Fig. 8).

mesenchymal cell   a cell that makes up part of mesenchyme; these cells, unlike epithelial cells, are not connected to one another; one of the two major types of cell arrangements in the embryo. *See:* **cell arrangements**.

mesenchymal stem cells (MSCs)   bone marrow-derived stem cells (BMDCs); are able to give rise to numerous bone, cartilage, muscle, and fat lineages.

mesenchyme   a loosely arranged type of embryonic tissue, with stellate cells embedded in substantial amounts of loose extracellular matrix, which can develop from any germ layer and is not necessarily of mesodermal origin.

mesenchyme-to-epithelium transitions   occur when mesenchymal cells collectively form an epithelium, e.g., formation of the kidney tubules. *See:* **epithelial-to-mesenchymal transitions**.

mesendoderm   embryonic tissue that differentiates into mesoderm and endoderm. *See:* **mesentoderm**.

mesenteries   membranes that suspend various parts of the digestive system in the abdominal (peritoneal) cavity (see Fig. 6).

mesenteron rudiments   in insects, are groups of cells closely associated with the stomodaeal and proctodaeal ingrowths; these cells multiply and grow toward each other and, finally, enclose the yolk in the form of a complete tube, the mesenteron or embryonic mid-intestine.

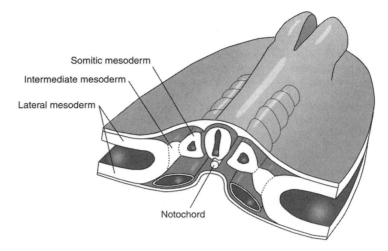

Somitic mesoderm

Intermediate mesoderm

Lateral mesoderm

Notochord

Figure 28. Regions of embryonic mesoderm. Reprinted from Frank J. Dye, *Human Life Before Birth*, Harwood Academic Publishers, 2000, fig. 8-11, p. 70.

mesentoblasts   in snail embryos, the cells derive from the 4d blastomere that give rise to both the mesodermal (heart) and endodermal (intestine) organs.

mesentoderm   an intermediary, composite germ layer formed during gastrulation in some animals, e.g., starfish, which subsequently separates into mesoderm and endoderm. *See:* **mesendoderm**.

mesocoel   the aqueduct of Sylvius.

mesoderm   the middle of the three primary germ layers; literally, "middle skin." In insects, the cells of the lower layer (the mesoderm of the two-layered germ band) of the embryo become arranged, for the most part, in two longitudinal bands that are shortly afterward marked off into segments corresponding with those bearing the appendages. Generally, most of these mesodermal segments acquire a pair of cavities or coelomic sacs; the dorsal or splanchnic wall of these sacs gives rise to the gonads, visceral muscles, and fat body, while the ventro-lateral or somatic wall produces the muscles of the body and the appendages (see Fig. 28).

mesoderm induction   in Xenopus, (1) a ventral signal is released throughout the vegetal region of the embryo, which induces the marginal cells to become mesoderm; (2) on the dorsal side (away from the point of sperm entry), a signal is released by the vegetal cells of the Nieuwkoop center (the dorsalmost vegetal cells); this dorsal signal induces the formation of the Spemann organizer in the overlying marginal zone cells; (3) the ventral mesoderm is induced by the other vegetal cells; (4) the dorsalization of the lateral marginal cells by the dorsal marginal cells (i.e., the organizer) produces intermediate mesoderm, while the other marginal cells adopt ventral fates.

mesodermal mantle   the mesoderm, not including the dorsal axial mesoderm, which enters the interior of the gastrula over the ventral and lateral blastopore lips. *See:* **dorsal axial mesoderm**.

mesogastrium   the part of the embryonic mesentery that suspends the stomach.

mesoglea   the gelatinous layer between the ectoderm and endoderm in the body wall of the Coelenterata.

mesolecithal eggs   eggs that have an intermediate amount of yolk; more than microlecithal and less than macrolecithal eggs; frogs have mesolecithal eggs (see Figs. 12 and 14).

mesomere   a bilateral portion of mesoderm in the embryo, also called nephrotome because of its role in the formation of the kidneys; also called intermediate mesoderm.

mesomeres   in those species (e.g., sea urchins) that have blastulas made up of various sized blastomeres, the intermediate size blastomeres are called mesomeres.

mesometencephalic fold   a fold in the embryonic brain, between the mesencephalon and the metencephalon, produced by flexure of the neural tube.

mesonephric ducts   ducts of the mesonephros that give rise to the epididymal ducts and vasa deferentia (sperm ducts) in males. In females, mesonephric ducts degenerate, but give rise to the canals of Gartner. *See:* **Wolffian ducts**.

mesonephric ridge   a fold of the dorsal wall of the coelom lateral to the mesentery formed by development of the mesonephros; also called mesonephric fold.

mesonephric tubules   the tubular components of the mesonephros. In males, give rise to the efferent ductules of the epididymis. In females, give rise to the epoophoron.

mesonephros   the chronologically intermediate, embryonic, nephroi (kidneys) of human development; chronologically between the earlier pronephros and the later metanephros.

mesorchium   the membranous support for the testis.

mesosalpinx   the upper part of the broad ligament that forms the mesentery of the uterine (Fallopian) tube.

mesovarium   the membranous support of the ovary in the abdominal cavity (see Fig. 33).

Mesozoic Era   started 225 million years ago; includes the Jurassic Period (135–197 mya), the Age of Reptiles begins.

messenger RNA (mRNA)   a subclass of nucleic acids that carries the genetic message from DNA to the ribosomes, where protein is made.

Metabola   all insects passing through a metamorphosis; includes the Hemimetabola and Holometabola. *See:* **Ametabola**.

metacentric chromosome   a chromosome in which the centromere is equidistant from the two ends of the chromosome; therefore, the chromosome's two arms are of equal length.

metacoel   that portion of ventricle IV of the brain that is the cavity of the metencephalon.

metadiscoidal placenta   a placenta in which the villi are scattered at first but later become restricted to a disc, as in humans and monkeys.

metafemales   human females with more than the normal number (two) of X chromosomes; exhibit varying degrees of mental retardation.

metagenesis   the phenomenon in which one generation of certain plants and animals reproduces asexually, followed by a sexually reproducing generation; also known as alternation of generations.

metalloproteinases   enzymes that digest the extracellular matrix and permit epithelial cells to separate from each other; e.g., in compensatory regeneration of mammalian liver, these enzymes (which may be activated by the trauma of partial hepatectomy) may permit the hepatocytes to separate and proliferate; in the formation of the salamander limb regeneration blastema, these enzymes (probably secreted by macrophages in the wound site) digest the extracellular matrices holding epithelial cells together.

metallothioneins   cysteine-rich, metal-binding proteins, which also have antioxidant activity. *See:* **glutathione, phytochelatins**.

metamere   segment; one of the linearly arranged similar segments of the body of metameric animals.

metamerism   the condition of an animal body characterized by the repetition of similar segments (metameres), exhibited especially by arthropods, annelids, and vertebrates in early embryonic stages and in certain adult structures; also known as segmentation.

metamorphic climax   that time when the visible changes of metamorphosis occur rapidly.

metamorphic molt   the molt undergone by the last instar (larval) stage of holometabolous insects to become a pupa.

metamorphosis   the transition stage in development between an immature stage and the adult stage of the life cycle; as undergone by, e.g., the frog tadpole and insect pupa (see Plate 12 in the color insert).

metanephric diverticulum   an evagination from the caudal end of each mesonephric duct that gives rise to a ureter, pelvis, calyces, and collecting tubules of the metanephric kidney.

metanephrogenous mesoderm   that portion of intermediate mesoderm that is induced by the metanephric diverticula to give rise to the functional units (nephrons) of the metanephric kidneys.

metanephros   the definitive human kidney; chronologically, the final nephros (kidney) of human development; following, sequentially, the pronephros and the mesonephros.

metaphase   the stage of mitosis or meiosis when the centromeres of the chromosomes are aligned on the equator of the spindle (see Fig. 29).

metaphase chromosome spread   a random spread of metaphase chromosomes, created on a slide, usually following treatment of dividing cells with a mitotic inhibitor (e.g., colchicine) and hypotonic solution; used to make a karyotype of the chromosomes. *See:* **karyotype, tetrad**.

metaphase plate   an imaginary plane perpendicular to the spindle fibers, and midway between the spindle poles, of a dividing cell, along which chromosomes align during metaphase.

metaplasia   or transdifferentiation; transformation of one form of adult tissue into another; observed, for example, in Wolffian regeneration, *lens* regeneration from the *iris* of the eye in salamanders.

metastasis   the invasion of a malignant cell into other tissues.

metatela   the thin roof of the myelencephalon.

Metatheria   pouched mammals, marsupials; an infraclass of the Theria (live-bearing mammals).

metazoa   multicellular animls, as opposed to protozoa; includes all animal phyla other than the protozoa, i.e., includes diploblasts and triploblasts, according to traditional classification of animals. Most metazoans belong to the Kingdom Animalia and a few belong to the Kingdom Protoctista, according to the five kingdoms classification of organisms. *See:* **protozoa**.

metazoan development   from work done in the latter half of the nineteenth century, a series of developmental steps, which became accepted as the basic procedure in the development of most metazoa, eventually became crystallized:

(1)   the blastula, typically a single-layered hollow structure, becomes converted into
(2)   the two-layered gastrula, by a process of invagination of one wall or delamination of one wall of the blastula, then,
(3)   by an outpouching of a part of the inner layer of the gastrula, or by an ingression of cells from this layer, or from the outside ectoderm, a third layer of cells, the mesoderm, comes to lie between the entoderm and ectoderm; and finally,
(4)   the inner layer of mesoderm eventually develops into a two-layered structure with a coelomic cavity between the layers.

The main chronological steps of metazoan development are (1) gametogenesis (preamble), (2) fertilization (create unique genotype), (3) cleavage (adds multicellularity), (4) gastrulation (formation of the germ layers), (5) organogenesis (groups of cells are assigned to organ formation), (6) histogenesis (primordia acquire specialization), (7) coordination (integration by the nervous and endocrine systems), (8) metamorphosis (a juvenile becomes an adult), (9) homeostatic physiology (includes regeneration and repair), as well as (10) aging and senescence.

Metchnikoff, Elie (1845–1916)   Russian naturalist; corecipient of a Nobel Prize in Physiology or Medicine, in 1908, in recognition of his work on immunity, for the demonstration of phagocytosis by white blood cells; appreciated that evolution consists of modifying embryonic organs, not adult ones. Metchnikoff and Aleksandr O. Kovalevsky, on the Bay of Naples (in 1864–1865), studied the development of germ layers in a range of invertebrates and lower chordates. He studied the development of insects, scorpions, and cephalopods; he was intent on establishing structural and functional homologies of the germ layers throughout the animal kingdom. Metchnikoff and Kovalevsky, in 1867, were jointly awarded the Karl Ernst von Baer Prize (by von Baer himself) for their substantial embryological accomplishments. In 1891, he wrote, "From the purely natural historical point of view, it would be possible to recognize man as an ape's "monster," with an enormously developed brain, face and hands." Metchnikoff attempted to make a phylogeny of all organisms on the basis of their germ layers, and he believed that all mesodermal cells could be characterized by their ability to phagocytize foreign substances. His discoveries in comparative embryology eventually allowed him to formulate the conceptual foundations of a new science, immunology. *See:* **hopeful monster; Kovalevsky, Aleksandr O**.

metencephalon   the fourth vesicle from the rostral end of the early five-vesicle brain; gives rise to the cerebellum and pons varolii; the roof of the metencephalon gives rise to a thickened transverse ridge, the cerebellum; gives rise to the cerebellum (coordination of complex muscular movements) and the pons (fiber tracts between the cerebrum and cerebellum in mammals) (see Fig. 8).

methoprene   a juvenile hormone mimic causing delayed metamorphosis; a major ingredient in flea collars and used to control ant and mosquito infestations.

Methuselah **gene**   flies with mutations in this gene live 30–40% longer than wild-type flies.

methylation   the chemical addition of a methyl group to a molecule, e.g., to DNA. Methylation "turns off" chromosomes or parts of chromosomes; i.e., histone deacetylases are recruited to methylated regions and will remove acetyl groups from the associated histones, making them more basic and causing them to bind more tightly to the DNA. At gametogenesis, chromosomes are equally methylated; after fertilization, some chromosomes/chromosome parts are turned off by methylation. Heritable changes in methylation state are epigenetic; i.e., these alterations are not mutations but are instructions that affect the expression, not the transmission, of genes. *See:* **DNA methylation**, **genomic imprinting**.

methylation erasure   in mice, there are two waves of methylation erasure; in the germ cell precursors and in the early embryo prior to implantation. *See:* **transgenerational continuity**.

microarray   an array is an orderly arrangement of samples. In general, arrays are described as macroarrays or microarrays, the difference being the size of the sample spots; macroarrays contain sample spot sizes of approximately 300 microns or larger and can be easily imaged by existing gel and blot scanners; the sample spot sizes in microarrays are typically less than 200 microns in diameter, and these arrays usually contains thousands of spots. Microarrays require specialized robotics and imaging equipment.

microbial consortia   structured arrangements of microbes that function together as a unit; found in numerous species; e.g., the oligochaete worm, *Olavius algarvensis* harbors an entire ecosystem of symbionts.

microcephaly   a congenitally smaller than normal head, often, but not always, associated with mental retardation; this condition may be attributed to congenital hypoplasia of the cerebrum.

microevolution   evolution resulting from changes in gene frequencies in a population over time; changes in gene frequencies and trait distributions that occur within populations and species. *See:* **macroevolution**.

microfilament   that component of the cytoskeleton of the cell made up of long, solid, filaments of actin protin (see Fig. 29).

microgamete   the smaller, or male, gamete produced by heterogametic species.

microgametogenesis   the formation of the pollen grain by a haploid microsopre undergoing mitotic division.

microgametophyte   the pollen grain of angiosperms.

microglossia   literally, "a small tongue," to the extent that it is abnormally small.

microinjection   a method for directly introducing substances (e.g., genes) into cells (e.g., newly fertilized eggs). *See:* **electroporation**, **transfection**.

microlecithal eggs   eggs that have a small amount of yolk; less than mesolecithal and macrolecithal eggs; humans have microlecithal eggs. Mammals in general have microlecithal eggs; notable exceptions are the duck-billed platypus and the spiny anteater that have macrolecithal eggs. A striking feature of mammalian development is the accelerated development of the cardiovascular mechanism and the trophic (food, nourishment) membranes, related to the paucity of yolk in mammalian ova. Echinoderms also have microlecithal eggs (see Figs. 12 and 14).

micromeres   in those species that have blastulas made up of various-sized blastomeres, the smallest blastomeres are called micromeres.

microphyll   literally, "small leaf." A leaf with a single, unbranched vein running through its length; typical of some mosses, primitive ferns, and horsetails. *See:* **megaphyll**.

micropyle   a passageway through the tough egg coats of some species by which the fertilizing spermatozoon may gain access to the egg itself; e.g., the tough chorion of fish and insect eggs has a micropyle; in plants, a passageway through the integuments, partially surrounding the embryo sac, through which the growing pollen tube passes to convey the sperm cells to the embryo sac.

microRNAs (miRNAs)   are post-transcriptional regulators that bind to complementary sequences of target messenger RNA transcripts (mRNAs), usually resulting in gene silencing; miRNAs are short RNA molecules, on average only 22 nucleotides long; the human genome may encode more than 1000 miRNAs, which may target approximately 60% of mammalian genes and are abundant in many human cell types. Each miRNA may repress hundreds of mRNAs. MiRNAs are well conserved in eukaryotic organisms and are thought to be a vital and evolutionarily ancient component of genetic regulation. *See:* **small-interfering RNA (siRNA)**.

microspikes   pointed filopodia that by elongation and contraction move the growth cone of an elongating axon.

microsporangium   the pollen sacs of the anthers of flowering plants.

microspore   a small spore, such as those produced by angiosperms, which give rise, by mitosis, to the male gametophyte, the pollen grain.

microspore mother cell   a cell in the sporophytes of angiosperms that, by undergoing meiosis, gives rise to four microspores.

microsporocyte   also called microspore mother cell.

microsporogenesis   the formation of microspores in the microsporangia (pollen sacs) of the anther.

microsporophyll   a leaflike structure bearing a microsporangium or microsporangia.

microstomia   an abnormally small mouth.

microsurgical epididymal sperm aspiration (MESA)   the withdrawal (by aspiration; suction) of sperm directly from any part of the epididymis by means of a tiny cannula (tube).

microtrabecular lattice   a controversial (as to its existence) part of the cytoplasm, revealed by high-voltage electron microscopes. It may be the physical basis of the classic cytosol or cytoplasmic matrix; it is regarded as an artifact by some cell biologists.

microtubule-associated proteins (MAPs)   a variety of proteins that regulate the dynamic behavior of microtubules, e.g., MAP-1, MAP-2, and tau proteins; apparently MAPs play a role in establishing cell shape and polarity; e.g., tau plays a role in the stabilization and organization of microtubules in nerve cell axons.

microtubule-organizing centers (MTOCs)   organelles in the cell that facilitate the polymerization (incorporation) of tubulin protein into microtubules, e.g., centrosomes.

microtubules   a type of proteinaceous, hollow, fibrillar structure found in the cytoplasm of eukaryotic cells; making up part of structures as diverse as the cytoskeleton, spindle, centrioles, cilia, and flagella (see Fig. 29).

micturition   the frequent urination of a small amount of urine by a pregnant woman because of pressure on the urinary bladder.

midblastua transition (MBT)   in the frog, *Xenopus laevis*, fertilization is followed by 12 synchronous mitotic divisions; within 7 hours, the number of cells goes from one cell (the fertilized egg) to 4000 cells (at the midblastula stage); during this time, there is no G_1 period, no G_2 period, and no RNA synthesis. After the 12 synchronous cleavages, the midblastula transition occurs, characterized by (1) slowing down of the cell cycle, (2) appearance of G_1 and G_2 phases, (3) initiation of transcription, (4) asynchrony of cell division, and (5) appearance of cell motility. The midblastula transition heralds a profound change in cell activity. The timing of the midblastua transition seems to be dependent on a nucleus-cytoplasm interaction.

midbrain   the middle of the early brain vesicles of the three-vesicle brain (see Plate 5 in the color insert).

midgut   the middle portion of the early embryonic gut, which gives rise to the small and large intestines. In the frog tadpole, the midgut begins to elongate after hatching; its length elongates to approximately nine times that of the body, but it is shortened by approximately one third during metamorphosis. In the chick embryo, the midgut is that part of the gut caudal to the foregut, where yolk (rather than endoderm) is still the only floor. Differentiation and local specializations appear in the digestive tract only in regions that have ceased to be midgut (see Plate 5 in the color insert and Fig. 6).

midgut loop   the part of the elongating midgut of the embryo that extends into the umbilical cord.

midpiece   the portion of a typical spermatozoon located between the head and the tail (flagellum) of the spermatozoon; the midpiece contains the mitochondria of the cell that provide the energy (ATP) for the cell's motility.

mifepristone   *See:* **RU486 (mifepristone)**.

migration   a basic morphogenetic movement; stationary cells begin to move away from the edges of a coherent tissue mass.

Mintz, Beatrice (1921–)   American developmental biologist; Mintz is especially well known for developing mouse chimeras. In one of her experiments, early embryos consisting of only a few cells were removed from pregnant mice and placed in close contact with similar cells of genetically unrelated embryos to form a composite, which were then implanted in a mouse uterus creating a cellular mosaic; this techniques enabled Mintz to trace the tissue site of specific genetic diseases.

***Minute* technique**   the cells of *Drosophila* carrying a mutation in the *Minute* gene grow more slowly than those in the wild type. By using flies, heterozygous for the *Minute* mutation, clones can be made in which a mitotic recombination event has generated a marked cell that is normal because it has lost the *Minute* mutation, and so it is wild type. This normal cell proliferates faster than the slower growing background *Minute* heterozygous cells, and thus, large clones of the marked cells are produced.

MIS   *See:* **Müllerian-inhibiting substance (MIS)**.

miscarriage   *See:* **spontaneous abortion**.

MITF   the microphthalmia protein, a transcription factor necessary for the development of pigment cells and their pigments.

mitochondria   cellular organelles with several functions, one of the most important of which is providing the cell with readily available biochemical energy in the form of adenosine triphosphate (ATP). In early mouse development, egg mitochondria undergo changes; up to the late four-cell stage, egg mitochondria are condensed; then up to the early morula stage, the mitochondria have their orthodox appearance; and from the late morula stage to the blastocyst stage, one observes large, bleb-like, uncoupled mitochondria. During early development, most energy comes from glycolysis and glycogen. Male mitochondria appear orthodox throughout, even in sperm, up to the blastocyst stage; also, they stay together all in one cell.

mitochondrial cloud   a mitochondria-rich, cytoplasmic region, formed during *Xenopus* oogenesis; it is the precursor of the germ plasm, and its location determines the future vegetal pole of the egg.

mitochondrial ribosomal RNA (mtrRNA)   a germ plasm component of *Drosophila* pole plasm; injection of mtrRNA into embryos formed from ultraviolet-irradiated eggs restores the ability of these embryos to form pole cells.

mitogen   an agent that stimulates cells to undergo mitosis, e.g., growth factors.

mitogen-activated protein kinase (MAPK)   a protein kinase that is activated by phosphorylation as a result of a signal transduction pathway initiated by a mitogen such as a growth factor.

mitogenic   inducing or stimulating mitosis.

mitosis   one of two general types of cell divisions undergone by eukaryotic cells. It is the more ubiquitous of the two types of cell divisions, playing a prominent role in the growth and maintenance of organisms (see Figs. 10 and 29).

mitosis-promoting factor (MFP)   *See:* **maturation-promoting factor (MPF)**.

mitotic recombination   recombination occurring in somatic cells rather than, as usually, in meiotic germ cells. Chromosome breaks induced with X-rays, e.g., in *Drosophila*, result in the exchange of material between homologous chromosomes, just after the chromosomes have replicated into chromatids. Such an event can generate a single cell with a unique genetic construction that will be inherited by all the cell's descendents, which in flies usually form a coherent patch of tissue. *See: Minute* **technique**.

mitral valve   the valve between the left atrium and left ventricle of the heart, formed from endocardial cushion tissue.

mittelschmerz   mild pain at the time of ovulation resulting from blood irritating the peritoneum.

mixed cranial nerves   cranial nerves composed of both sensory and motor components.

mixed nerves   nerves composed of both sensory and motor components.

model organisms   species that are easily studied in the laboratory and have special properties that allow their mechanisms of development to be readily observed, e.g., sea urchins, snails, mice, and birds.

Modern Synthesis   sometimes called neo-Darwinism; the synthesis of evolutionary biology and Mendelian genetics; evolution was redefined to mean changes in gene frequencies in a population over time; the mechanisms of evolution observed as constituting problems of population genetics, the developmental approach to evolution was excluded from the synthesis; the broad-based effort, accomplished during the 1930s and 1940s, to unite Mendelian genetics with the theory of evolution by natural selection; also called the Evolutionary Synthesis. The change in allelic frequencies predicted by the mathematics of population genetics was confirmed by DNA sequencing, making the Modern Synthesis one of the most successful and important explanatory theories in science.

modularity   the concept that animals are organized into developmentally and anatomically distinct parts and that gene regulatory regions are organized into discrete *cis*-regulatory elements.

molar pregnancy   *See:* **hydatidiform mole**.

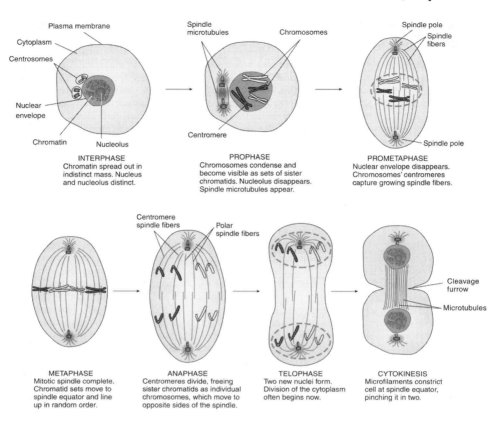

Figure 29. The five phases of mitosis. Interphase is not part of mitosis but represents a nondividing cell. Reprinted from Frank J. Dye, *Human Life Before Birth*, Harwood Academic Publishers, 2000, fig. 3-6, p. 15.

mold   a cultural, rather than a scientific, term, used to refer to a subset of fungi, e.g., bread mold to refer to various fungi that grow on bread, such as the black bread mold, *Rhizopus stolonifer*. Also used to refer to organisms that are not members of the *Kingdom Fungi*, such as slime molds, which are members of the *Kingdom Protoctista*.

molecular clock   differences in sequences of RNA, DNA, and protein molecules accumulate as species diverge. In theory, geological information can be used to calibrate the rate of sequence changes between lineages. The relative sequence divergence between species can be used to infer phylogenetic relationships.

molecular cloning   amplification of single genes to a chemically useful quantity. *See:* **polymerase chain reaction**.

molecular parsimony   although development differs enormously from lineage to lineage, development within all lineages uses the same types of molecules; also called the small tool kit.

molting   every insect, during its growth, sheds (molts) its skin one or more times; also called ecdysis; provides the only means for any increase in surface area or change of form of the exoskeleton. The molting process is initiated in the brain of the larva, where neurosecretory cells release prothoracicotropic hormone (PTTH), which stimulates the production of ecdysone by the prothoracic gland.

monocarpic plants   plants that flower once and then senesce. *See:* **polycarpic plants**.

monocots   *See:* **monocotyledonous plants**.

monocotyledonous plants   plants whose seeds contain one seed leaf or cotyledon.

Monod, Jacques Lucien (1910–1976)   French molecular biologist, co-recipient of the 1965 Nobel Prize in Physiology or Medicine, for work on the genetic control of enzyme and virus synthesis. In 1960, Monod and Francois Jacob introduced the term "operon" for a closely linked group of genes, each of which controls a different step in a given biochemical pathway. In 1961, Monod and Jacob postulated the existence of a molecule, messenger RNA, that carries the genetic information necessary for protein synthesis from the operon to ribosomes, where proteins are made.

monoecious   literally, "one house"; refers to those flowering plants where male and female structures (stamens and pistils) are found on the same plant. *See:* **dioecious**.

monomer   the single unit of a polymer; e.g., amino acids are monomers of polymeric protein molecules.

monophyletic   *See:* **clade**.

monophyletic group   the set of species or populations descended from a common ancestor. *See:* **clade**.

monosomy   one body; as it pertains to chromosomes, a single copy of normally paired chromosomes; although a solitary human X chromosome is found in Turner's syndrome, *autosomal* monosomies are essentially nonexistent among live-born infants.

monospermy   the typical condition where only one sperm enters the egg. *See:* **polyspermy**.

monotremes   egg-laying mammals, represented by the extant duck-billed platypus and the spiny anteater.

monozygotic twins   *See:* **identical twins**.

monster   a now archaic term of only historical interest, once used to describe birth defects; i.e., to describe malformed infants.

Morgan, Thomas Hunt (1866–1945)   American embryologist; fellow graduate of Johns Hopkins and friend of Edmund Beecher Wilson; recipient of a Nobel Prize in Physiology or Medicine, in 1933, for his discoveries concerning the role played by the chromosome in heredity.

Following the experimental approach of Eduard Pflüger, Wilhelm Roux, Hans Driesch, and Laurent Chabry, Morgan began his own line of research on teleost fish. Studies with isolated blastomeres in various species suggested that external conditions alone do not direct development.

Morgan, as a result of his own studies of echinoderm eggs, did not agree with Theodor Boveri's interpretation and his emphasis on nuclear inheritance; Morgan thought it likely that instead of the nucleus and the chromosomes carrying the important material for development, a simple mechanical explanation was probably at the root of the matter. Additional experiments showed that the sea urchin, as opposed to Driesch's work establishing the early isotropy of the egg and its cleavage products, is already cytoplasmically differentiated by the two-cell stage and probably even before.

He also investigated frog development and regeneration. He postulated a gradient of anterior-producing materials concentrated in the head region of flatworms.

morning sickness   daily episodes of nausea, often in the morning, during early pregnancy.

morph   phenotypic variation caused by environmental differences.

morphallactic regeneration   a type of regeneration in which the new individual is *not* produced by the addition of parts to the remaining piece, as in salamnder limb regeneration, but, rather, a type of regeneration in which reorganization of the remaining part of the body of the animal is involved, as in regeneration of *Hydra* from a tiny piece of the animal. *See:* **compensatory regeneration, epimorphic regeneration**.

morphallaxis   See **morphallactic regeneration**.

morphogen   a substance that can direct the differentiation of cells; as along a gradient of morphogenetic substances; cells respond differently at different concentrations. *See:* **morphogenetic determinants**.

morphogen receptor gradient   a gradient of those molecules that recognize the morphogen; the expression of the *Brachyury* and *goosecoid* genes has been correlated with the number of activin receptors on each cell that are bound by activin.

morphogenesis   literally, "the origin of form"; refers to those processes during development that give rise to the form or shape of the organism or a structure. Cell movement plays a significant role in animal morphogenesis; plant morphogenesis does not involve the translocation of cells.

morphogenetic constraints   developmental constraints involving morphogenetic construction rules; e.g., the use of reaction-diffusion mechanisms to construct limbs may constrain the possibilities that can be generated during development because only certain types of limbs are possible under these rules. *See:* **reaction-diffusion model**.

morphogenetic determinants cytoplasmic substances that cause the determination of blastomeres; they appear to work by selectively activating specific genes. *See:* **bicoid protein**.

morphogenetic fields transient embryonic units that are self-differentiating as a whole but regulative within their boundary. A discrete region of an embryo that will give rise to a structure and within which pattern formation is largely independent of other developing structures. A group of cells whose position and fate are specified with respect to the same set of boundaries; the general fate of a morphogenetic field is determined (i.e., a particular field will give rise to a particular organ, e.g., a forelimb); however, the individual cells within the field are not committed and the cells of the field can regulate their fates to make up for missing cells in the field. *See:* **modularity**.

morphogenetic furrow in *Drosophila* eye development, the furrow that moves across the eye disc and initiates the development of the ommatidia.

morphogenetic gradient a concentration gradient of morphogens, such as the concentration gradients of bicoid protein and nanos protein in the early embryo (syncytial stage) of *Drosophila*.

morphogenetic law the concept that developmental sequences tend to be constant, and the earlier the ontogenetic stage, the greater the degree of constancy. Developmental sequences tend to be preserved simply because they are interrelated at all stages with other events going on at the same time; thus, it is not so readily possible to change a developmental sequence as it is to add a further step at the end, for a change along the sequence requires adjustments in all related sequences.

morphogenetic movements cellular movements carried out by groups of cells, which result in the origin of or change in form of the developing organism or part of it, e.g., involution, convergent extension, and invagination.

morpholino an antisense oligonucleotide against a mRNA, used to inhibit protein synthesis.

morphology the study of structure.

morphospecies populations that are designated as separate species based on morphological differences.

morula an early stage in the development of animals, including humans, when cleavage is going on and the embryo consists of a solid ball of cells (blastomeres).

mosaic an organism made up of distinct cells or cell populations, e.g., XX/XY mosaics. The expression "mosaic development" pertains to those kinds of embryos not able to compensate for a missing part of the embryo. *See:* **regulative development**.

mosaic behavior surgical removal of parts of an embryo causes a defect in the final anatomy corresponding exactly to the fate map.

mosaic determination the type of determination where there is a qualitative separation of organ-forming materials into the early blastomeres; also referred to as determination by cytoplasmic specification or immediate determination. *See:* **progressive determination**.

mosaic development that type of development where the early embryo is not able to compensate for the loss of part of the embryo; typically exhibited by embryos of molluscs and annelids. Prospective potency is equal to prospective fate. Widespread throughout the animal kingdom, mosaic development is especially observed among ctenophores, annelids, nematodes, and molluscs. Note that regulative development and mosaic development are the poles of a continuum. *See:* **regulative development**.

mosaic pleiotropy a type of pleiotropy where the affected organ systems are separately affected by the abnomal gene function, e.g., Waardenburg syndrome, caused by heterozygosity for *MITF*, where four different parts of the body (skin pigment, iris, inner ear tissue, and mast cells) independently use the MITF protein as a transcription factor. *See:* **relational pleiotrophy**.

motifs parts of proteins consisting of simple combinations of a few secondary structural elements with a specific geometric arrangement. Examples of protein motifs are the helix-loop-helix motif, the Greek key motif, and the beta-alpha-beta motif. Some motifs are known to have specific functions; e.g., the zinc finger motif functions in DNA binding by proteins such as the estrogen receptor and the progesterone receptor. Several motifs usually combine to form compact globular structures called domains. *See:* **domain**.

motor cranial nerves cranial nerves that innervate muscles but lack sensory components.

motor proteins proteins that carry cargo (e.g., mitochondria) along tracks (e.g., microtubules) within cells; dynein, kinesin, and myosin are families of motor proteins.

movement proteins   proteins that increase the effective pore size of plasmodesmata, allowing molecules as big as 40,000 D to pass through.

mucopolysaccharides   chemical conjugates of protein and carbohydrate; an example of which is released, beneath the vitelline membrane, by the cortical reaction of sea urchin eggs; by causing an osmotic influx of water, it causes the vitelline membrane to rise away from the plasma membrane of the egg.

mucous gland   a gland that secretes mucus. *See:* **oral sucker**.

mucus   a viscid fluid secreted by mucous glands.

Muller, Fritz (1822–1897)   German naturalist and zoologist; noted defender of Charles Darwin and Darwinism; first clearly enunciated the view, in his well-known work, *Fur Darwin* (Leipzig, 1864; English edition, *Facts for Darwin,* 1869), that a knowledge of embryonic and larval histories would lay bare the secrets of race-history and enable the course of evolution to be traced, and so lead to the discovery of the natural system of classification, gave a powerful stimulus to morphological study in general and to embryological investigation in particular.

Müller, Johannes Peter (1801–1858)   German physiologist and comparative anatomist; professor of anatomy and physiology at the University of Berlin; his pupils included Rudolf Virchow and Ernst Haeckel. Although he made his major contribution to science in neurophysiology, he worked with Robert Remak on embryological problems and was the first to describe what later became known as the Müllerian ducts. He discovered fetal pronephric ducts; explained the nature of hermaphroditism; studied embryology and metamorphosis of echinoderms.

Müllerian ducts   a pair of ducts that develops from intermediate mesoderm and that gives rise to the fallopian tubes, the uterus, and at least, part of the vagina in females. Müllerian ducts develop in males as well as in females, but they remain rudimentary or later disappear in males. In males, the degenerated Müllerian ducts give rise to the appendices of the testes. The Müllerian ducts of placental mammals give rise to oviducts, uteri, and vaginas; embryonic Müllerian ducts typically unite at their caudal ends; thus, the adult, human, female tract is paired anteriorly and unpaired posteriorly, terminating as an unpaired vagina. In the remaining placental mammals, there are varying degrees of fusion of the caudal ends of the mullerian ducts. Also called paramesonephric ducts. *See:* **bicornuate uterus**, **duplex uterus**, **simplex uterus**.

Müllerian-inhibiting substance (MIS)   another name for anti-Müllerian duct hormone.

multicellular   consisting of more than a single cell and, generally, of a large number of cells.

multicellularity   exhibiting a multicellular composition, which, during animal development, is initially provided by cleavage.

multifactorial inheritance   the inheritance of a characteristic (e.g., intelligence) that depends on more than one gene and on environmental factors as well.

multigravida   a woman who has been pregnant more than one time.

multinucleated   possessing more than one nucleus (generally, many) within the same mass of cytoplasm; e.g., syncytiotrophoblast and skeletal muscle fibers are multinucleated.

multiple allelism   the existence of more than two different allelic forms for a given gene in a given species; e.g., in humans, the ABO blood typing system exhibits three different allelic forms (alleles), I^A, I^B, and i.

multiple pregnancy   in humans, a pregnancy in which more than one conceptus is carried at a time.

multipotent cardiac progenitor cells   progenitor cells of the heart field that form cardiomyocytes, endocardium, epicardium, and the Purkinje neural fibers of the heart.

multipotent stem cells   adult stem cells commited to give rise to only a small subset of all the cell types found in the body; e.g., the hematopoietic stem cell can give rise to the granulocyte, platelet, and red blood cell lineages.

mural trophoblast   that portion of the trophoblast opposite the inner cell mass. *See:* **polar trophoblast**.

muscle spindles   mechanoreceptors in skeletal muscle; also called neuromuscular spindles.

mushroom   a cultural, rather than a scientific, term; used to refer to a subset of fungi, which often, but not always, have a visible cap and stalk and a subterranean mycelium.

mutagenesis screens   the examination of thousands of mutations affecting development.

mutagenized   caused to have mutations, as by treatment with a chemical mutagen.

mutagens   agents that cause mutations; agents in the environment responsible for genetic damage that results in malformations.

mutation   a genetic change; an alteration in a genome compared with some reference state.

mutualism   symbiosis where the relationship benefits both partners; an example of a mutualistic developmental symbiosis is that between photosynthetic algae and salamander egg masses. *See:* **leghemoglobin**.

mutualistic consortia   *See:* **microbial consortia**, **mutualism**.

mycelium   a mat of hyphae, such as that produced by mushrooms.

mycorrhiza   "fungus root"; a symbiotic relationship between seed plants and fungi; mycorrhizal fungi induce changes in gene expression in the plant, just as the plant changes gene expression in the fungus; all orchids need a mycorrhizal partner for their early development. *See:* **leghemoglobin**.

mycosporines   potent sunscreens used by marine embryos; have high absorption in the ultraviolet part of the spectrum; animals cannot synthesize these compounds, but must get them from plants that they eat, e.g., sea urchins from algae that they eat. *See:* **test cells**.

myelencephalon   the caudal-most vesicle of the early five-vesicle brain; gives rise to the medulla; the thin roof becomes folded and vascularized to form the posterior choroid plexus, whereas the floor and ventrolateral walls become thickened with the formation of nerve tracts. The medulla is the reflex center of involuntary activities (see Fig. 8).

myelin   a lipid substance that ensheathes some nerves, which are then referred to as myelinated nerves.

myelination   the process of ensheathing nerves in a covering of myelin, carried out by glial cells; by oligodendrocytes in the central nervous system and by Schwann cells in the peripheral nervous system.

myelocoel   that portion of ventricle IV of the brain that is the cavity of the myelencephalon.

myeloid precursor cell   the pluripotent stem cell that gives rise to three kinds of lineage-restricted stem cells: (1) CFU-GM, (2) CFC-Meg, and (3) BFU-E.

myeloschisis   a complete or partial failure of the neural plate to form a neural tube, resulting in a cleft spinal cord.

myoblasts   cells that are the precursors of muscle cells or muscle fibers.

myocardium   the muscular wall of the heart.

myocoel   cavity formed in the somites of some species, e.g., Amphioxus, the frog.

MyoD   a transcription factor of the MyoD family, which, in addition to other functions, also directly activates its own gene; i.e., once the *myoD* gene is activated, its protein product binds to the DNA immediately upstream of the *myoD* gene and keeps it active.

MyoD family   *See:* **myogenic bHLH**.

myoepithelial cells   smooth muscle cells derived from ectoderm and found in lacrimal, mammary, salivary, and sweat glands.

myofibrils   bundles of myofilaments (elongated aggregates of contractile protein) found in muscle fibers.

myogenesis   the development of muscle.

myogenic bHLH   a family of transcription factors, also called the MyoD family, the members of which all bind to similar sites on DNA and activate muscle-specific genes, e.g., MyoD and Myf5. These proteins are so powerful that they can turn nearly any cell into a muscle cell.

myomere   a muscle segment differentiated from the myotome, which divides to from the epimere and the hypomere.

myoplasm   special cytoplasm in ascidian eggs involved in the specification of muscle cells.

myosin   a ubiquitous protein in eukaryotic cells, occurring in especially abundant and organized form in skeletal muscle (where it is organized into thick filaments). *See:* **motor proteins**.

myotome   the portion of each somite that gives rise to muscle; the muscle plate that differentiates into myomeres. In the frog, muscle fibrillae develop, within myotomes, which are to form the muscles of the back; from the outer ventral edges of the myotomes and from the ventral edges of cutis plates (dermatomes) develop ventral body and limb musculature.

myotubes   cylindrical syncytia (multinucleated cells), derived from the fusion of unicellular myoblasts, which give rise to skeletal muscle fibers.

N

N-acetylglucosaminadase is an enzyme, released by the zona reaction of the mouse egg, which breaks down the N-acetylglucosamine of ZP3 of the mouse zona pellucida; this prevents additional spermatozoa from attaching to the zona pellucida.

N-acetylglucosamine:galactosyltransferase is an enzyme exposed on the surface of the head of the mouse spermatozoon, by the acrosome reaction; it binds to N-acetylglucosamine of ZP3 of the zona pellucida of the mouse egg, thereby attaching the spermatozoon to the zona pellucida.

N-cadherin also called neural cadherin, is a cell adhesion molecule first observed on mesodermal cells in the gastrulating embryo as they lose their E-cadherin expression; it is also highly expressed on the cells of the developing central nervous system. A factor in the initiation of neural crest cell migration is the loss of the N-cadherin that had linked them together; migrating neural crest cells have no N-cadherin on their surfaces, but they begin to express it again as they aggregate to form the dorsal root and sympathetic ganglia.

nanos a key posterior group maternal gene in early *Drosophila* development.

***nanos* mRNA** in *Drosophila*, *nanos* mRNA is localized at the extreme posterior pole of the unfertilized egg; this mRNA is not translated until after fertilization. Nanos is essential for germ cell formation; pole cells lacking nanos do not migrate into the gonads and fail to become gametes.

nanos protein *nanos* mRNA is translated, after fertilization of the egg of *Drosophila*, to give a concentration gradient of nanos protein, with the highest level at the posterior end of the embryo. Unlike the bicoid protein, the nanos protein does not act directly as a morphogen; its function is to suppress, in a graded way, the translation of the posterior mRNA of another maternal gene, maternal *hunchback*.

nares nostrils.

nasal passages communications between the nares and the nasopharynx.

nasal pits paired depressions on the head of the early embryo, which develop into parts of the nasal cavities.

nasal placodes paired ectodermal thickenings on the head of the early embryo, which will sink beneath the surface of the head to form the nasal pits.

nasal septum the septum (partition) between the two nasal cavities.

nasofrontal process *See:* **frontonasal process**.

nasolacrimal groove the furrow that separates the maxillary and lateral nasal processes of the embryo.

nasolateral processes also lateral nasal processes; paired thickenings, on the head of the embryo, lateral to and partially surrounding the nasal pits.

nasomedial processes also median nasal processes; paired thickenings, on the head of the embryo, medial to and partially surrounding the nasal pits.

nasopalatine duct a canal between the oral and nasal cavities of the embryo at the point of fusion of the maxillary and palatine processes.

nasopharynx the portion of the pharynx above the soft palate.

native DNA DNA in its natural state, that of a two-stranded double helix. *See:* **denatured DNA**.

native protein protein in its natural state; where protein molecules have their weaker chemical bonds intact and, therefore, have their natural shape (e.g., the albumin protein in raw egg white) and are able to carry out their normal functions. *See:* **denatured protein**, **egg chemistry**.

Dictionary of Developmental Biology and Embryology, Second Edition. Frank J. Dye.
© 2012 Wiley-Blackwell. Published 2012 by John Wiley & Sons, Inc.

natural selection   the sum total of those factors in the environment that favor the reproduction of some organisms and work against the reproduction of others; the driving force behind organic evolution.

nebenkern   the fused mitochondria of fly sperm; sperm mitochondria are often highly modified to fit the streamlined cell.

neck of the latebra   a narrow strand of white yolk, extending from the latebra to the surface of the oocyte, formed by the peripherad migration of the oocyte nucleus.

neck neural crest   *See:* **vagal neural crest**.

necrosis   cell death resulting from a force outside of the cell; conventionally, any nonsuicidal cell death. *See:* **apoptosis**.

Needham, Joseph (1900–1995)   English biochemist; he was a university demonstrator in biochemistry at Cambridge (in 1928–1933), working mainly to discover the process underlying the development of the fertilized egg into a differentiated and complex organism. In his *Chemical Embryology* (1931), he concluded that embryonic development is controlled chemically. He published *A History of Embryology* (1934).

Nematoda   a pseudocoelomate phylum of animals; the round worms.

Nemertea   an acoelomate phylum of animals; the ribbon worms.

neoblasts   a special type of cell, found in annelids, which serve for the formation of regenerating parts; these cells migrate to and proliferate at sites of repair and regeneration.

neo-Darwinism   *See:* **Modern Synthesis**.

neoplasia   formation of new tissue; formation of tumors or neoplasms.

neoplasm   a tumor; an aberrant new growth of abnormal cells or tissues.

neoteny   somatic retardation; a type of paedomorphosis where attainment of sexual maturity is not speeded up, but the terminal stages in development of other systems are retarded; a phenomenon peculiar to some salamanders, in which large larvae become sexually mature while still retaining gills and other larval features; the retention of the juvenile form owing to retardation of body development relative to the germ cells and gonads, which achieve maturity at the normal time, e.g., as occurs in the Mexican axolotl, *Ambystoma mexicanum*, a salamander, and in species of the salamanders *Necturus* and *Siren*.

nephrocoel   cavity of the nephrotome.

nephrogenic cord   the longitudinal cordlike mass of mesenchyme derived from the mesomere or nephrostomal plate of the mesoderm, from which develop the functional parts of the pronephros, mesonephros, and metanephros.

nephrogenic tissue   the tissue of the nephrogenic cord derived from the nephrotome plate that forms the blastema or primordium from which the embryonic and definitive kidneys develop.

nephron   functional unit of the kidney.

nephrostomes   openings of pronephric and primitive mesonephric tubules into the body cavity.

nephrotome   that part of the embryonic mesoderm that gives rise to the excretory system.

nephrotomic plate   *See:* **nephrotome**.

nerve   a group of nerve cell processes contained within a single sheath.

nerve growth factor (NGF)   a protein growth factor; promotes survival of and nerve process outgrowth of specific classes of neurons.

nerve tracts   bundles of nerve cell processes within the central nervous system; equivalent to nerves in the peripheral nervous system.

netrins   proteins that provide chemotactic cues that guide axons in the developing nervous system; netrins can serve as chemoattractants.

network   two or more regulatory circuits that are linked by regulatory interactions between components of each circuit.

neural cadherin   *See:* **N-cadherin**.

neural canal   the lumen of the neural tube. Also called neurocoel.

neural cell adhesion molecule (N-CAM)   a cell adhesion molecule; its structure resembles that of the immunoglobulin molecule; therefore, it is an example of an immunoglobulin superfamily CAM. N-CAM is needed for the proper attachment of axons to target muscle cells and appears to be critical in bundling axons together so that they travel as a unit.

neural crescent   the fate maps of amphibian blastulas shows that the presumptive neural ectoderm maps to a crescent-shaped region, the neural crescent, on the surface of the blastula.

neural crest   a group of cells derived from the edges of the neural plate during neurulation, which initially occupies a position dorsal or dorsolateral to the neural tube. Neural crest cells migrate throughout the developing embryo and give rise to such an array of derivatives that the neural crest is sometimes regarded as a fourth germ layer. Examples of neural crest derivatives are spinal ganglia, cranial ganglia, autonomic ganglia, the adrenal medulla, and chromatophores. The development of neural crest cells (and the epidermal placodes that give rise to the sensory nerves of the face) distinguish the vertebrates from the protochordates; i.e., the neural crest is a vertebrate invention (see Figs. 6 and 31). *See:* **neurulation**.

neural-derived mitotic factors   mitosis-stimulating factors released from neurons that increase the proliferation of regeneration blastema cells.

neural ectoderm   *See:* **neuroectoderm**.

neural folds   the elevated margins of the neural plate, which develop early during the process of neurulation (see Fig. 30).

neural groove   the medial depression of the neural plate during neurulation, which brings together the neural folds in anticipation of neural tube formation (see Figs. 19 and 31).

neural plate   the thickened neuroectoderm on the dorsal side of the developing embryo, the formation of which initiates the process of neurulation. In frogs, at 40 hours at 25 °C. The human embryo shows a relatively larger neural plate, than chick or pig embryos, in its future forebrain region (see Figs. 30 and 31).

neural retina   that portion of the retina derived from the inner wall of the two-walled optic cup; made up of glia, ganglion cells, interneurons, and light-sensitive photoreceptor neurons.

neural ridges   in insects, the central nervous system develops as a pair of longitudinal neural ridges of the ectoderm that are separated by a median neural groove.

neural tube   the tubular precursor of the central nervous system of chordates, the formation of which completes the process of neurulation; in frogs, at 56 hours at 25 °C. Differentiation of the neural tube takes place at three levels: (1) gross anatomical level, where bulges and constrictions give rise to the chambers of the brain and spinal cord; (2) tissue level, where cell populations rearrange themselves into different functional regions; and (3) cellular level, where neuroepithelial cells differentiate into neurons and glial cells (see Figs. 6, 30, and 31).

neuralizing activity   "activity" that came from the amphibian organizer and induced the ectoderm to be neural; chordin and noggin, for example, constitute the neuralizing factors secreted by the organizer. *See:* **posteriorizing activity**.

neurenteric canal   the passageway between the neurocoel and the archenteron in amphioxus embryos. In frog embryos, the passageway between the neurocoel and the archenteron, formed when the neural folds fuse and roof over the dorsal portion of the blastopore; the neurenteric canal becomes severed before hatching and the nerve cord continues straight out into the tail. *See:* **postanal gut**.

neurites   nerve cell processes; that is, axons and dendrites.

neuroblasts   cellular precursors of neurons (nerve cells), derived from the neural tube or neural crest (see Fig. 18).

neurocoel   the lumen of the neural tube; also called neural canal.

neurocranium   that portion of the cranium that houses the brain. *See:* **viscerocranium**.

neuroectoderm   that portion of the ectoderm of the early embryo that normally gives rise to the neural tube and neural crest. *See:* **epidermal ectoderm**.

neuroepithelium   the epithelium derived from the neuroectoderm (see Fig. 18).

neurogenesis   origin of the nervous system.

neurogenin   the key protein involved in activating the neural phenotype in the ectoderm; a transcription factor that activates a series of genes whose products are responsible for the neural phenotype; specifies sensory neurons from the neural crest.

neurohypophysis   that portion (posterior lobe) of the pituitary gland derived from the infundibulum of the brain.

neuromeres   a series of enlargements of the neural plate marked off by constrictions In the chick embryo, at 24 hours, there are 11 neuromeres: the 3 most rostral give rise to the prosencephalon, 4 and 5 to the mesencephalon, and 6–11 to the rhombencephalon. Rostrally, the constrictions disappear except for one between the prosencephalon and the mesencephalon and one between the mesencephalon and the rhombencephalon. Rhombencephalic neuromeres persist for a longer time; the two anterior neuromeres of the rhombencephalon give rise to the metencephalon, and the four posterior neuromeres of the rhombencephalon give rise to the myelencephalon. In insects, neural ridges become segmentally constricted into neuromeres or primitive nerve ganglia, while their intersegmental portions give rise to the connectives. The first three neuromeres, the protocerebrum, the deutocerebrum, and the tritocerebrum, amalgamate to form the brain, while the succeeding three cephalic neuromeres fuse to become the suboesophageal ganglion; the neuromeres that follow develop into the thoracic and abdominal ganglia.

neurons   nerve cells.

neuropores   openings at both ends of the newly formed neural tube, which in human embryos places the neurocoel in continuity with the amniotic cavity until the neuropores close (see Fig. 30).

neurotrophins   proteins released from potential target tissues, of axons, which work at short ranges as either chemotactic factors or chemorepulsive factors, e.g., nerve growth factor (NGF), brain-derived neurotrophic factor (BDNF), neurotrophin 3 (NT-3), and NT-4/5.

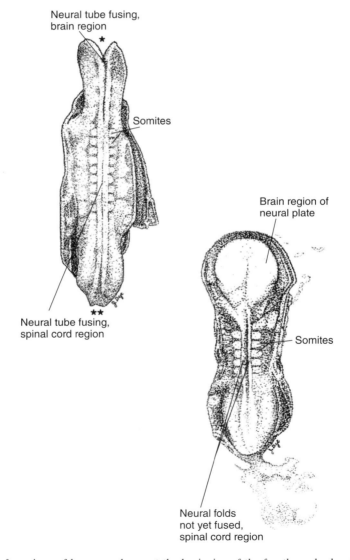

Figure 30. Surface views of human embryos at the beginning of the fourth week, showing the neural tube progressively closing, with open anterior (*) and posterior (**) neuropores. Reprinted from Frank J. Dye, *Human Life Before Birth*, Harwood Academic Publishers, 2000, fig. 8-9BC, p. 68.

neurula   the stage in embryonic development during which the formation of the neural plate and its closure to form the neural tube are the dominant processes (see Plate 5 in the color insert).

neurulation   the embryonic process, following or overlapping gastrulation, beginning with the appearance of the neural plate and ending with the completion of the neural tube. The two major types of neurulation are primary neurulation and secondary neurulation. In insects, the whole of the central nervous system and the sense organs are ectodermal in origin (see Plate 5 in the color insert and Fig. 31).

neurulation, primary   *See:* **primary neurulation**.

neurulation, secondary   *See:* **secondary neurulation**.

NF-kB   a vertebrate transcription factor with considerable homology to the dorsal protein transcription factor of *Drosophila*. I-kB is a protein that binds NF-kB and prevents it from entering the nucleus before the cell has received the appropriate signal that dissociates the complex (I-kB to NF-kB); when the NF-kB protein enters the nucleus, it effects the transcription of several lymphocyte-specific genes involved in the inflammatory response. I-kB has homology with the *Drosophila* cactus protein. *See:* ***dorsal***.

NGF   *See:* **nerve growth factor (NGF)**.

niche construction   when a developing organism induces changes in its physical environment in ways that make the environment more fit for the organism, e.g., the induction of a gall by a goldenrod gall fly larva; a contribution of eco-devo. An inductive phenomenon. *See:* **adaptive developmental plasticity**.

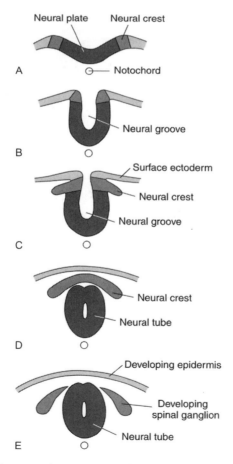

Figure 31. Neurulation. Diagrammatic transverse sections through 3-week human embryos undergoing neurulation. Reprinted from Frank J. Dye, *Human Life Before Birth*, Harwood Academic Publishers, 2000, fig. 8-9A, p. 68.

Nicolaus Kleinenberg (1842-1897) homologized the two germ layers of *Hydra* with the initial layers found in the development of more complex invertebrates and vertebrates.

nidation implantation.

Nieuwkoop center the dorsalmost vegetal cells of the *Xenopus* blastula, which releases a signal that, in turn, induces the formation of the Spemann organizer in the overlying marginal zone cells.

Nieuwkoop, Pieter (1917–1996) Dutch embryologist who worked on development of the embryo in its early stages after fertilization; after whom the Nieuwkoop center is named to honor his discovery of mesoderm induction.

NIMZ *See:* **noninvoluting marginal zone (NIMZ) cells.**

nitrogenous waste waste containing nitrogen-containing chemical compounds, e.g., urea.

Nkx2-5 a transcription factor critical in instructing the mesoderm to become heart tissue; it activates the synthesis of other transcription factors; in turn, these transcription factors activate the expression of genes encoding cardiac muscle-specific proteins. Nkx2-5 is homologous to the Tinman transcription factor of *Drosophila*. *See:* **Tinman**.

Nobel Prizes three Nobel Prizes for Physiology or Medicine have been awarded to embryologists/developmental biologists: (1) to Hans Spemann (in 1935) for his discovery of the organizer effect in embryonic development; (2) to Edward B. Lewis, Christiane Nüsslein-Volhard, and Eric F. Wieschaus (in 1995) for their discoveries concerning the genetic control of early embryonic development; and (3) to Robert Edwards (in 2010) for the development of human *in vitro* fertilization (IVF) therapy.

nocodazole a microtubule-depolymerizing drug that has been used experimentally to prevent cortical rotation of amphibian eggs.

***nodal* gene** in all vertebrates studied so far, the critical event in left-right axis formation is the expression of a *nodal* gene in the lateral plate mesoderm on the left side of the embryo.

node Hensen's node in mammalian development; one of two signaling centers in the mammalian embryo; responsible for creation of all of the body; produces chordin and noggin; works together with the anterior visceral endoderm to form the forebrain. *See:* **anterior visceral endoderm.**

nodes those sites along the length of a stem where lateral buds are found; the union of a leaf and the stem.

noggin the first of the soluble organizer molecules to be isolated; it can accomplish two of the major functions of the Spemann–Mangold organizer: (1) It can induce dorsal ectoderm to form neural tissue, and (2) it can dorsalize mesoderm cells that would otherwise contribute to the ventral mesoderm. Secreted by the organizer and prevents the bone morphogenetic protein (BMP) from binding to the ectoderm and mesoderm near the organizer. *See:* **chordin, follistatin.**

non-cell autonomous genes genes that encode signaling ligands that act on other cells. *See:* **cell autonomous genes.**

nondisjunction (1) mitotic: the failure of sister chromatids to be distributed to opposite cell poles during the mitotic anaphase resulting in aneuploid products of mitosis; (2) meiotic: the failure of chromosomes contained in pairing configurations to separate regularly at first anaphase, or the distribution of unpaired homologues to the same pole at first anaphase, both cases resulting in aneuploid products of meiosis (gametes or meiospores).

noninvoluting marginal zone (NIMZ) cells those cells of the marginal zone of the amphibian embryo that do not undergo involution during gastrulation.

nonsister chromatids those chromatids of different chromosomes in a pair of homologous chromosomes. *See:* **sister chromatids.**

nonsurgical sperm aspiration the withdrawal (by aspiration; suction) of sperm directly from the testis by means of a tiny cannula (tube).

norgestrel one form of a variety of progestogen hormones found in birth control pills; has activity similar to the natural hormone progesterone.

Northern blot an RNA blot; different RNAs separated by electrophoresis and transferred to nitrocellulose paper are incubated in a solution containing a radioactive DNA fragment, from a particular gene, which binds to the region of the filter where the complementary RNA is located; if the mRNA for the gene is present, the labeled DNA will bind to it and can be detected by autoradiography. Used to determine the temporal and spatial locations where RNAs are expressed. *See:* **developmental Northern blot.**

nostrils the openings of the nasal cavities to the outside of the body.

notochord a dorsal rod of mesoderm, formed during gastrulation in all chordates, which gives some structural support to the early embryo. In chordates, it runs from head to tail and lies beneath the developing central nervous system. In human development, the notochord is replaced by the vertebral column and only gives rise to the nuclei pulposi of the intervertebral discs of the adult (see Plates 1 and 4 in the color insert as well as Figs. 6 and 28).

notochordal canal a canal formed by a continuation of the primitive pit into the head process (notochord) of mammalian embryos.

novelty new morphological characters of adaptive value.

NSA *See:* **nonsurgical sperm aspiration**.

nucellus another name for megasporangium; more specifically, the diploid tissue surrounding the megaspore or the sporophytic tissue of the ovule that surrounds the embryo sac.

nuclear envelope the nuclear membrane when viewed with an electron microscope, consisting of two membranes separated by a space (perinuclear cisterna) and pierced by numerous nuclear pores. Intimately associated with the endoplasmic reticulum, the outer membrane may have attached ribosomes on its cytoplasmic side.

nuclear equivalence the concept that members of a group of nuclei are genetically equivalent to each other, e.g., that all the nuclei in the group are totipotent, as is the case with the four blastomeres of the four-cell sea urchin embryo or the two blastomeres of the two-cell frog embryo. As development continues, the members of the group of nuclei cease to be equivalent and, instead, they become determined to follow different paths of cell lineages; e.g., pairs of blastomeres of eight-cell tunicate embryos are determined to give rise to different parts of the organism. *See:* **mosaic development, regulative development**.

nuclear membrane the membrane found at the periphery of the cell nucleus, separating the nuclear contents from the cytoplasm; not found in prokaryotic cells and breaks down during cell division in most eukaryotic cells, with some protists being important exceptions. *See:* **nuclear envelope**.

nuclear pores passageways through the nuclear envelope between the cytoplasm and the interior of the nucleus.

nuclear receptor superfamily transcription factors; intracelluar receptors; stimulated by lipophilic ligands, including steroids, thyroid hormones, and retinoic acid; have a hormone binding domain, a DNA-binding domain, and a transcription activation domain. The receptor is normally complexed with a heat shock protein called hsp90 and thereby retained in the cell cytoplasm; on binding the hormone, the hsp90 is displaced and the factors can form active dimers that move to the nucleus and bind to the target genes. Examples include the retinoic acid receptors and the ecdysone receptors.

nuclear RNA (nRNA) the primary transcript from a gene; also called heterogeneous nuclear RNA (hnRNA) or pre-messenger RNA (pre-mRNA).

nuclei clusters into which neurons of the brain are organized.

nucleocytoplasmic ratio (N/P ratio) the volumetric ratio between the nucleus and the cytoplasm; during cleavage, this ratio is decreased until the species-specific ratio is reached. In the sea urchin, the N/P ratio of the fertilized egg is 1:550; at the end of cleavage, the N/P ratio is 1:6, the ratio specific for the mature sea urchin.

nucleolar organizer the region of a chromosome concerned with transcribing ribosomal RNA that associates with other such regions on other chromosomes to form the nucleolus; it is a region of a chromosome containing a cluster of ribosomal RNA genes.

nucleolus (pl. nucleoli) a cellular organelle, found in the nucleus of eukaryotic cells, concerned with the formation of ribosomal subunits; it is where ribosomal RNA is transcribed.

nucleoplasm the protoplasm of the nucleus; the nonstaining or slightly chromophilic ground substance of the interphase nucleus that fills the nuclear space around the chromosomes and nucleoli.

nucleosome the basic unit of chromatin; made up of a nucleosome core particle, consisting of an octomer of histone proteins enveloped by two loops of DNA (approximately 146 basepairs), linker DNA (approximately 50 basepairs), and H1 histone attached to the linker DNA.

nucleosome positioning the establishment of specific positions of nucleosomes with respect to a gene's promoter.

nucleus (pl. nuclei) (1) cells: a cellular organelle found in almost all cells. In eukaryotic cells, it has a limiting membrane, the nuclear membrane; in prokaryotic cells, no such membrane exists and the term "nucleoid" refers to the comparable DNA-containing region; (2) nervous system: an aggregate of nerve cell bodies in the central nervous system; the equivalent of a ganglion in the peripheral nervous system.

nucleus of Pander   in chicken eggs, a whitish disc of white yolk, found at the end of the neck of the latebra at the oocyte surface; on the surface of the which is found the germinal disc of active cytoplasm and the germinal vesicle.

nucleus pulposus   (pl. nuclei pulposi) the center of intervertebral discs derived from the notochord.

null mutation   a complete loss of function corresponding to a complete lack of active gene product.

nurse cells   of the *Drosophila* ovary, are produced by stem cells undergoing four mitotic divisions (one of the resulting 16 cells, per stem cell, becomes the oocyte); the nurse cells produce large quantities of proteins and RNAs that are exported into the egg through cytoplasmic bridges. *Bicoid* mRNA is made by nurse cells located next to the anterior end of the developing oocyte and is transferred from them into the egg.

Nüsslein-Volhard, Christiane   (1941–) (Germany) corecipient, with Eric F. Wieschaus (United States) and Edward B. Lewis (United States), of the 1995 Nobel Prize for Physiology or Medicine for discovery of genes that control the early stages of the body's development.

nutritional polyphenism   polyphenism controlled by diet; e.g., the caterpillar morphs of the moth, *Nemoria arizonaria*, differ in appearance depending on whether they eat oak flowers or oak leaves.

nymph   a stage in the life cycle of insects exhibiting the hemimetabolous type of metamorphosis; as a result of the presence of abundant yolk, the immature insects are hatched from eggs in a relatively advanced phase of developement and are called nymphs. A nymph generally resembles the adult insect in its general structure and body form. The nymph usually adopts the same mode of life as its parents, frequenting the same habitat and feeding on similar food; however, in the Ephemeroptera, Odonata, and Plecoptera, the nymphs are exceptional in being aquatic, while the adults lead an aerial life. In the Hemimetabola, the nymphal instars are preparatory to the development of the imago.

O

obesigen   a substance that stimulates the body both to produce fat cells and to accumulate fat within these cells.

obstetrician   a physician who specializes in the care of pregnant women.

odontoblasts   the mesenchymal cells, of a tooth germ, derived from neural crest, which secrete dentin.

odontogenesis   formation of teeth.

olfactory cells   cells that act as receptors for the sense of smell.

olfactory epithelium   a sheet of olfactory cells found in a nasal cavity.

olfactory lobes   the constricted anterior ends of the cerebral hemispheres.

olfactory nerves   the first pair of cranial nerves, going from the olfactory epithelium in the nose to the olfactory lobes of the brain; concerned with the sense of smell.

olfactory pits   a bilateral pair of depressions that forms on the rostrad end of the head, which give rise to the sensory linings of the nasal passages.

olfactory placodes   a pair of epidermal thickenings, the most anterior of the cranial ectodermal placodes, that form the ganglia for the olfactory nerves.

oligodendrocytes   glial cells found in the central nervous system, where they myelinate nerves (see Fig. 18).

oligohydramnios   a deficieny of amniotic fluid. *See:* **polyhydramnios**.

oligolecithal   containing very little yolk.

omental bursa   the space dorsal to the stomach and lesser omentum, lined with peritoneum, and communicating with the general abdominal cavity.

omphalocele   a result of a failure of the developing gut to return completely to the abdominal cavity from its embryonic location in the umbilical cord; it is a congenital umbilical hernia into which intestine and peritoneum protrude.

omphalogenesis   development of the yolk sac.

omphalomesenteric arteries   vascular channels out to the yolk sac. In the chick embryo, the omphalomesenteric arteries develop at approximately 40 hours; the proximal portions develop as branches of the dorsal aortae and extend peripherally, while the distal portions develop in the extra-embryonic vascular area and extend toward the embryo. At approximately 40 hours, the omphalomesenteric arteries communicate with the vitelline vascular plexus. Therefore, at approximately 40 hours, blood cells formed in the yolk sac are carried into the embryonic body for the first time.

omphalomesenteric duct   *See:* **vitelline duct**.

omphalomesenteric veins   return vascular channels from the yolk sac. In the chick, the omphalomesenteric veins develop as caudal extensions of the endocardial primordia; when the vascular plexus on the yolk sac connects with the omphalomesenteric veins, at approximately 33—35 hours, the vascular channels are established from the vitelline plexus to the heart. The omphalomesenteric veins, which are at first paired, are brought together by closure of the ventral body wall and become fused to form a median vessel within the embryonic body. *See:* **vitelline veins**.

oncogene   literally, "a cancer gene"; a mutated proto-oncogene that no longer normally regulates some aspect of cell proliferation or differentiation; examples of oncogenes are *myc* and *ras*. *See:* **proto-oncogene, tumor suppressor gene**.

oncosis   (1) cell death by swelling; (2) formation of a tumor or tumors.

one gene–one enzyme hypothesis   George Wells Beadle and Edward Lawrie Tatum's idea that specific genes control the production of specific enzymes (one gene controls the production of one enzyme), which virtually created the science of biochemical genetics. *See:* **one gene–one polypeptide**.

Dictionary of Developmental Biology and Embryology, Second Edition. Frank J. Dye.
© 2012 Wiley-Blackwell. Published 2012 by John Wiley & Sons, Inc.

Oogenesis

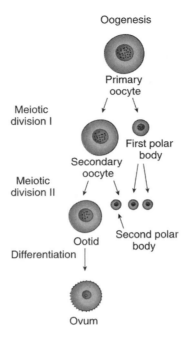

Primary
oocyte

Meiotic
division I

Secondary
oocyte

First polar
body

Meiotic
division II

Ootid

Second polar
body

Differentiation

Ovum

Figure 32. Oogenesis. Reprinted from Frank J. Dye, *Human Life Before Birth*, Harwood Academic Publishers, 2000, fig. 6-6, p. 49.

one gene–one polypeptide   the modification of George Wells Beadle and Edward Lawrie Tatum's one gene-one enzyme hypothesis, to reflect the reality that many enzymes consist of more than one kind of polypeptide, each of which is under the control of a different gene.

one gene–one family of proteins   differential RNA processing by the splicing of the mRNA precursors into messages for different proteins by using different combinations of potential exons, resulting in one gene creating a family of related proteins.

ontogenetic hypertrophy   the principle that developmental sequences tend to become longer or more complex in evolution. This is a logical sequel to the principle of hypermorphosis. In vertebrates, in general, the trend toward increase in length of the developmental period is fairly clear.

ontogenetic inconstancy   evolutionary change in developmental sequences is continuous; the change is inexorable simply because of the inconstancy of the environment.

ontogeny   the development of the individual organism.

ontogeny recapitulates phylogeny   *See:* **biogenetic law.**

Onychophora   a phylum of soft-bodied animals possessing unjointed appendages; closely related to the arthropods.

oocyte   a female germ cell after it has developed from an oogonium and begins its cell differentiation toward becoming an ovum (egg); in addition to growth and differentiation, the oocyte undergoes meiosis as it progresses toward becoming an egg (see Plate 8 in the color insert as well as Figs. 32 and 33).

oogamy   sexual reproduction by the union of gametes that are not similar in size; i.e., the egg is large and nonmotile, while the spermatozoon is smaller and motile. *See:* **anisogamy, isogamy.**

oogenesis   the type of gametogenesis that occurs in females and results in the production of ova (eggs) (see Fig. 32).

oogonia in adult mammalian ovaries   The doctrine that female mammals are born with a finite number of oocytes fated to be exhausted with age has been challenged by recent results claiming that new oocytes can be continuously formed in the postnatal mouse ovary.

oogonium   (pl. oogonia) (1) a type of female germ cell that functions as a stem cell until it begins to undergo cell differentiation and becomes an oocyte. In humans, all oogonia have differentiated into primary oocytes before the birth of a female. In frogs, oogonia persist throughout reproductive life, as a stem cell population, multiply and produce eggs for the next breeding system. (2) a type of unicellular reproductive structure in, for example, algae, which contains the female gamete or ovum (see Fig. 32).

oolema   plasma membrane of the egg.

ooplasm   cytoplasm of the egg.

ooplast   an enucleated oocyte.

ootid   a female germ cell that results from meiosis II (see Fig. 32).

oozooid   the individual developing from an egg. *See:* **blastozooid.**

operculum   gill cover; in the frog, between hatching and metamorphosis, a fold of integument, the operculum, grows posteriad, from the posterior border of each hyoid arch, over the gills.

operon   a unit of genetic regulation in bacteria consisting of an operator gene and one or more structural genes. The operator gene encodes a gene product that may, depending on the specific operon, inhibit or activate transcription of the structural gene(s).

Operon Theory   first described by Francois Jacob and Jacques Monod in the 1950s. *See:* **operon.**

optic bulges (protuberances)   a bilateral pair of surface bulges that appear on the developing heads of many species of embryos, indicative of underlying eye development; e.g., these are found on the surface of the head of the 5-mm frog tadpole, on the dorsal aspects of the sense plate.

optic chiasma   the structure in the floor of the vertebrate forebrain that contains crossed fibers from the two optic stalks going to opposite sides of the brain.

optic cups   the bilaterally symmetrical, double-layered structures, resulting from invaginations of the optic vesicles, which give rise to the retinas of the eyes (see Fig. 7).

optic lobes   the anterior (superior) pair of colliculi of the mammalian corpora quadrigemina; in the frog also known as corpora bigemina; develop as a pair of dorsolateral swellings of the mesencephalon.

optic nerves   the second pair of cranial nerves; they go from the retinas of the eyes to the superior (anterior) colliculi of the midbrain.

optic recess   a recess that extends downward and forward above the optic chiasma.

optic stalk   the constricted portion of the original optic vesicle; found between the optic cup and the lateral wall of the diencephalon.

optic tectum   the region of the brain in amphibians and birds where the axons from the retina terminate.

optic vesicles   paired, lateral evaginations of the wall of the forebrain that give rise to the optic cups and optic stalks (see Plate 2b in the color insert and Fig. 8).

oral-aboral axis   the axis perpendicular to the pentaradial arms of echinoderms.

oral evagination   an outpocketing of the foregut endoderm toward the stomodaeal invagination of ectoderm.

oral plate   a double layer of endoderm, from the oral evagination of the foregut, and ectoderm, from the stomodaeal invagination; when the oral plate ruptures the mouth of the embryo is opened. In the frog, the oral plate becomes perforated a few days after hatching.

oral sucker   an invagination at the ventral end of each mandibular arch formed from the sense plate of the frog embryo; the two oral suckers eventually fuse to form a single mucous gland. The mucous gland allows the frog tadpole, after hatching, to attach to objects. Also called the mucous gland.

oral surface   the surface on which the mouth is found; in the sea urchin, this is the lowermost surface when the animal is in its normal, upright position. *See:* **aboral surface.**

Ordovician   the geological period from approximately 490 to 439 million years ago, which followed the Cambrian and preceded the Devonian.

Organ of Jacobson (Jacobson's organ)   an olfactory canal in the nasal mucosa that ends in a blind pouch; highly developed in reptiles and vestigial in humans.

organelles   subcellular components specialized for specific functions, such as mitochondria and lysosomes.

organizer   a signaling center in a developing embryo or field that induces the development of surrounding tissues; examples include the Spemann organizer of amphibian embryos and the focus at the center of the butterfly eyespot field. The self-differentiating tissue of the dorsal lip of the amphibian blastopore; Hensen's node is similarly effective in organizing secondary embryos in birds and mammals. *See:* **anterior tip**; **embryonic induction**; **Mangold, Hilde**; **mesodermal induction**; **Spemann, Hans**.

organizer graft   the most famous experiment in embryology, first performed by Hans Spemann and Hilde Mangold in 1924; a piece of tissue from above the dorsal blastopore lip is implanted into the ventral marginal zone, leading to the formation of a double dorsal embryo.

organogenesis   the formation of organs. In vertebrate embryos, organogenesis is initiated by neurulation.

oropharyngeal membrane   the double-layered membrane composed of stomodeal ectoderm and pharyngeal endoderm; when this membrane breaks down, the mouth is open for the first time (see Fig. 6).

oropharynx   the oral pharynx, located between the lower border of the soft palate and the larynx.

orphan embryos   most developing organisms, with little or parental protection during their development. *See:* **embryo defenses**.

orthologous regulatory genes   e.g., *Pax6* genes are important for eye development throughout the animal kingdom; *tinman* genes are responsible for heart development in flies, worms, and mammals; and *Hox* genes are critical for specifying the axes in all animals studied so far. *See:* **orthologs**, **regulatory genes**.

orthologs   homologous genes in *different* species that originated from a single gene in the last common ancestor of these species; i.e., genes that are homologous between species, e.g., the *Hox* genes. *See:* **paralogs**.

orthotopic graft   a graft to the same position of another embryo; one of the usual methods of fate mapping. *See:* **heterotopic graft**.

os   *See:* **external os**.

oskar **mRNA**   specifies the egg posterior germplasm that gives rise to the germ cells in *Drosophila*; it is delivered into the oocyte by nurse cells and moved to the posterior end through its interaction with the oocyte's microtubule array. *See:* **nurse cells**.

ossicles   small bones, e.g., the incus, malleus, and stapes of the middle ear cavity.

ossification   the formation of bone.

ossification center   a region from which bone formation spreads outward.

Osteichthyes   the class of vertebrates composed of bony fish; ray-finned fish and lobe-finned fish.

osteoblasts   cells derived from mesodermal mesenchyme cells, which are the precursors of osteocytes and are responsible for bone deposition.

osteoclasts   cells that have the ability to break down already formed bone; play an important role in bone remodeling. Osteoclasts come from the CFU-GM, the same stem cell as macrophages and granulocytes; the growth factor interleukin 6 (IL-6) stimulates the production of osteoclasts; the production of IL-6 is inhibited by estrogen and testosterone.

osteocytes   bone cells derived from osteoblasts, which reside in already formed bone.

osteogenesis   literally, "bone formation."

osteoid matrix   a collagen-proteoglycan secreted by osteoblasts that can bind calcium.

osteopetrosis   a rare developmental error characterized chiefly by excessive radiographic density of most or all bones. If not enough osteoclasts are produced, the bones are not hollowed out for the marrow and osteopetrosis results. *See:* **osteoporosis**.

osteoporosis   deossification with an absolute decrease in bone tissue; resulting in enlargement of marrow spaces, decreased thickness of cortex and trabeculae, and structural weakness. If there are too many active osteoclasts, too much bone will be dissolved and osteoporosis will result. *See:* **osteopetrosis**.

osteoprotegerin   osteoclastogenesis inhibitory factor, is a member of the tumor necrosis factor (TNF) receptor superfamily; together with its ligand, regulates conversion of a macrophage stem cell into an osteoclast.

ostium   an opening, e.g., the opening of the fallopian tube near the ovary.

ostracoderms   the oldest known vertebrates.

otic capsule   a cartilaginous capsule surrounding the auditory vesicle during development.

otic placodes   *See:* **auditory placodes**.

otic vesicles   the bilateral pair of vesicles that develop by invagination from the surface of the head at the level of the hindbrain and give rise to the inner ears.

otocyst otic vesicle or auditory vesicle (see Plate 13 in the color insert).

otolith a calcareous concretion on the end of a sensory hair cell in the vertebrate ear and in some invertebrates (see Plate 13 in the color insert).

outgroup a taxon that diverged from a group of other taxa before those taxa subsequently diverged from one another.

ovarian follicle a follicle found in the ovary; consisting of a single oocyte and numerous follicle cells (and their products) (see Fig. 22).

ovaries the female gonads. Mammals have a compact type of ovary. In frogs, the ovaries are hollow and lobed; the lobes are connected by an internal cavity. Frog ovaries are also speckled because their contained eggs are easily visible through the transparent, hollow ovaries and the accumulation of dark pigment beneath the surface of the eggs' animal hemispheres and the accumulation of light-colored yolk beneath the surfaces of the eggs' vegetal hemispheres. In hens, two ovaries are formed, but one degenerates leaving the left one as the only functional ovary. Nicolaus Steno first introduced the term "ovary" in 1667 for the female reproductive gland.

ovariole the essential feature of the insect ovary is that it is divided into separate egg-tubes or ovarioles. A typical ovariole consists of a terminal filament, a germarium, and a vitellarium. There are three types of ovarioles: panoistic, polytrophic, and acrotrophic.

ovary the lowermost part of the pistil, the female reproductive structure found in flowers, which contains ovules with embryo sacs; with maturity, the embryo sacs produce embryos and endosperm, the ovules produce seeds, and the ovary produces fruit (see Figs. 16, 21, 22, 27, 33, and 34). *See:* **ovaries**.

oviducts generally, paired tubes that convey eggs or embryos from the ovaries to the uterine cavity. Many vertebrates (birds and snakes) have one oviduct. The cephalic end of each oviduct is, generally, open and near the ovary, but it is not united with it directly, except in teleost fishes. The cephalic opening of the oviduct is referred to as the infundibulum, *ostium tubae abdominale*, or when understood in context, ostium. In frogs, the anterior lengths of the oviducts are glandular and produce the gelatinous coverings of the eggs, whereas the posterior lengths of the oviducts are thin-walled, dilated, and referred to as uteri, which

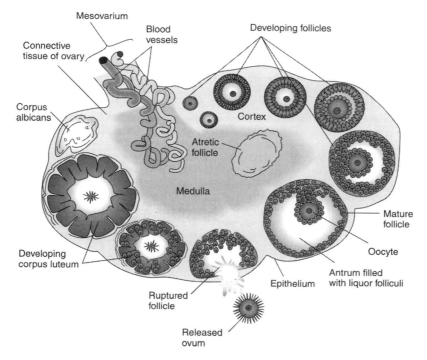

Figure 33. The ovary and ovulation. Reprinted from Frank J. Dye, *Human Life Before Birth*, Harwood Academic Publishers, 2000, fig. 5-7, p. 39.

store eggs. The frog oviduct interior is ciliated and propels the eggs along its length (ciliary action on the peritoneum of the frog conveys the eggs to the ciliated infundibulum of each oviduct that draws the eggs into the oviducts). In chickens, during its passage through the oviduct, the zygote becomes surrounded with albumin ("white of the egg") and accessory membranes. As it takes 16–24 hours before the egg reaches the most posterior part of the oviduct, when the chicken egg nears the cloaca, it is in an advanced stage of cleavage. The hen oviduct consists of the following parts: (1) infundibulum, a funnel-like region where fertilization takes place; (2) magnum, where layers of albumin are deposited around the egg (takes 4 hours); (3) isthmus, where the shell membranes are deposited; and (4) the shell gland or uterus, where "plumping" (5 hours), shell calcification (15–16 hours), and deposition of the cuticle take place. Mammillary knobs are laid down in the isthmo-uterus junction and provide nuclei for the growth of calcite crystals. In insects, two lateral oviducts join a secondary median oviduct that usually opens by a single primary gonopore into a wider passage or vagina that, in turn, opens by a secondary gonopore to the outside; there are variations on this theme in different species of insects.

oviparity   the birth of young, as opposed to the laying of eggs.

oviparous   *See:* **oviparity**.

ovist   a preformationist who believed that the new individual is already preformed in the egg and that development involves essentially the growth of the preformed organism. Albrecht von Haller (1708–1777) and Charles Bonnet (1720–1793) were among the foremost ovists. Haller considered parthenogenesis (as in aphids) to be one of the most powerful arguments in favor of ovism. Bonnet endeavored to explain the phenomenon of regeneration on the basis of preformation. Karl Ernst von Baer's celebrated work, *Development of Animals* (1828), added the final stroke against the preformation doctrine; he observed the egg and the actual developmental stages and showed differentiation to be a progressive process.

ovothiol   modified histidine, a sulfhydryl-containing antioxidant, found in sea urchin eggs, and eggs of several other invertebrates; deals with the sudden bolus of reactive oxygen species (ROS) produced at fertilization. *See:* **reactive oxygen species (ROS) theory of aging**.

ovoviviparity   the eggs do not have supplementation of the food supply of the egg (yolk) by food from the mother; these eggs simply develop entirely in the uterus instead of in some nest outside of the body.

ovoviviparous   *See:* **ovoviviparity**.

ovulation   release of an egg(s) from the surface of the ovary. Generally, vertebrate oocytes complete the first maturation division at the time of ovulation. In hens, ovulation is caused by hormone release (between 12:00 am and 8:00 am); 6–8 hours later, ovulation occurs (therefore in daylight); because it takes 24 hours to complete the egg, the egg is laid in daylight. At the time of human ovulation, (1) there is a temperature increase (0.6 °F), as a result of the thermogenic action of progesterone, and (2) there may be *mittelschmerz* (middle pain) (see Plate 8 in the color insert as well as Figs. 16 and 27). *See: mittelschmerz*.

ovule   the structure, consisting of nucellus and integuments, found in the ovaries of pistils that contains the embryo sac; with maturity, the embryo sac gives rise to an embryo and endosperm and the ovule gives rise to a seed (see Plate 11 in the color insert).

ovum   a mature female germ cell, the female gamete (see Fig. 32).

oxycephaly   *See:* **acrocephaly**.

oxytocin   a hormone produced by the posterior lobe of the pituitary gland, which stimulates contraction of the smooth muscles of the uterus at the time of labor and is used clinically to induce labor.

P

p53   a transcription factor that is one of the most important regulators of cell division.

P-cadherin   also called placental cadherin, a cell adhesion molecule that is apparently expressed primarily on those placental cells (trophoblast cells) that contact the uterine wall and the uterine wall epithelium; it is possible that P-cadherin facilitates the connection of the trophoblast to the uterus.

P element   *See:* **transposable element**.

P-granules   ribonucleoprotein complexes, found in early embryos of *Caenorhabditis elegans*, which probably function in specifying the germ cells.

pacemaker   the sinoatrial node that functions to stimulate the heart and to control the rate of heartbeat; that which controls the rate of heartbeat.

pachytene   the third stage of prophase I of meiosis, by which time the chromosomes have completed synapsis.

Paedogenesis   reproduction by the immature organism; an example is in the Gall-midge *Miastor*, whose larvae give rise asexually to daughter larvae, thus involving parthenogenesis and viviparity; the progeny eat their way out of the parent larva and reproduce in a similar manner; sexual acceleration; a type of paedomorphosis where sexual development is accelerated and terminal stages of development in other systems are omitted after reaching sexual maturity.

paedomorphosis   where late developmental sequences in certain organs or systems are simply omitted and formerly juvenile characteristics thus appear in the terminal (adult) or subterminal (subadult) stages. The character that was embryonic or juvenile in the ancestor becomes adult in the descendant; the neotenic salamanders (e.g., *Necturus*) are excellent examples; phylogenetic change in which adults retain juvenile characters.

pair-rule genes   *Drosophila* genes (e.g., *even-skipped* and *fushi tarazu*), expressed in a series of seven transverse stripes in the blastoderm; each pair-rule gene being expressed in alternate parasegments. The parasegments of the embryo are delimited by the activity of the pair-rule genes. Pair-rule gene expression is revealed by staining for the pair-rule proteins, which gives a striking zebra-striped embryo. The positions of the stripes of pair-rule gene expression are determined by the pattern of gap gene expression; a nonrepeating pattern of gap gene activity is converted into repeating stripes of pair-rule gene expression. Some pair-rule genes (e.g., *even-skipped*) define odd-numbered parasegments, whereas others (e.g., *fushi tarazu*) define even-numbered parasegments; however, the domains of expression of some pair-rule genes cross parasegment boundaries. Each stripe is specified independently. The independent localization of each of the stripes by the gap gene transcription factors requires that in each stripe, pair-rule genes respond to different concentrations and combinations of the gap gene transcription factors; i.e., the pair-rule genes require complex control regions with multiple binding sites for each of the different factors. Gap genes regulate pair-rule gene expression in each parasegment. With the initiation of pair-rule gene expression, the embryo becomes segmented; it is now divided into several unique regions, characterized mainly by the combinations of transcription factors being expressed in each; these include proteins encoded by the gap genes, the pair-rule genes, and the genes expressed along the dorsoventral axis. Most pair-rule genes code for transcription factors. Like gap genes, the activity of pair-rule genes is only temporary. During pair-rule gene expression, the blastoderm becomes cellularized. *See:* **even-skipped**.

palate   the roof of the mouth; partially hard (with bone) and partially soft (without bone).

palatine shelf   *See:* **medial palatine processes**.

Paleocene period   54–65 million years ago; most present-day angiosperm families develop.

Dictionary of Developmental Biology and Embryology, Second Edition. Frank J. Dye.
© 2012 Wiley-Blackwell. Published 2012 by John Wiley & Sons, Inc.

paleogenesis the concept that descendent ontogenies tend to recapitulate ancestral ontogenies; first clearly conceived by Walter Garstang in 1922.

paleomorphic lacking symmetry.

Paleozoic Era started 570 million years ago; includes the Carboniferous Period (280–345 mya), the Age of Amphibians, and the Silurian Period (405–425 mya), the Age of Fish.

palmar erythema redness of the palms of the hands often experienced by pregnant women.

pampiniform plexus a network of multiple small veins originating from veins leading out of the testis and surrounding the testicular artery; this complex vascular system constitutes a highly efficient countercurrent heat exchanger in which the arterial blood is precooled before it reaches the testis to the temperature of the scrotal skin and the venous blood is warmed to body temperature before returning to the abdomen.

pancreas an organ of the digestive system, found in the abdominal cavity; the pancreas produces both digestive enzymes (e.g., trypsin) and hormones (e.g., insulin). In the frog, at the posterior margin of the bile duct opening into the foregut, a pair of outgrowths develop, connected by a single piece of tissue—the future pancreatic duct. The free ends of the outgrowths grow anteriad and fuse anterior to the bile duct; later, these are joined by a mass of tissue that originated from the dorsal wall of the gut (its duct eventually regresses). The three fused elements give rise to the pancreas; eventually the pancreatic duct comes to open into the bile duct instead of directly into the gut itself. *See:* **digestive system**.

pancreatic diverticulum one of two diverticula (dorsal and ventral) from the embryonic duodenum or hepatic diverticulum that form the pancreas or its ducts. *See:* **digestive system**.

Pander, Christian (1794–1865) Russian zoologist, regarded as a founder of the science of embryology; together with Karl Ernst von Baer (in 1817), formulated the germ layer concept as a structural fact for vertebrate embryology. By observations on the chick, he extended the knowledge of Wolff's leaf-like layers and elaborated the conception of Wolff. Pander recognized the presence of three primary layers, an outer, a middle, and an inner, out of which the tissues of the body are formed. In his published *Beiträge* (1817), a monograph of 42 pages on chick development, he observed the blastoderm separating into two germ layers, a *Serosenblatt* (a "serous layer") and a *Schleimblatt* (a "mucous layer"); he later included the appearance of a third layer, the *Gefässblatt* (the "vascular layer"), from which the vascular system arose. He made his most valuable contribution not in the factual details but in his conception of the germ layers, which he believed were the antecedents to later embryonic structures and through their growth and interaction embryonic form came into being. Pander's *Beiträge* was illustrated by Eduard J. d'Alton, with it descriptive embryology was to become heavily dependent on its pictorial presentations. According to Cleveland P. Hickman (1966), Pander made the first description of three germ layers, the description was first made on the chick, and later the concept was extended by von Baer to include all the vertebrates.

Pangenesis the oldest theory of generation; according to this theory, minute granules develop in particular areas of the body and represent the parts in which they originate; at the time of sexual maturity, they come together in the reproductive organs and carry information that is transmitted in inheritance. Charles Darwin resuscitated the theory of pangenesis in his *The Variation of Animals and Plants under Domestication, Volume 2* (1868), to account for the origin of variation. He theorized that body cells secrete minute corpuscles or "gemmules" that record growth patterns for the area they represent. These gemmules were believed to be carried by body fluids and the bloodstream to the reproductive organs and there packed into the eggs or sperm. In the new individual, they were considered to determine the visible characteristics and the growth pattern.

panoistic ovariole the type of ovariole that is devoid of trophocytes; the oocytes are nourished by products elaborated from the blood by the follicular epithelium. This is the primitive type of ovariole, found in the Apterygota, Orthoptera, and other of the lower Pterygota.

Panspermy the theory that generation depended on a primordial, indestructible, and unorganized substance or principle comparable with air, water, and earth, but endowed with life.

papilla a short, rounded nipple-like bump or projection.

parabiosis experimental joining of two individuals, so that their blood circulations are continuous, to study the effects of one partner on the other.

Table 2. Examples of the Foru Major Paracrine Factor Families

Families	Examples	Functions
Fibroblast growth Factor	Fgf8	Lens induction
	Fgf7	Skin development
Hedgehog	Desert Hedgehog	Spermatogenesis
	Indian Hedgehog	Bone growth
Wnt	Wnt Proteins	Cell morphology
	Wnt Proteins	Cell movement
TGF-ß Superfamily	BMP4	Bone formation
	BPM7	Kidney development

paracrine factors   protein molecules used in paracrine signaling; these factors are the inducing factors of the classical experimental embryologists. Most paracrine factors fall into one of four major families: Hedgehog, Wnt, TGF-ß, and Fibroblast Growth Factor (Table 2).

paracrine signaling   a mode of cell-cell communication in which signaling molecules (paracrine factors) act as local mediators and only affect cells in the immediate environment of the signaling cell. *See:* **autocrine signaling, endocrine signaling, juxtacrine signaling**.

paradidymis   the atrophic remains of the mesonephric tubules, which separate from the mesonephric duct and lie near the convolutions of the epididymal duct.

paradigm   an outstandingly clear or typical example or archetype.

***ParaHox* genes**   a small cluster of genes that developed from an early duplication of a primitive *Proto-Hox* gene cluster. *ParaHox* genes function in endoderm development and may be involved in the evolution of the through gut in bilaterians. Considered to be a sister complex to *the Hox* genes.

paralogous genes   genes within a species that have developed by duplication and divergence.

paralogous group   a term applied to genes having similar structures, the same relative positions on each of several chromosomes, and similar expression patterns; e.g., the genes *Hoxa-1*, *Hoxb-1*, and *Hoxd-1*, in the mouse genome, make up a paralogous group.

paralogs   homologous genes, *within a species*, that are related by duplication of an ancestral gene; e.g., the *Antennapedia* gene of *Drosophila* is a paralogue of the *Ultrabithorax* gene of *Drosophila*. *See:* **orthologs**.

paramesonephric ducts   *See:* **Müllerian ducts**.

paraptosis   a form of cell death accompanied by cells swelling and developing internal fluid-filled vacuoles, and lacking membrane blebing, nuclear fragmentation, and cell breakup characteristic of cells undergoing apoptosis.

parasegments   units of the *Drosophila* embryo composed of the posterior part of one segment and the anterior part of the adjacent segment. Independent developmental units, under the control of a particular set of genes, that give rise to the segments of the larva and adult of *Drosophila*. The 14 parasegments first appear as transient grooves on the surface of the embryo after gastrulation; they are the fundamental units in the segmentation of the embryo. The parasegments are initially similar, but each eventually acquires its own unique identity. The parasegments are out of register with the final segments by half a segment. The parasegments are delimited by the activity of the pair-rule genes.

parasitism   symbiosis when one partner benefits at the expense of the other; an example of a developmental parasitic symbiosis is parasitoid wasp species laying their eggs inside the larvae of other insects.

parathyroid glands   two pairs of glands that develop as evaginations of the third and fourth pharyngeal pouches; they play a role in the regulation of calcium levels in the blood. In the chick embryo, one pair is budded off the caudal faces of the third pharyngeal pouches and the other pair in a similar manner from the fourth pouches.

paraxial mesoderm   that portion of the mesoderm that bilaterally flanks the embryonic axis and gives rise to somites. Also called somitic or segmental or dorsal mesoderm (see Fig. 41).

paraxial protocadherin   one of two cell adhesion molecules that apparently direct the adhesive changes driving convergent extension. *See:* **axial protocadherin**.

Paraxis protein   a transcription factor encoded by the *Paraxis* gene; which is an essential part of the conversion of mesenchyme to epithelium, as in the epithelialization of somites.

parazoa   a name proposed for a subkingdom of animals that includes the sponges.

parenchymal cells   those cells of an organ that make up the specialized parts of the organ, as opposed to the organ's connective tissue.

parietal   positioned along the edges or wall, rather than on the axis; as in parietal pericardium or parietal peritoneum.

parietal mesoderm   *See:* **somatic mesoderm**.

parietal placentation   in plants, ovules attached to the walls of the ovary.

parthenocarpy   development of a fruit without fertilization or seed production.

parthenogenesis   of virgin origin; development without fertilization of the egg, which may occur naturally (e.g., some insects, such as aphids) or by artificial means (e.g., parthenogenetic activation of the frog egg by means of a needle prick). Parthenogenetic activation of the egg demonstrates that the egg has the necessary apparatus to begin development, but it needs a mechanism to trigger it.

parthenogenesis induction   turning haploid embryos that normally would become males into diploid females; e.g., *Wolbachia* (bacteria) induction of parthenogenesis, in mites and parasitoid wasps, by doubling the chromosome number in unfertilized eggs to create a diploid embryo with two identical sets of chromosomes.

parthenomerogony   development of a nucleated fragment of an unfertilized egg after parthenogenetic stimulation.

partial birth abortions   these are generally called "D&X" procedures, an abbreviation of "dilate and extract," as well as "intact D&E" or "intrauterine cranial decompression" abortions. The terms "partial birth abortion" and "D&X" were created by pro-life groups when the procedure became actively discussed at a political and religious level. The procedure is performed during the fifth month of gestation or later. The woman's cervix is dilated, and the fetus is partially removed from the womb, feet first. The surgeon inserts a sharp object into the back of the fetus' head, removes it, and inserts a vacuum tube through which the brains are extracted. The head of the fetus contracts at this point and allows the fetus to be more easily removed from the womb.

partial cleavage   cleavage in which only part of the egg divides into blastomeres.

partial hepatectomy   removal of part of the liver; procedure used to study compensatory regeneration in mammalian liver.

parturition   childbirth.

passive immunity   immunity passively acquired, as in the form of antibodies coming through the placenta or in the form of antibodies received through an injection. *See:* **active immunity**.

Patella   genus of limpet (snails) used for studies on cell determination. *See:* **Wilson, Edmund Beecher**.

patent ductus arteriosus   a ductus arteriosus that remains open after birth.

paternal epigenetic effects   epigenetic changes in the male germline that can be passed from one generation to the next, e.g., as caused by environmental agents, such as the endocrine disruptors, BPA (bisphenol A), or vinclozolin.

pathway   in cell-cell signaling, the components required for the sending, receiving, and transduction of a signal, including one or more ligands, membrane-associated receptors, intracellular signal transducers, and transcription factors.

pattern formation (patterning)   the process by which cells in the embryo acquire identities resulting in an ordered spatial pattern of cell types and activities. In *Drosophila*, the sequential activities of maternal and zygotic genes pattern the embryo in a series of steps; broad regional differences are established first, and these are then resolved to produce a larger number of smaller developmental domains.

pattern of cleavage   the manner of cleavage; the way in which the embryos of a given species undergo cleavage; humans undergo a holoblastic, equal, rotational, cleavage pattern.

patterning   *See:* **pattern formation (patterning)**.

Pax **genes**   a subfamily of homeobox genes in vertebrates that contain a homeobox and another conserved domain known as paired; these genes encode transcription factors with various functions in development.

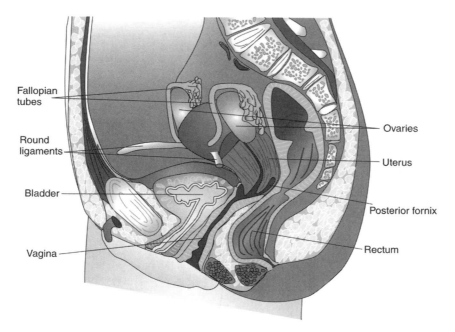

Fallopian tubes

Round ligaments

Bladder

Vagina

Ovaries

Uterus

Posterior fornix

Rectum

Figure 34. The female pelvis. A sagittal section of the female pelvis. Reprinted from Frank J. Dye, *Human Life Before Birth*, Harwood Academic Publishers, 2000, fig. 5-12, p. 43.

Pax proteins   transcription factors; characterized by a DNA-binding region called a paired domain; many Pax proteins also contain a homeodomain, e.g., Pax6 in the eye and Pax3 in the developing somite.

Pax6   a protein, a transcription factor belonging to the homeodomain family; necessary for mammalian eye, nervous system, and pancreas development.

PDA   protostome-deuterostome ancestor. *See:* **Urbilateria**.

PDGF   *See:* **platelet-derived growth factor (PDGF)**.

pelvic cavity   the lowermost portion of the abdominal cavity. In females, it contains the nongravid, internal, reproductive organs—specifically the uterus and ovaries.

pelvic inflammatory disease (PID)   infection of any female reproductive organ.

pelvic kidney   a kidney abnormally located in the pelvis.

pelvis   the space in the kidney that receives urine from the calyces of the kidney and from which the ureter receives urine. Also, that part of the female body that contains the female reproductive organs (see Fig. 34).

penetrance   the frequency (per cent) with which a gene or gene combination manifests itself in the phenotype. *See:* **expressivity**.

penis   the male organ of intercourse, which, together with the scrotum, make up the male external genitalia.

perennial   a plant that lives 3 or more years.

perfect flower   a flower possessing both male and female parts.

perianth   the calyx and corolla of a flower, collectively, especially when they are similar in appearance.

periblast   the nucleated layer of cytoplasm that surrounds the blastodisc of an egg undergoing discoidal cleavage; the syncytial region associated with the expanding blastoderm of macrolecithal, heavily telolecital, cleaving eggs. Two types are delineated: (1) marginal periblast, peripheral to the expanding blastoderm, and (2) central periblast, beneath the blastoderm and its underlying segmentation cavity; both types are populated by yolk nuclei.

periblastula   the blastula of a centrolecithal egg, formed by superficial segmentation (cleavage).

pericardial cavity   the cavity, derived from the paired amniocardiac vesicle portions of the coelom, which contains the heart and is lined by the pericardium (see Fig. 19).

pericarp   the fruit wall and the seed coat together constitute the pericarp, as in the case of the corn kernel, which is actually a fruit.

perichondrium   the connective tissue covering of cartilage.

periclinal divisions   cell divisions that take place in a plane parallel to the surface of the structure of which they are a part; these divisions give rise to new tissue layers. *See:* **anticlinal divisions**.

pericyte   a mesenchymal cell found around a capillary; it may or may not be contractile.

periderm   the outer layer of the two-layered skin of the embryo at the end of the embryonic period. *See:* **basal layer**.

periosteal bone   bone that adds thickness to long bones and is derived from mesoderm via intramembranous ossification.

periosteal buds   ingrowths of the periosteum into the cartilage of growing bone; vascular osteogenic tissue from the periosteum.

periosteum   the connective tissue covering of bones.

peripheral   at the periphery; not at the center.

peripheral nervous system (PNS)   that portion of the nervous system outside the central nervous system (brain and spinal cord).

peripheral zone   region of a meristem surrounding the central zone of a meristem; with higher rates of cell division; divisions in the peripheral zone contribute more directly to new organs; the rate of division here is roughly equivalent to the division rate of cells in newly established organs.

periplasm   the outer layer of protoplasm that surrounds the centrally located yolk of the insect egg.

peristalsis   the spontaneous contraction of the smooth muscle of the wall of the gut, which moves food along the digestive tract.

peristome   the membrane surrounding the mouth of the sea urchin, through which, and into the perivisceral cavity, a solution of KCl is injected when causing the artificial shedding of gametes by the sea urchin.

peritoneal cavity   cavity derived from the coelom and containing most of the viscera; its lining is the peritoneum. Also called the abdominal cavity.

peritoneum   lining of the peritoneal (abdominal) cavity and covering of the abdominal internal organs.

perivisceral cavity   the cavity of the sea urchin that contains the viscera (internal organs), including the gonads; when 0.5 M KCl is injected into this cavity, the consequential gonadal muscle contraction will cause the sea urchin to shed its gametes (see Plate 6 in the color insert).

perivitelline space   the fluid-filled space between the plasma membrane of the egg and the zona pellucida in mammals and between the plasma membrane of the egg and the fertilization membrane in many other animals, such as amphibians and echinoderms (see Plate 10 in the color insert).

permissive interaction   occurs when the responding tissue contains all the potentials that are to be expressed and needs only an environment that allows the expression of these traits; e.g., a solid substrate containing fibronectin or laminin may enable what has been determined to be expressed. *See:* **instructive interaction**.

peroxidase   an enzyme released by the cortical reaction of sea urchin eggs, which hardens the fertilization membrane.

petal   the outermost floral part of a flower. *See:* **corolla**.

Pfeffer, Wilhelm Friedrich Philipp (1845–1920)   German botanist; is usually given credit for having introduced the word and concept of induction into biology, in his 1881 *Pflanzenphysiologie*. However, according to Jane M. Oppenheimer (1994), Pfeffer used the word 10 years before, in 1871, and he was speaking of external, not internal, inducing agents. Discoved that alpha-malic acid is a chemoattractant, playing a role in fern fertilization (in 1884). Through the two volumes of his *Textbook of Plant Physiology* (*Lehrbuch der Pflanzenphysiologie*), which was translated into English and French, he rose to an international teacher of botany. Today he is thought to be one of the founders of modern plant physiology.

Pflüger, Eduard (1829–1910)   offered a new combination of the *methods* of physiology and the *problems* of morphological embryology; he, for example, concentrated on manipulating and controlling the environment external to the developing individual frog to ascertain what determines the sex of a frog embryo. Pflüger carried out a series of experiments to test the effects of orientation within the gravitational field on development; he concluded that gravity determines the direction of the cleavage plane and, consequently, the body orientation; thus, he concluded that external conditions can direct development. Pflüger's results meant that the embryo could not already be lying within the egg, but neither is the egg a mass of undifferentiated material *driven by ancestral heredity*, as Ernst Haeckel maintained. Pflüger's work stimulated experiments on frog embryos by several researchers, including Gustav Born and Wilhelm Roux. His work suggesting that an experimental attack on embryological problems might be productive is what inspired others.

PGCs   *See:* **primordial germ cells (PGCs)**.

pH   a numerical scale for expressing hydrogen ion concentration; the negative log of the hydrogen ion activity.

phallus   an undifferentiated embryonic structure derived from the genital tubercle that differentiates into the penis in males and the clitoris in females (see Fig. 49).

phanerogam   a plant that produces seeds. *See:* **cryptogam**.

pharyngeal arches   *See:* **branchial arches**.

pharyngeal bursa   a small pit caudal to the pharyngeal tonsil, resulting from the ingrowth of epithelium along the course of the degenerating tip of the notochord of the vertebrate embryo.

pharyngeal pouches   endodermal diverticula from the lateral walls of the pharynx.

pharyngobranchial duct   the narrow medial part of a pharyngeal pouch.

pharyngula stage   in chordates, the phylotypic stage is the pharyngula. In amphibian (frog) development, all the cells originating from the fertilized egg give rise to the pharyngula stage; in avian (chick) development, of the approximately 60,000 cells present when the egg is laid, only approximately 500 will actually construct the pharyngula; the other cells form the extraembryonic membranes; in mammalian (mouse) development, the inner cell mass gives rise to the pharyngula proper. *See:* **phylotypic stage**.

pharynx   the cavity between the oral (mouth) cavity and the esophagus (see Plates 1 and 13 in the color insert as well as Fig. 6).

phenocopy   a mimic of a genetic mutant phenotype, usually caused by environmental changes.

phenotype   the outward, frequently visible, expression of the genotype.

phenotypic accommodation   the adaptive mutual adjustment, without genetic change, *among variable aspects of the phenotype* following a novel input during development; may facilitate the evolution of a novel morphology; e.g., the origin of human bipedalism has possibly been contributed to by phenotypic accommodation. *See:* **genetic accommodation**.

phenotypic heterogeneity   the phenomenon whereby the same mutation can produce a different phenotype in different individuals.

phenotypic plasticity   the ability of an individual to express one phenotype under one set of circumstances and another phenotype under another set; e.g., upon the death of the male in its cohort, a female blue-headed wrasse (fish) becomes a male . Also, the ability of an organism to react to an environmental input with a change in form, state, movement, or rate of activity. *See:* **developmental plasticity**.

phenotypic sex   the apparent sex of an individual; having the characteristics of a male or a female. *See:* **genetic sex**.

phenotypic suppression   the phenomenon by which more posterior HOX-cluster genes inactivate the more anterior without affecting transcription or translation; e.g., *ems* can confer identity to trunk segments when other HOX-cluster gene activities are absent, in trunk segments of wild-type embryos; however, *ems* activity is prevented by phenotypic suppression; i.e., HOM gene products normally expressed in anterior regions are suppressed by more posterior products. *See:* **posterior prevalence**.

philtrum   the groove on the medial surface of the upper lip, directly beneath the nose.

phloem   the component of plant vascular tissue that is made up of nutrient-conducting cells. *See:* **cambium, procambium, tracheophytes, vascular cambium**.

phocomelia   refers to birth defects in which the hands or feet are attached directly to the shoulders or hips, respectively, the intervening parts being absent; a type of birth defect caused by the drug thalidomide.

-phora   carry, bear, e.g., Ctenophora and comb bearer.

phosphatidylinositol bisphosphate (PIP₂)   *See:* **phosphoinositol pathway**.

phosphoinositol pathway　a common intracellular signal transduction pathway. During sea urchin fertilization, when bindin (a ligand on the surface of the spermatozoon) binds to the bindin receptor on the vitelline envelope of the egg, a G protein is activated. The G protein, in turn, activates the enzyme phospholipase C, which hydrolyzes its substrate, phosphatidylinositol bisphosphate (PIP_2), into two products, diacylglycerol (DAG) and inositol triphosphate (IP_3). The DAG activates protein kinase C, which in turn activates a H^+/Na^+ exchanger, which by simultaneously transporting H^+ out of the egg and Na^+ into the egg causes alkalinization of the egg's cytoplasm. Meanwhile, the IP_3 binds to a receptor on the membrane of the endoplasmic reticulum (ER), causing the ER to release Ca^{++} into the cytosol of the egg. Together, alkalinization of the cytoplasm and elevated Ca^{++} concentration of the egg cytoplasm cause activation of the egg.

phospholipase C　*See:* **phosphoinositol pathway**.

phosvitin　a heavily phosphorylated yolk protein found in frog oocytes, derived from the splitting of vitellogenin and packaged into yolk platelets.

photolyase　a flavoprotein enzyme; a repair system for ultraviolet (UV) damge to DNA; absorbs light in the yellow range of the visible spectrum and uses the energy to reverse thymidine dimerization in DNA caused by UV. *See:* **mycosporines**.

photomorphogenesis　refers to the influence that light has on the morphogenesis of a plant, e.g., the influence exerted by blue light on the transition from one- to two-dimensional growth in the development of the fern prothallus (see Plate 15 in the color insert).

photoperiod　duration of daily exposure to light.

photoperiodism　the physiological responses of an organism to the length of night or day or both.

photoreceptor cells　highly specialized, light-sensitive cells containing photopigments, e.g., sensory cells, found in the retinas of the eyes, that are light receptors for the sense of vision.

phototropin　a blue light receptor; functions in phototropism, blue light-induced opening of stomata, and the migration of chloroplasts within a cell. *See:* **cryptochromes, daylength**.

phototropism　bending toward the light.

phthalates　endocrine disruptors; ubiquitous in industrialized societies, e.g., used in vinyl plastics; have been shown to induce experimentally many components of testicular dysgenesis syndrome.

phyletic constraints　historical restrictions based on the genetics of an organism's development; e.g., once a structure comes to be generated by inductive interactions, it is difficult to start over again. *See:* **polycomb proteins**.

phyletic transformation　the evolution of a new morphospecies by the gradual transformation of an ancestral species, without a speciation or splitting event taking place; also called anagenesis.

phyllotaxy　the pattern of leaf primordia initiation; the positioning of leaves on the stem; involves communication among existing and newly formed leaf primordia; the two basic patterns are spiral phyllotaxy and whorled phyllotaxy. *See:* **plastochron**.

phylogenetic tree　a depiction of the evolutionary relationships among species and their order of "branching" from common ancestors; a diagram of the relationships of ancestry and descent among a group of species or populations; also called an evolutionary tree.

phylogeny　the study of the evolutionary history of a group of organisms.

phylotypic stage　a stage, the tailbud stage, in the development of vertebrate embryos, during which the embryos of different vertebrate classes closely resemble each other; that stage when all of these embryos possess a distinct head, notochord, neural tube, and somites. The embryonic stage of chordates when pharyngeal gill slits are present; when features distinguishing the phylum Chordata are just recognizable—the notochord, the dorsal hollow nerve cord, gill slits, and a post-anal tail. This stage was recognized by Karl Ernst von Baer in 1828 when he observed the similarity of early embryos of chick, rat, and salamander just after the completion of neurulation. In insects, the mature germ band stages of both long-germ and short-germ insect embryos look similar (they all show six head segments, three appendage-bearing segments, and a variable number of abdominal segments); this, therefore, is a common stage, the phylotypic stage, through which all insect embryos develop. At this stage, the homeotic gene expression pattern of the *Hox*/HOM-C genes is observed most clearly and is remarkably similar in all animals so far investigated. The homeobox genes of fungi and plants are not homologous with those in animals. Development both preceding and following the phylotypic stage is extraordinarily diverse. *See:* **animal**.

phylum   a taxon; a classification division just beneath the taxon kingdom; each phylum corresponds to a body plan, e.g., Chordata and Echinodermata.

physical constraints   the laws of physics, e.g., diffusion, hydraulics, and physical support, allow only certain mechanisms of development to occur.

physiology   the study of function.

phytochelatins   polymers of glutamate-cysteine that bind heavy metals in plants, and perhaps animals as well. *See:* **glutathione, metallothioneins.**

phytochromes   plant pigments that transduce photoperiodic signals from the external environment; absorb in the red and far-red parts of the spectrum; daylight is rich in red light, moonlight in far-red light, shade under the leaf canopy is mostly far-red, and the light at the bottom of a lake is mostly red; phytochrome responses to light include seed germination, greening, stem elongation, flowering, and the induction of gene expression, especially genes involved in photosynthesis. *See:* **photoperiodism, signal transduction.**

PiED   *See:* **preimplantation embryo diagnosis (PiED).**

pigment cells   cells specialized for the production of pigment; e.g., melanocytes are pigment cells specialized for the production of the pigment melanin.

pigmentation   having pigment.

pigmented retina   that portion of the retina derived from the outer wall of the two-walled optic cup; its cells produce melanin pigment.

pineal gland   an endocrine gland derived from the roof of the diencephalon; secretes melatonin, which may play some role in the onset of puberty.

pinnae   the external flaps of the ears found on the sides of the human head.

pioneer nerve fibers   axons that go ahead of other axons and serve as guides for them.

pioneer transcription factors   transcription factors that can penetrate repressed chromatin and bind to their enhancer DNA sequences, a step critical to establishing certain cell lineages, e.g., FoxA1, Pax7, and pbx.

pistil   the innermost, female, part of a flower, consisting of the stigma, style, and ovary; the term also refers to an individual carpel or a group of fused carpels.

pistillate flowers   imperfect flowers lacking stamens; also called carpellate flowers.

pituitary gland   an endocrine gland; has a dual origin; part of it (neurohypophysis) is derived from the floor of the diencephalon (infundibulum), and part (adenohypophysis) is derived from the roof of the stomodaeum. Although under the control of the hypothalamus of the brain, the pituitary gland secretes hormones that play central roles in reproductive physiology. Also called hypophysis. In frogs, increased secretion of pituitary hormone in the spring brings about ovulation.

placenta   (1) in animals, an organ, partially fetal and partially maternal in origin, through which the mother and fetus exchange vital substances such as oxygen, food, and waste products. The intervillous space of the mature placenta contains 150 mL of blood, which is replenished three to four times per minute. The rate of uteroplacental blood flow is (i) 50 mL/minute at 10 weeks and (ii) 500–600 mL/minute at full term. Generally, the full-term placenta is a flat, circular, or oval disc, with a diameter of 15–20 cm, a thickness of 2–3 cm, and a weight of 500–600 g. The fetal surface of the placenta is smooth with the umbilical cord usually attached near the center. The maternal surface of the placenta is rough and raised into 10–38 cotyledons. Grooves between the cotyledons mark the sites of the placental septa. The amnion and chorion are continuous with the edges of the placenta; at parturition, the umbilical cord, placenta, amnion, and chorion follow the fetus as the afterbirth. (2) in plants, the part of the ovary wall to which the ovules or seeds are attached (see Figs. 35–37).

placenta accreta   a placenta that has partially grown into the myometrium of the uterus.

placenta fetalis   the chorion frondosum or bushy chorion. The influence of the embryo causes striking changes in the endometrium. Because of the postpartum shedding and replacement of much of the endometrium, the term "decidua" (to shed) is used for the endometrium of pregnancy. As the chorionic vesicle grows, the overlying portion of the endometrium is stretched out over it; this part of the decidua is called the decidua capsularis. The portion of the endometrium elsewhere, other than at the site of attachment of the chorionic vesicle, is called the decidua parietalis. The part of the endometrium directly underlying the chorionic vesicle is the decidua basalis. The maternal blood supply is most direct and abundant in the decidua basalis. The chorionic vesicle, at first uniformly villated over its entire surface, has by the end of the fourth month become denuded of its villi everywhere except where they lie in the decidua basalis. The part of the chorionic vesicle

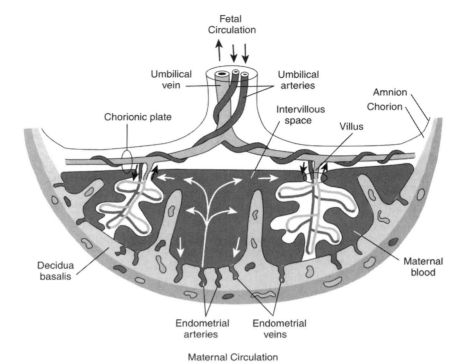

Maternal Circulation

Figure 35. The placenta. A diagrammatic section through the fully developed placenta. Reprinted from Frank J. Dye, *Human Life Before Birth*, Harwood Academic Publishers, 2000, fig. 10-2, p. 82.

under the decidua capsularis, which has lost its villi, becomes known as the chorion laevae (smooth) and that part of the chorion next to the decidua basalis, where the villi are highly developed, is termed the chorion frondosum (bushy). It is the interlocked chorion frondo-sum and decidua basalis that together constitute the placenta. The dual origin of the placenta is emphasized by the terms placenta fetalis (chorion frondosum) and placenta materna (decidua basalis). During the last half of pregnancy, the chorion laeve is pushed (by the growing embryo) tight against the uterine walls and the decidua capsularis is squeezed against the decidua parietalis. Adherent to the inner face of the chorion laeve is the amnion, which during the third month expands to fill the entire chorionic sac. Placental attachment normally occurs relatively high up in the body of the uterus; the decidua capsularis, chorion laeve, and adherent amnion, only, stretch over the cervical outlet. With the beginning of muscular contractions, at the onset of labor, amniotic fluid is squeezed into this thin part of the chorionic sac, which is the preliminary dilator of the cervical canal (see Plate 9a in the color insert and Fig. 36).

placenta materna the decidua basalis, which is the maternal contribution to the placenta (see Plate 9b in the color insert and Fig. 37). *See:* **placenta fetalis**.

placenta previa an abnormal position of the placenta in which it blocks the internal os, that is, the opening from the uterine cavity into the cervical canal that makes up part of the birth canal; this would obstruct the fetus' exiting of the uterine cavity during parturition.

placental activities: endocrine secretion (1) protein hormones; human chorionic gonadotro-pin (HCG), human placental lactogen (HPL), and thyrotropin; (2) steroid hormones: estrogens and progesterone.

placental activities: metabolism (1) synthesizes glycogen, cholesterol, and fatty acid; (2) probably serves as a source of nutrients and energy for the embryo.

placental activities: transfer (1) mechanisms; most materials are transferred by simple dif-fusion, facilitated diffusion, active transport, and pinocytosis; others, red blood corpuscles from fetus to mother as a result of microscopic breaks in the placental membrane, maternal lymphocytes and syphilis organisms on their own power (fetal lymphocytes also enter the

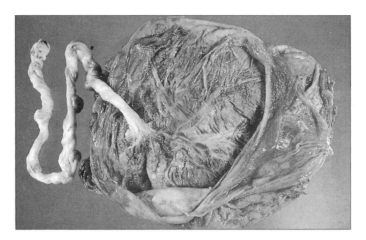

Figure 36. The fetal side of the placenta and most of the umbilical cord, including its attachment to the placenta (See Figure 37). (Courtesy of Jack Wolk, MD, the Danbury Hospital.) Reprinted from Frank J. Dye, *Human Life Before Birth*, Harwood Academic Publishers, 2000, fig. 23-3A, p. 207.

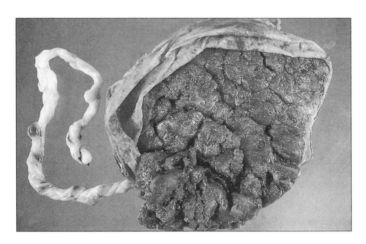

Figure 37. The maternal side of the same placenta (See Figure 36). (Courtesy of Jack Wolk, MD, The Danbury Hospital.) Reprinted from Frank J. Dye, *Human Life Before Birth*, Harwood Academic Publishers, 2000, fig. 23-3B, p. 207.

maternal circulation), some bacteria and protozoa as a result of infection and placental lesions; (2) gases; oxygen, carbon dioxide, and carbon monoxide, by simple diffusion; (3) nutrients; water (freely and rapidly), maternal cholesterol, triglycerides, and phospholipids (little or none), fatty acids (little), vitamins (yes and essential), and glucose (yes and quickly).; (4) hormones; protein hormones (not significant amounts) and steroid hormones (freely); (5) electrolytes; freely exchanged; (6) antibodies; some passive immunity to the fetus for dipththeria, smallpox, and measles; (7) wastes; carbon dioxide, urea, uric acid, and bilirubin; (8) drugs; most drugs cross the placenta freely; (9) infectious agents; many viruses may pass.

placental barrier the tissues intervening between the maternal and fetal blood of the placenta, which prevent or hinder certain substances or organisms from passing from mother to fetus.

placental cadherin *See:* **cadherins, P-cadherin**.

placental circulation that portion of prenatal circulation that carries blood from the fetus to the placenta and back, by way of the umbilical cord; that is, that portion of the circulation that leaves the fetus by way of the umbilical arteries and returns by way of the umbilical vein. The course of FETAL placental circulation is as follows: fetus-umbilical arteries (deoxygenated blood in umbilical cord)-radial vessels of the chorionic plate- arterial, capillary, venous system of villi-chorionic plate-umbilical vein (oxygenated blood in umbilical cord)-fetus. The course of MATERNAL placental circulation is as follows: mother-decidua basalis-maternal arteries (80–100 spiral endometrial arteries)-intervillous spaces-spurts toward chorionic plate-flows over villi-maternal endometrial veins-decidua basalis-mother (see Figs. 35 and 47).

placental classification   placentas may be classified on the basis of (1) the distribution of villi: e.g., mice and humans have discoid placentas, while pigs have diffuse placentas, and (2) the number of intervening layers between fetal and maternal blood; e.g., humans have hemochorial placentas with three intervening layers, while pigs have epitheliochorial placentas with six intervening layers.

placental membrane   the tissues separating the maternal and fetal blood; e.g., in humans, with a hemochorial placenta, in which the chorionic villi are bathed in maternal blood, these are three fetal (villus) tissues: capillary endothelium, connective tissue, and trophoblast.

placental septa   the placental septa are projections of the maternal side of the placenta into the intervillous space, which divide the fetal part of the placenta into 15–30 irregular areas called cotyledons.

placental stage   that stage of labor extending from the birth of the baby to the birth of the placenta.

placentation   the arrangement of ovules and placenta in the ovary of a plant; the formation and fusion of the placenta to the uterine wall in animals.

Placodermi   an extinct class of vertebrates, made up of armored fish.

placoderms   the first gnathostomes.

placodes   thickenings of head ectoderm that contribute to the formation of some sense organs (e.g., olfactory placodes, lens placodes, auditory placodes, and cranial placodes) and some cranial ganglia.

plagiocephaly   a type of strongly asymmetric cranial deformation, as a result of a number of causes, such as the disordered sequence of suture closure.

planaria   a group of flatworms that has been extensively used for regeneration research and teaching.

planes   (see Fig. 38). *See:* **frontal plane, sagittal plane, transverse plane**.

plant   any organism belonging to the kingdom Plantae.

-plasia   a word root that denotes formation, development, or growth; used with several prefixes to form the names of numerous birth defects having to do with abnormal growth, e.g., cleidocranial dysplasia.

plasma   *See:* **blood plasma**.

plasma cell differentiation   differentiation of plasma cells; this type of cell differentiation involves permanent alteration of the cell's genome.

plasma cells   antibody-secreting cells derived from B cells that bind antigen.

plasma membrane   the part of every cell found at its periphery and separating the contents of the cell from the cell's environment. Its function is vital because it determines what does and does not enter and leave the cell. It makes the interior of the cell different from its surroundings, which is necessary for life.

plasmodesmata   *See:* **cytoplasmic bridges, movement proteins**.

plasmodium   a stage in the life cycle of slime molds, of two general types: (1) the true, i.e., syncytial plasmodium, and (2) the pseudoplasmodium, a multicellular structure.

plastochron   the time interval between successive leaf initiations, is used to describe leaves at different stages; i.e., a leaf that has just initiated is at plastochron 1 (P_1), this primordium becomes P_2 when a new primordium forms, etc. *See:* **phyllotaxy**.

platelet-derived growth factor (PDGF)   a protein growth factor; stimulates proliferation of connective tissue cells and some neuroglial cells; in angiogenesis, it is necessary for the recruitment of pericyte cells. *See:* **pericyte**.

Platyhelminthes   a phylum of animals; the flatworms.

pleiotropic   as applied to genes and proteins; genes or proteins with multiple functions.

pleiotropy   the production of several effects by one gene. *See:* **mosaic pleiotropy, relational pleiotropy**.

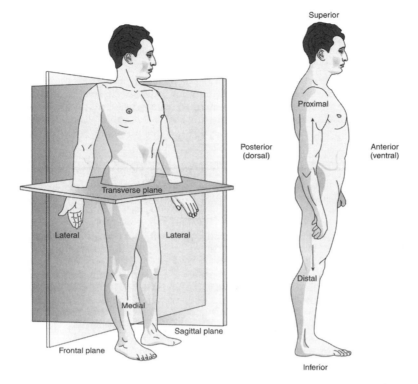

Figure 38. Relative positions and planes of the body. Relative positions are reference points. Note that the elbow is closer (proximal) to the shoulder than is the wrist (distal to the shoulder). Three important planes—transverse, sagittal, and frontal—divide the body. Reprinted from Frank J. Dye, *Human Life Before Birth*, Harwood Academic Publishers, 2000, fig. 1-2, p. 3.

plesiomorphic   a trait that occurred in the common ancestor of the taxon being considered and its sister group.

pleural cavities   contain the lungs and are lined by pleurae; derived from the coelom.

-ploid   suffix; having several chromosomes that have a specified relationship to, or are a multiple of, the basic number of chromosomes of a group, e.g., haploid, diploid, or heteroploid.

plumule   the stem with several leaf primordia, of grass embryos; actually the shoot apical meristem and several embryonic leaves.

pluripotent   describes a cell that is capable of giving rise to several cell types of an organism; e.g., the pluripotent stem cells of the bone marrow give rise to all classes of blood cells, and neural crest cells can differentiate into several different cell types, such as cholinergic neurons and adrenergic neurons, depending on their location; also used to refer to stem cells that have the ability to become all the cell types of the embryo except trophoblast.

pluripotent epidermal stem cell   a stem cell whose offspring become epidermis, sebaceous gland, or hair shaft.

pluripotent neural stem cell   a stem cell whose offspring become neural and glial cells.

pluripotential hematopoietic stem cell   the common precursor cell shared by erythrocytes, basophils, eosinophils, mast cells, monocytes, tissue macrophages, osteoclasts, neutrophils, platelets, and T and B lymphocytes.

pluteus larva   echinoderm larva of the class Echinoidea (sea urchins); a motile (by cilia) larval stage in the early development of sea urchins.

PMZ   *See:* **posterior marginal zone (PMZ)**.

PNS   peripheral nervous system.

poikilothermic   unable to maintain a constant body temperature independent of the environment; poikilothermic vertebrates include fish, amphibians, and reptiles. *See:* **homoiothermic**.

polar body   a cell formed during oogenesis, from the division of an oocyte, that normally plays no role in development (see Fig. 32). *See:* **polyploidy**.

polar coordinate model   a model that attempts to explain the numerous phenomena of limb regeneration, where each cell has a circumferential value specifying the anterior-posterior axis and a radial value specifying the proximal-to-distal axis.

polar granule component (Pgc)   a component of *Drosophila* pole plasm; a nontranslatable RNA, the exact function of which is unknown; evidence indicates it has a role in the migration of pole cells to the gonads.

polar granules   granules found at the posterior pole of the cellularizing blastoderm of *Drosophila* embryos; work with mutants of *Drosophila* suggest that the polar granules are directly concerned with germ cell production.

polar lobe   an extrusion of cytoplasm, contained within the plasma membrane, from one of the early blastomeres of certain spirally cleaving embryos, e.g., mollusks (see Plate 18 in the color insert).

polar nuclei   the two nuclei of the central cell, found at the center of an embryo sac, which, when fertilized by a spermatozoon, give rise to the triploid, nutritive, tissue called endosperm. *See:* **central cell**.

polar trophoblast   that portion of the trophoblast associated with the inner cell mass. *See:* **mural trophoblast**.

polarization   the first stage of cell migration, wherein a cell defines its front and its back.

pole cells   the first intact cells to develop, at the 512-nuclei stage, from the syncytial *Drosophila* blastoderm; these cells give rise to the germ cells. One of the first events of insect development is to separate the future germ cells from the rest of the embryo.

pole plasm   that portion of the cytoplasm of the *Drosophila* egg that is incorporated into the germ cells; components of *Drosophila* pole plasm include the mRNA of the *germ cellless* gene, proteins such as oskar and nanos, mitochondrial ribosomal RNA, and polar granule component; the vegetal region of amphibian eggs also appears to contain pole plasm, which plays a role in germ cell determination; there is no obvious germ plasm in mammals. *See:* **germ plasm**.

poles   the two ends of the spindle found in dividing cells; the two ends of the primary axis of the egg (animal pole and vegetal pole).

pollen   *See:* **pollen grain**.

pollen germination   the emergence of a pollen tube from a pollen grain, which occurs if a pollen grain and the stigma on which it lands are compatible; this is preceded by the pollen taking up water (hydrating). Pollen germination may also occur under artificial conditions if pollen is exposed to a sugar solution.

pollen grain   the male gametophyte of angiosperms, consisting of, initially, two *nuclei*: the tube nucleus and the generative nucleus. A mature pollen grain consists of two *cells*, one within the other; the tube cell contains a generative cell within it (the generative cell divides to produce two sperm). Gymnosperms also produce pollen grains.

pollen tube   the tube formed under the influence of the tube nucleus of the pollen grain, which conveys the spermatozoa to the female embryo sac.

pollination   occurs when a pollen grain lands on the stigma of a pistil; refers to the landing and subsequent germination of the pollen on the stigma.

polocyte   *See:* **polar body**.

poly-   word root meaning "many," e.g., polyspermy and polyploid.

polyandry   a mating system in which at least some females mate with more than one male.

polycarpic plants   plants that flower repeatedly. *See:* **monocarpic plants**.

polychlorinated biphenyls (PCBs)   ubiquitous chemicals introduced into the environment in 1928; associated in mammals with reproductive abnormalities, neurological and cognitive deficits, thyroid deficits, cancers, and immune dysfunctions.

polycomb proteins   proteins conserved throughout evolution that silence *Hox* genes at specific times and places.

polydactyly   the existence of supernumerary (extra) fingers or toes.

polyembryony   is the production of two or more individuals, instead of one, from a single egg. It happens in cases of identical twins in humans, in certain armadillos, and in Polyzoa, but assumes its extreme development among parasitic Hymenoptera; the essential feature of the process is the separation of the blastomeres of the egg into groups of cells or morulae, each of which grows into an adult insect.

polygamous   with unisexual and bisexual flowers on the same plant.

polygenic   involving many genes.

polygyny   a mating system in which at least some males mate with more than one female.

polyhydramnios   an excessive volume of amniotic fluid. *See:* **oligohydramnios**.

polyinvagination islands   *See:* **primary hypoblast, secondary hypoblast**.

polymastia   the presence of more than two breasts.

polymer   a large molecule made of small, repeating subunits (monomers).

polymerase chain reaction (PCR)   a highly efficient, *in vitro*, enzymatic method of cloning (making many identical copies of) specific pieces of DNA.

polymorphic   variable; with many forms.

polymorphism   (1) a region of the genome that varies between individual members of a population; (2) the existence within a population of more than one variant for a phenotypic trait or of more than one allele.

polyphenism   discontinuous ("either/or") phenotypes elicited by the environment, e.g., short-winged and long-winged forms of migratory locusts. *See:* **nutritional polyphenism, reaction norm, seasonal polyphenism, sexual polyphenism**.

polyphyletic group   a set of species that are grouped by similarity but not by descent from a common ancestor; a group is said to be polyphyletic if it does not contain a singular most recent common ancestor; e.g., it was once thought that all pachydermous mammals (e.g., rhinos, hippos, and elephants) were descended from a single large ancestor; it is now known that each of these animals evolved from a seperate small ancestor, and the common ancestor of all of them was small and slightly built, with presumably thin skin and fur.

polyploidy   having many (more than two) multiples of the haploid number of chromosomes; may result from polyspermy or restitution nucleus formation as a result of abortive polar body formation (e.g., abortive separation of the first polar body from the secondary oocyte during oogenesis).

polyspermy   the fertilization of a single egg by more than one sperm. Polyspermy is abnormal, in general, e.g., in sea urchins, frogs, and mammals. Polyspermy is common in chicken fertilization, but only one spermatozoon nucleus eventually undergoes karyogamy with the female pronucleus; the abortive nuclei of the supernumerary spermatozoa eventually disintegrate.

polytene chromosomes   literally, "many-thread chromosomes" (where each thread is a chromatid); giant chromosomes found in various organisms as diverse as plants and insects, but especially well developed in the larval salivary glands and other larval organs of dipteran insects. These chromosomes are unique because they are giant, interphase chromosomes resulting from the uncoupling of DNA replication and mitosis; each chromosome may consist of up to 1000 or more chromatids. Especially in *Drosophila*, hundreds of genes have been mapped to specific bands along the lengths of these chromosomes; it is even possible to visualize active genes, because when actively undergoing transcription, a given gene locus will result in the formation of a microscopically visible chromosome puff. *See:* **chromosome puff**.

polytrophic ovariole   the type of insect ovariole that shows successive chambers containing oocytes and trophocytes; found in most Endopterygota.

pons   a part of the brain derived from the metencephalon, which has fibers connecting the cerebral cortex with the cerebellum.

pontine   relating to the pons of the brain (see Fig. 8).

pontine flexure   a flexure in the embryonic brain concave dorsally, occurring in the region of the myelencephalon.

position   one of four descriptions of the fetus' alignment in the uterus; the position of the fetus relates a chosen part of the fetus to the right or left side of the mother, e.g, face left or face right (see Fig. 3). *See:* **attitude, lie, presentation**.

positional gene cloning   a method used to find genes that are involved in normal human development and whose mutations cause congenital developmental malformations; pedigree analysis highlights a region of the genome where a particular mutant gene is thought to reside; by making a DNA map of that area, one may find a region of DNA that differs between people who have the mutation and people who lack it; then by sequencing that region of the genome the gene can be located, e.g., the discovery of the human *Aniridia* gene. *See:* **candidate gene mapping**.

positional information information that tells the cell where it is within the context of the organism; because plant cells are constrained by their fixed position, positional information is the guiding force in plant development. *See:* **lineage-based information**.

positional value the identity acquired by a cell that is related to its position; e.g., its position along a line, where it is exposed to a specific concentration of a morphogen along a concentration gradient of the morphogen.

postanal gut in the frog, as the tail region of the tadpole develops, the neurenteric canal is drawn out into a small tube beneath the posterior end of the notochord. Before hatching, the neurenteric canal breaks away from the neural tube and persists for a brief period as the postanal gut.

posteriad toward the posterior.

posterior dominance the process whereby the most posteriorly expressed *Hox* genes can inhibit the action of more anteriorly expressed *Hox* genes when they are expressed in the same region.

posterior fornix the deepest and most posterior of the four parts of the vault of the vagina; the vault of the vagina is subdivided by the uterine cervix, which projects into the blind end of the vagina, into fornices (see Fig. 34).

posterior group *Hox* genes *Hox* genes expressed in the posterior region of bilaterans and located at the 5′ end of *Hox* clusters; include the *Abd-B*-related and *Hox9-13* genes.

posterior intestinal portal in the chick embryo, the opening from the midgut into the hindgut.

posterior marginal zone (PMZ) the region of the chick blastoderm where hypoblast formation begins; these posterior marginal cells contribute the inducing cells of the hypoblast and prevent other regions of the margin from inducing their own hypoblasts. In bird and mammalian embryos, the PMZ may be the equivalent of the amphibian Nieuwkoop center; the chick homologue of *Vg1* is transcribed in this region.

posterior neuropore the temporary opening at the caudal end of the neural tube. In human development, an improperly closed posterior neuropore gives rise to the birth defect known as spina bifida (see Fig. 30). *See:* **anencephaly**.

posterior organizing center classic embryological experiments demonstrated that there are at least two "organizing centers" in the insect egg, one is the anterior organizing center, and the other the posterior organizing center. In *Drosophila* embryos, the posterior organizing center is defined by the activities of the *nanos* gene. *See:* **anterior organizing center, terminal gene group**.

posterior prevalence meaning that when moved to an anterior location, a posterior gene had priority over its anterior collegues; e.g., HOM gene products normally expressed in anterior regions are suppressed by more posterior products. *See:* **phenotypic suppression**.

posteriorizing activity or "mesodermalizing" activity; originates in the posterior of the amphibian embryo and weakens anteriorly; candidates for the posteriorizing factor include eFGF, retinoic acid, and Wnt3a. *See:* **neuralizing activity**.

postpartum after birth.

posttranslational regulation protein modifications that determine whether a protein will be active.

postzygotic isolation reproductive isolation between populations caused by dysfunctional development or sterility in hybrid forms; postzygotic isolation refers to the inability of a zygote resulting from cross-species fertilization to develop into a viable organism; note that viable organisms must be able to reproduce. Mules are an example of postzygotic isolation even though they are able to complete development successfully. *See:* **prezygotic isolation**.

potassium chloride (KCl) used as a 0.5 M solution, which, when injected into the perivisceral cavity of sea uchins, causes contraction of the gonadal muscles and, consequently, the release of gametes through the genital pores.

potency the range of possible cell type or structures into which a particluar cell can develop. *See:* **pluripotent, totipotent**.

prechordal plate also prochordal plate; the precursor of the head mesoderm.

precipitation the theory of precipitation was based on the notion that the embryo is formed suddenly at the moment of fecundation, by precipitation of materials already present in the ovum.

precocenes compounds isolated from herbaceous composites (sunflower family) that cause the death of insect herbivores by eliciting metamorphosis too early; cause the selective death of corpus allatum cells responsible for synthesizing juvenile hormone in immature insects.

precocious   developing or appearing very early; with the flowers developing before the leaves. *See:* **serotinous**.

predator-induced defense   *See:* **predator-induced polyphenism**.

predator-induced polyphenism   a phenotypic change caused by a predator where the phenotypic modification increases the fitness of its bearers when the predator is present; e.g., the snail *Thais lamellosa* develops a thickened shell and a "tooth" in its aperature when exposed to the effluent of the crab species that preys on it; alteration of prey rotifer morphology in the presence of predator rotifers. *See:* **kairomones**.

predictive adaptive responses   early environmental cues shift the developmental pathway to modify the phenotype in expectation of the later environment, thereby benefitting the organism later in life; e.g., arctic animals whose coats are brown in summer but white in winter—the processes were initiated long before the first snows of winter fell. *See:* **immediate adaptive responses**.

preformationism   the doctrine that the new individual is already preformed in the spermatozoon or egg and that development involves essentially the growth of the preformed organism; historically, it seemed to explain best the continuity between generations; however, it could not explain variations or the generation of "monstrosities" when both parents were normal. Jan Swammerdam visualized the process of embryology as a simple enlargement from minute but preformed organisms to adults. Marcello Malpighi believed that the chick embryo was preformed. During the controversy between preformationism and epigenesis, from the point of view of preformationist doctrine, it was only a question of whether the miniature organism was in the egg or in the sperm (ovism or animalculism). *See:* **epigenesis**.

pregnancy   the condition of a woman who is carrying a conceptus (the product of conception or fertilization).

pregnancy reduction   selective abortion of some fetuses of a multiple pregnancy.

preimplantation embryo diagnosis (PiED)   removal of a cell from an early embryo for very early prenatal diagnosis; the embryo is the result of *in vitro* fertilization (IVF), and the procedure is carried out in a dish.

premature   in the context of human development, the birth of a child before completion of the normal, full term of a pregnancy.

premature chromosome condensation   when a cell whose chromosomes are in metaphase is fused with a cell in interphase, the nuclear membrane of the interphase nucleus dissolves and the chromosomes condense; the appearance of the prematurely condensed chromosomes depends on the nuclear stage (G_1, S, G_2) of the interphase cell at the time of the fusion.

premaxilla   incisive bone; in humans is fused with the maxilla.

pre-messenger RNA (pre-mRNA)   *See:* **nuclear RNA**.

prenatal   before birth.

prenatal diagnosis   the evaluation of the condition of the embryo or fetus before birth.

preoral gut   *See:* **foregut**.

prepupa   a stage in the life cycle of holometabolous insects, marked by cessation of feeding; among Diptera, it is the stage in which the puparium is formed.

presentation   one of four descriptions of the fetus' alignment in the uterus; presentation of the fetus refers to the fetal part leading the way down the birth canal (see Fig. 3). *See:* **attitude, lie, position**.

presumptive   that which is expected during normal development; e.g., it is expected that presumptive neural plate will give rise to the neural plate.

prezygotic isolation   reproductive isolation between populations caused by differences in mate choice or timing of breeding, so that no hybrid zygotes are formed; caused by mechanisms that come into play before or during fertilization, e.g., temporal isolation (as a result of differences in timing of reproduction; 17-year cicadas and 13-year cicadas will release gametes at the same time only once in 221 years), ecological isolation (populations have adapted to different microenvironments; in marine marsh areas, salt-tolerant plants grow well, but just a few feet away, on higher ground, salt-intolerant species will dominate); and behavioral isolation (courtship rituals are recognized by members of a given species but not by members of other species). *See:* **postzygotic isolation**.

primary axis   of the embryo, the animal-vegetal axis. *See:* **animal-vegetal axis**.

primary embryonic induction   classic experiments by Hans Spemann and Hilde Mangold in 1924 provided evidence that the dorsal lip of the amphibian blastopore was the "organizer" of the embryo, responsible for primary embryonic induction, i.e., the induction of the dorsal axis of the embryo. Recent work provides evidence that the organizer is not the *primary* inducer in amphibian development; rather, the three most dorsal vegetal blastomeres of 64-cell *Xenopus* embryos can induce the formation of the dorsal lip of the blastopore. This small cluster of vegetal blastomeres is responsible for allowing their adjacent marginal cells to invaginate and form the dorsal axis of the embryo. *See:* **Nieuwkoop center**.

primary follicle   an early ovarian follicle consisting of a single oocyte and a single layer of follicle cells.

primary hypoblast   the hypoblast formed, from the chick blastoderm, when certain cells migrate individually into the subgerminal cavity; also called polyinvagination islands. *See:* **secondary hypoblast**.

primary inducer   Hans Spemann thought that searching for an inducer molecule was folly; it was discovered that a huge variety of things could induce a neural plate; furthermore, whatever is responsible for induction does not necessarily demand physical contact between the ectoderm and the chordamesoderm.

primary larvae   larvae with body plans and morphology that are distinct from that of the adult, e.g., sea urchin larvae, which are bilaterally symmetrical, but the adults are not. *See:* **larva, secondary larvae**.

primary mesenchyme cells   cells that undergo ingression into the blastocoel of the sea urchin during the first stage of gastrulation; these cells, by forming syncytial cables, give rise to the spicules of the skeleton of the pluteus larva.

primary nephric ducts   *See:* **pronephric ducts**.

primary neurulation   the chordamesoderm induces the ectoderm to form the hollow neural tube; birds, amphibians, and mice undergo both primary (anteriorly) and secondary (posteriorly) neurulation. *See:* **secondary neurulation**.

primary oocyte   an oocyte that has begun meiosis but has not yet completed meiosis I.

primary ossification centers   locations in the forming fetal skull where bone formation begins.

primary sex determination   the determination of the gonads.

primary tissues   plant tissues produced directly by the activity of apical meristems. A root has three primary meristematic regions that give rise to three primary tissues: (1) the procambium gives rise to the primary xylem, vascular cambium (a lateral meristem), and primary phloem; (2) the ground meristem gives rise to the pericycle, endodermis, and cortex; and (3) the protoderm gives rise to the epidermis.

primary transcript   the initial, unprocessed RNA product of a gene.

primary vesicles   the three (in human embryos, form during the fourth week.), then five, (in human embryos, form during the fifth week), early subdivisions of the embryonic brain. The three primary vesicles: prosencephalon, mesencephalon, and rhombencephalon. The five primary vesicles: telencephalon, diencephalon, mesencephalon, metencephalon, and myelencephalon.

primigravida   a woman who is pregnant for the first time.

primitive alveoli   terminal air sacs of the lungs before their linings have become as thin as they will later in lung development.

primitive blood cells   the first blood cells to form; appear first in the blood islands of the yolk sac.

primitive groove   in the chick embryo, the central furrow running down the length of the primitive streak.

primitive gut   the tubular structure in embryos that differentiates into the alimentary canal; in chick embryos, the cavity between the endoderm and the yolk. At 18 hours, the primitive gut is a cavity with a flat roof of endoderm, a floor of yolk, and is bound peripherally on all sides by the germ wall. By 20 hours, the region of the primitive gut underlying the cephalic part of the embryo consists of a pocket of endoderm; this is the first part of the gut to acquire a floor of endoderm (versus yolk) and is called the foregut.

primitive knot   alternative name for Hensen's node.

primitive node   the mound of cells at the cephalic end of the primitive streak. Cells passing through the primitive node during gastrulation give rise to the notochord. Also called Hensen's node.

primitive pit   in the early chick embryo, a depression at the cepahlic end of the primitive groove and just behind Hensen's node.

primitive ridges   the thickened margins of the primitive streak that flank the primitive groove in the chick embryo.

primitive streak   a thickened streak of cells found in the upper layer of the two-layered human embryonic disc. Cells pass through it during gastrulation, giving rise to the germ layers. Other higher vertebrates also develop primitive streaks early in their development. On frog embryos, the primitive streak is a vertical groove formed with the fusion of the lateral lips of the blastopore. On chick embryos, the primitive streak is a thickened region that forms on the area pellucida; at 3–4 hours of incubation, it involves one quadrant of the area pellucida; by 16 hours, this thickened region has undergone cephalocaudal elongation and is a prominent feature of the embryo. At 16 hours, the features of the primitive streak itself include the primitive groove, two primitive ridges, and Hensen's node. The cephalic end of the chick primitive streak is within the area pellucida and is marked by Hensen's node; the caudal end of the streak lies near the area opaca, so one can relate the orientation of the primitive streak to the longitudinal axis of the future embryonic body. The primitive streak of the chick embryo is considered to be the homologue of the fused lips of the blastopore in the amphibian embryo (see Plate 2b in the color insert).

primitive streak stage   the 16-hour chick embryo; when the primitive streak is the most prominent feature of the embryo.

primordial follicle   an early ovarian follicle consisting of a single oocyte and an incomplete layer of follicle cells.

primordial germ cells (PGCs)   migratory germ cells first observed on the yolk sac of mammalian embryos. Once they enter the developing gonad, they are known as oogonia or spermatogonia. In insects before the blastoderm is complete, some of the dividing nuclei pass to the posterior pole of the early embryo to form the primordial germ cells.

primordium   a discrete field of cells that will give rise to a particular organ, appendage, or tissue type. *See:* **anlage**.

Pringsheim, Nathaniel (1823–1894)   German botanist, one of the founders of the scientific study of algae; was the first to observe actual cellular fertilization in plants; saw in 1856 a spermatozoon enter the egg cell of *Oedogonium*; was the first to see a sperm enter a female cell.

proacrosin   a mammalian sperm protein that adheres to the inner membrane of the acrosome and binds to sulfated carbohydrate groups on the zona pellucida glycoproteins.

proamnion   a misnomer, refers to a mesoderm-free area in the cephalic part of the blastoderm of the chick; as mesoderm grows into this area, the proamnion decreases in size.

procambium   the primary meristem that gives rise to the vascular tissue of plants; begins to become visible in the heart-shaped embryo of angiosperms; one of three tissue systems that emerge at the globular stage of angiosperm embryogenesis; the procambium forms the xylem and phloem. *See:* **dermal tissue, ground tissue**.

procaryotes   *See:* **prokaryotes**.

prochordal plate   *See:* **prechordal plate**.

proctodaeum   also called proctodeum; the posterior part of the alimentary canal in embryos, formed by invagination of the anus; in frogs, a pit found at the ventral end of the primitive streak that eventually perforates to form the anus. In insects, a posterior ingrowth of ectoderm that becomes the embryonic hind-intestine.

proctodeum   *See:* **proctodaeum**.

proerythroblast   the first recognizable member of the erythrocyte lineage.

progenesis   the retention of the juvenile form as a result of the gonads and germ line developing at a faster rate than normal, e.g., as occurs in the tropical tree salamander, *Bolitoglossa occidentalis*.

progenitor   an ancestor in the direct line. *See:* **lineage**.

progenitor cell   a current term used for any undifferentiated cell (retained in the adult) that has the potential to replace differentiated cells; progenitor cells, although related to stem cells, are not capable of unlimited self-renewal; they have the capacity to divide only a few times before differentiating; e.g., such cells are found in hematopoiesis, spermatogenesis, and differentiation of neurons.

progeria   *See:* **Hutchinson–Gilford progeria syndrome**.

progesterone the steroid hormone responsible for the maintenance of pregnancy by maintaining the endometrial lining of the uterus.

programmed cell death also called apoptosis; cell death as a result of a genetic program for it and not as a result of a traumatic occurrence.

progress zone a region of cell division, approximately 200 μm in from the apical ectodermal ridge (AER) of the limb bud; the limb bud elongates by means of the proliferation of the mesenchyme cells of this region underneath the AER.

progressive determination cell determination involving cells interacting with one another to determine the fates of one or both of the participants; also called embryonic induction. *See:* **mosaic determination**.

prokaryotes are organisms that are made up of prokaryotic cells, i.e., cells that lack true (membrane-bound) nuclei. Also called procaryotes.

prolactin this anterior pituitary hormone aids in the development of mammary glands and in the formation of milk.

proliferation a basic morphogenetic movement; occurs when a small population of cells multiplies at a rate much greater than neighboring cells.

proliferin (PLF) a factor secreted by the placenta to induce angiogenesis. *See:* **proliferin-related protein (PRP)**.

proliferin-related protein (PRP) a peptide secreted by the placenta to inhibit angiogenesis, when the placental blood vessels have become established. *See:* **proliferin (PLF)**.

promeristem the central region of a meristem containing the initials; the cells capable of continued division.

prometaphase the stage of mitosis or meiosis when the chromosomes have been released from the nuclear membrane but are not yet aligned on the equator of the spindle (see Fig. 29).

promoter the region of a gene near the start site of transcription to which the general transcription machinery binds; the nucleotide sequence in DNA to which RNA polymerase binds to initiate transcription; generally located before (upstream from) structural genes.

pronephric ducts also called primary nephric ducts; the paired ducts of the pronephros, derived from intermediate mesoderm.

pronephric swellings bilateral, and posterior to the dorsal ends of the gill plates, swellings found on the surface of the 5-mm tadpole (*Rana*) embryo; these are indicative of underlying pronephros development.

pronephric tubules the tubular components of the pronephros.

pronephros the most primitive type of kidney that develops during vertebrate embryonic development.

proneural cluster a group of cells from which one or more neural precursor cells will segregate, divide, and differentiate.

proneural genes in *Drosophila*, the genes activated in the neural ectoderm that enable a cell to become a neuroblast; these genes encode the transcription factors Achaete and Scute, among others. *See: MASH1* **gene**.

pronuclei fusion generally, a fertilized egg will contain two haploid pronuclei, one of maternal and one of paternal origin; fusion of these pronuclei marks the completion of fertilization. In the sea urchin, *Arbacia*, actual fusion of the pronuclei occurs; in the nematode, *Ascaris*, actual fusion of the pronuclei does not occur; rather, separate groups of chromosomes are deposited on the first division spindle.

pronucleoli in the mouse, 13 pronucleoli are found in each nucleus at the two-cell stage; by the morula stage, there are 2 definitive nucleoli; in trangenic experiments, it is necessary to avoid pronucleoli when injecting into the male pronucleus.

pronucleus when gametes are formed, the number of chromosomes is reduced to the haploid number; if fertilization occurs, the egg and sperm nuclei, within the cytoplasm of the fertilized egg, will have only the haploid number of chromosomes; such nuclei are called pronuclei. At the time of the second maturation division of oogenesis in the female, the haploid, female, pronucleus is formed; when the nucleus of the fertilizing spermatozoon is in the cytoplasm of the egg, it decondenses and gives rise to the haploid, male pronucleus. In mammals, both the male and female pronuclei are required for normal development. *See:* **genomic imprinting**.

pronymph stage a stage in development of some (e.g., ametabolous) insects, immediately after hatching, bearing the structures that enabled it to get out of the egg; but after this stage, the insect begins to look like a small adult.

prophase   the stage of mitosis or meiosis when the chromosomes are becoming visible but are still contained within the nuclear membrane (see Fig. 29).

prosencephalon   the embryonic forebrain; the most rostral of the three primary brain vesicles. In the frog tadpole, it is possible to subdivide the prosencephalon by passing a plane, transversely, from the anterior side of the anterior choroid plexus to the anterior side of the optic recess; the anterior portion is the telencephalon, and the posterior portion is the diencephalon (see Plate 5 in the color insert).

prosocoel   the cavity of the forebrain in a primitive vertebrate embryo.

prospective fate   what a cell or group of cells will give rise to during normal development.

prospective potency   what a cell or group of cells is capable of giving rise to, revealed by experimental manipulation, often involving transplantation of the cells in question. During development, prospective potency is reduced to prospective fate; i.e., although during early development a group of cells may be able to give rise to a variety of structures under experimental conditions (prospective potency), later in development the same group of cells will be capable of giving rise only to what it would give rise to under normal conditions (prospective fate). *See:* **pluripotent, totipotent.**

prostaglandins   fatty acids that belong to a diverse group of hormones, known as eicosanoids, because they contain 20 carbon atoms; they are chemical signals that have numerous activities in the body, including stimulating contraction of smooth muscle of the female reproductive system.

prostate gland   one of four types of male auxiliary sex glands, the prostate gland makes a major contribution to the liquid portion (seminal plasma) of semen. Prostatic fluid contains citric acid, acid phosphatase, and the metals zinc and magnesium (see Fig. 4).

protamines   basic proteins, tightly compacted through disulfide bonds, that package the DNA of the sperm nucleus.

proteases   are protein-digesting enzymes, e.g., trypsin and metalloproteinases; a protease released by the cortical reaction of sea urchin eggs, breaks the links between the plasma membrane and the vitelline membrane of the egg.

protective cloning   prey cloning in response to a predator; i.e., sand dollar larvae clone themselves in response to predator cues (fish mucus). *See:* **embryo defenses.**

protein kinase C   *See:* **phosphoinositol pathway.**

protein–protein domain   that domain of a transcription factor by which it interacts with another protein.

proteins   the class of organic molecules that plays an important role in almost every cellular activity.

proteoglycans   high-molecular-weight complexes of protein and polysaccharide; major components of the extracellular matrix.

proteome   the complete set of proteins encoded by the genome; the structure and activities of all the proteins an organism can make in its lifetime. *See:* **genome, transcriptome.**

proteomics   the study of proteomes; analysis of the total complement of proteins that make up cells.

Proterozoic   the geological era from approximately 2500 to 544 million years ago; the later of two divisions of Precambian time, during which sponges, sea worms, and other forms of sea life appeared.

prothallial cell   one of the initial cells produced when a fern spore undergoes germination, which gives rise to the fern prothallus; the other cell is the rhizoidal cell.

prothallium   *See:* **prothallus.**

prothallus   also called prothallium; the small, usually flat, thallus-like growth germinating from a spore; the gametophyte generation in the alternation of generations; the free-living, haploid, photosynthetic, gametophyte of fern development (see Plate 15 in the color insert).

prothoracic gland   one of the paired glands in the prothorax of insects; the source of ecdysone, in insect larvae, which is the prohormone precursor of 20-hydroxyecdysone, the active molting hormone. *See:* **corpora allata.**

prothoracicotropic hormone (PTTH)   a peptide hormone, it stimulates the production of ecdysone by the prothoracic gland. *See:* **molting.**

protoblast   the single-cell stage of an embryo; a blastomere from which a definite organ or part is developed.

protocadherins   a class of cadherins that lack the attachment to the actin skeleton through catenins; important in separating the notochord from surrounding tissues during its formation.

protocerebrum *See:* **neuromeres**.

protochordates members of the phylum Chordata that are not vertebrates; namely, urochordates and cephalochordates; formerly, but no longer, included the hemichordates. The development of neural crest cells (and the epidermal placodes that give rise to the sensory nerves of the face) distinguish the vertebrates from the protochordates.

protoderm a primary meristematic tissue that gives rise to epidermis in angiosperm development; already visible as the outer layer of the globular angiosperm embryo.

Proto-Hox gene complex the pre-bilaterian complex of three to four genes, duplicated early in metazoan history, which gave rise to the ancestral *Hox* complex and a sister *ParaHox* complex.

protonema the early, filamentous or platelike gametophyte of mosses and some liverworts, which grows from the germinated spore.

proto-oncogene normal genes that regulate cell proliferation and differentiation; mutations of these genes often result in what are called oncogenes (cancer genes) that play a role in the development of cancer.

protoplasm a heuristic, if somewhat archaic, term referring to the living substance of cells, as opposed to inclusions such as oil droplets, yolk, and pigment granules.

protostomes one of two clades of bilaterians; those animals in which the blastopore gives rise to the mouth and anus formation is secondary; includes most animal phyla other than the phyla Echinodermata and Chordata.

Protostomia a, "so-called," superphylum containing coelomate phyla characterized by spiral cleavage, early acting cytoplasmic determinants, and schizocoely. *See:* **Deuterostomia**.

Prototheria a subclass of mammals, the egg-laying mammals, monotremes; composed of the spiny anteater and the duck-billed platypus.

protozoa animal organisms made up of a single cell; a phylum of unicellular animls, as opposed to the multicellular metazoa, according to traditional classification of animals. Protozoans belong to the kingdom Protoctista, according to the five kingdom classification of organisms. *See:* **metazoa**.

proximal an anatomic term referring to a structure that is close to an anatomic reference point (see Fig. 38). *See:* **distal**.

pseudocoelom a body cavity only partly lined by mesoderm.

pseudocoelomates animals having a body cavity only partially lined by mesoderm, e.g., nematodes (round worms).

pseudogene the remnant of a gene that has been rendered nonfunctional through the accumulation of mutations; a region of DNA that shows extensive similarity to a known gene but that cannot itself function, either because it has lost the signal required for transcription (the promoter sequence) or because it carries mutations that prevent it from being translated into protein.

pseudoglandular period a period during lung development when the lungs resemble an endocrine gland.

pseudohermaphroditism a genetic disorder in which the external genitalia resemble one sex while the gonads are those of the opposite sex.

pseudoplasmodium *See:* **plasmodium**.

pseudopregnancy a condition resembling pregnancy that occurs in some mammals, marked by persistence of the corpus luteum and usually following infertile copulation. In some animals may be induced without copulation.

Pteridophyta ferns and fern allies; the principal groups include the ferns, club mosses, and horsetails.

pterospermous with winged seeds.

Pterygota one of two subclasses of insects; winged insects that are sometimes secondarily apterous; metamorphosis very varied, rarely slight, or wanting. Includes the divisions Exopterygota (= Hemimetabola) and Endopterygota (= Holometabola). *See:* **Apterygota**.

puberty the time period during which changes occur that transform a juvenile into a sexually mature individual.

pubescence (1) puberty, or the coming on of puberty; (2) the presence of fine soft hairs.

pulmonary circulation the portion of the circulatory system that carries blood from the heart to the lungs and back again; that is, the portion of the circulation from the pulmonary trunk leaving the right ventricle to the pulmonary veins returning to the left ventricle.

pulmonary stenosis   a narrowing of the pulmonary trunk, coming from the right ventricle of the heart, as a result of unequal partitioning of the truncus arteriosus.

pulmonary trunk   the main arterial blood vessel of pulmonary circulation that leaves the right ventricle of the heart.

punctuated equilibrium theory   an extreme alternative to gradualism; the two main ideas of this theory are that (1) species change very little most of the time and (2) most anatomical or other evolutionary change in individual species occurs during a geologically brief period at the time of speciation; speciation events punctuate an otherwise stable equilibrium. *See:* **gradualism**.

pupa   an insect during the immature, pupal, stage of development; at this stage, the wings and appendages become everted from their epidermal pouches and become evident externally.

puparium   in the Diptera, the last larval exuviae persists and, undergoing contraction and other changes, becomes a hardened, barrel-like case or puparium that protects the pupa within; the pupal case.

pupation   the process of becoming a pupa.

Purkinje, Jan Evangelista (1787–1869)   Czech physiologist (other than in German and English papers, he spelled his name "Purkyne"); discovered the germinal vesicle in birds' eggs; proposed the word "protoplasm" for formative material of young animal embryos. In 1825, he dedicated one of his first papers to Johann Friedrich Blumenbach: his remarkable study on the origin of the avian egg, demonstrating a "germinal vesicle" in the yolk, later shown to be the nucleus of the cell. Purkyne, in 1834, in his extensive speculations on generation, mentioned the transfer of traits from parents to offspring, although without using the term "heredity" (Orel et al., 1987). He supposed that a process of "involution" reduced the parental traits to a mere quality in the germs of the parents and that, after their fusion, a process of evolution produced the embryo of a new individual, bearing the traits of the parents. Purkyne also studied animal tissue under the microscope, and in 1837, he pointed to the analogy between plant cells and "globules" in the tissue of animals. Two years later, this analogy was generalized by Matthias Schwann (1810–1882) as cell theory. The idea of the cell as a common unit of structure and function in animals and plants became the starting point for new efforts to explain the enigma of generation and, later, that of fertilization.

Purkinje fibers   modified heart muscle fibers that form the terminal part of the heart's conducting system.

Purkyne, Johannes (1787–1869)   *See:* **Purkinje, Jan Evangelista**.

pyloric stenosis   unlike stenosis as a result of abnomal partitioning of a cavity, the narrowing of the pyloric region of the stomach is attributed to thickening (hyperplasia) of the wall of the pylorus.

Q

quadruplets the offspring of a single pregnancy simultaneously carrying four conceptuses; four siblings that happen to occupy the same uterus (womb) at the same time.

quantitative trait a character that exhibits continuous variation in a population.

quantitative trait loci (QTL) genetic loci that contribute to the variation in a quantitative trait in a population.

quiescent inactive, as an insect before a molt.

quiescent cell a cell that is not progressing through the cell cycle, e.g., a cell in G_o.

quiescent center a region of mitotically inactive cells in the center of the apical meristem; quiescent center formation precedes the organization of a root meristem, and it has been suggested that a quiescent center is necessary for meristem organization.

Dictionary of Developmental Biology and Embryology, Second Edition. Frank J. Dye.
© 2012 Wiley-Blackwell. Published 2012 by John Wiley & Sons, Inc.

R

rachischisis also called spina bifida, a condition characterized by fissure, incomplete closure, or absence of vertebrae of the spinal column.

radial cleavage a cleavage pattern that exhibits radial symmetry; at the eight-cell stage, the animal tier of blastomeres is directly above the vegetal tier of blastomeres. This cleavage pattern is a consequence of the longitudinal axes of the mitotic spindles being parallel to or at perpendicular angles to the animal-vegetal axis of the egg. Characteristic of echinoderms. *See:* **spiral cleavage**.

radial patterning in plants, produces three tissue systems (protoderm, ground meristem, and procambium); begins in the globular stage of angiosperm embryogenesis.

radial symmetry that form of symmetry where many planes passing through a given body axis will divide the organism into two mirror-image halves. *See:* **bilateral symmetry**.

Radiata metazoans exhibiting primary radial symmetry.

radiation (1) energy emitted and propagated through space or matter, e.g., X-rays and ultrasound; (2) the evolutionary divergence of a lineage into a variety of forms, often used to describe rapid diversification.

radicle the primary embryonic root of the embryo of a flowering plant.

Ramón y Cajal, Santiago (1852–1934) Spanish neuroanatomist, speculated that in order to orchestrate the connections between the 10^{14} neurons of the human nervous system there had to be specific chemoattractants to help guide the developing neurons; co-recipient of the Nobel Prize in Physiology or Medicine in 1906, "in recognition of his work on the structure of the nervous system."

Rana a genus of anuran amphibians (frogs); used extensively in embryology and developmental biology research and teaching, e.g., *Rana pipiens* (Northern Leopard Frog), *Rana sylvatica* (Wood Frog), and so on.

random epigenetic drift our DNA-methylating enzymes are prone to errors resulting in random epigenetic drift; may have profound effects on physiology; e.g., methylation of the promoter regions of estrogen receptors increases linearly with age, thought to result in "hardening of the arteries;" a characteristic of aging. *See:* **epigenetic progenitor model**.

random segregation of chromosomes during anaphase I of meiosis, members of pairs of homologous chromosomes randomly separate (segregate) from each other. A consequence is that a given gonad in a given individual will produce genetically different gametes (eggs or sperm); increases genetic diversity of gametes and, therefore, of the species. *See:* **crossing over**.

Ras a G protein in the RTK pathway; mutations in the *Ras* gene account for a large proportion of cancerous human tumors.

Rathke, Heinrich (1793–1860) German anatomist; practiced medicine; by the mid-1820s, he discovered the gill clefts in mammals and birds, which emphasized the affinity between embryos of higher and lower vertebrates. He published more than 125 monographs. Rathke's lasting contributions to classic descriptive embryology include (1) his discovery of the transitory gill clefts and associated vascular arches in amniotes; (2) the working out of early elements in the formation of the vertebrate skull; and (3) a clarification of certain fundamental relationships in the vertebrate urogenital system. According to Frederick Churchill (1994), Rathke recognized that in males, of the adder and other amniotes, the efferent ducts of the mesonephros each becomes transformed into a vas deferens coming from a testis and, that as the mesonephros degenerates, some Wolffian tubules develop into an epididymis; he recognized that in males, the Müllerian ducts serve no purpose, but they

Dictionary of Developmental Biology and Embryology, Second Edition. Frank J. Dye.
© 2012 Wiley-Blackwell. Published 2012 by John Wiley & Sons, Inc.

degenerate and disappear; his work clarified the contrasting origins of the vasa deferentia and oviducts and helped place in perspective the temporary nature of the mesonephros for all amniotes. However, Rathke failed to understand that the uterus develops out of a fusion of the two oviducts and remained uncertain regarding the independent origins of the metanephros or true kidney.

Rathke's pocket an evagination from the dorsal surface of the stomodeum that gives rise to the anterior lobe (adenohypophysis) of the pituitary gland. Also called Rathke's pouch and craniobuccal pouch.

Rb *See:* **retinoblastoma protein**.

reaction-diffusion model system of pattern generation; the Turing model; generation of periodic spatial heterogeneity can come about spontaneously when two reactants, S and P, are mixed together under the conditions that S inhibits P, P catalyzes production of both S and P, and S diffuses faster than P. *See:* **Turing, Alan Mathison**.

reaction norm the continuous range of phenotypes expressed by a single genotype across a range of environmental conditions.

reactive oxygen species (ROS) theory of aging this theory considers our metabolism the cause of our aging; i.e., aging is a by-product of normal metabolism. Approximately 2–3% of the oxygen atoms taken up by mitochondria are reduced insufficiently to ROS, which include the superoxide ion, the hydroxyl radical, and hydrogen peroxide, which can oxidize and damage cell membranes, proteins, and nucleic acids.

realisator genes those, as yet undiscovered, genes that are the targets of the homeotic gene proteins, whose function it is to form the specified tissue or organ primordia; the genes that respond to the regulatory genes to produce the body structures.

reawakening of the morphogenetic process there are two modes to reawaken the morphogenetic process at an advanced stage of the ontogenetic process: metamorphosis and regeneration.

Recapitulation Theory one of the widest generalizations of embryology; that animals are supposed, in their individual development, to recapitulate to a considerable degree phases of their ancestral history. It was suggested in the writings of Karl Ernst von Baer and Louis Agassiz, but it received its first clear and complete expression in 1863, in the writings of Fritz Muller, according to William A. Locy (1908). Although the course of events in development is a record, it is, at best, only an imperfect one; stages have dropped out, others are prolonged or abbreviated, or they appear out of chronological order; additionally, some structures have developed from adaptation of a given organism to its conditions of development and, therefore, rather than being ancestral, are recent additions. Again, according to Locy, recapitulation theory was a dominant note in all of Francis Balfour's speculations and received its most sweeping application in the works or Ernst Haeckel.

receptacle the portion of the stalk upon which the flower parts are borne.

receptor a protein molecule, embedded in either the plasma membrane or the cytoplasm of a cell, to which one or more specific kinds of signaling molecules may attach.

receptor gradient *See:* **morphogen receptor gradient**.

receptor proteins proteins that bind other molecules or ions and, as a result of the binding, influence cellular activity, e.g., insulin receptors in the cell surface and progesterone receptors in the cytoplasm.

receptor tyrosine kinase *See:* **RTK (receptor tyrosine kinase) signal transduction pathway**.

recessive descriptive of the allele, of a nonidentical pair of alleles, that is not expressed.

reciprocal embryonic induction the reciprocal nature of many inductive reactions; e.g., the optic vesicle induces lens formation, and the lens, in turn, induces the formation of the optic cup from the optic vesicle.

reciprocity of epigenetic and genetic cancer causation the two mechanisms appear to augment each other; indeed the large number of mutations that accumulate in cancer cells may have an epigenetic cause. *See:* **epigenetic progenitor model, random epigenetic drift**.

recombinant DNA any DNA molecule formed by joining DNA fragments from different sources, e.g., insertion of DNA restriction fragments into replicating DNA.

recombination the appearance in offspring of gene combinations not found in the parents; the process by which DNA is exchanged between pairs of homologous chromosomes during egg and sperm formation.

recombination experiment involves observing the development of the embryo after replacing an original part with a part from a different region of the embryo.

reconstitution   the reaggregation of isolated cells into a new whole animal; a phenomenon discovered in sponges. It has nothing in common with embryonic development as the individual cells into which the organism (sponge) is broken up each retain their specific character; the whole process rests mainly on a rearrangement of the cells in space rather than on a progressive differentiation.

rectovaginal fistula   a fistula between two internal structures, the vagina and the rectum.

recumbent   leaning back.

red blood cells   cells in the circulatory system that lack nuclei, contain hemoglobin, and are responsible for oxygenation of the tissues of the body.

reductionism   the concept that the properties of the whole can be known if all the properties of the parts are known. *See:* **wholist organicism**.

redundancy   refers to redundant, i.e., serving as a duplicate for preventing failure of an entire system upon failure of a single component; as the results of knockout techniques have demonstrated large numbers of null effects, the consensus has developed that they clearly indicate the existence of widespread functional redundancy in genetic pathways. *See:* **knockout, null mutation**.

regeneration   the replacement of a lost part, as, for example, in salamanders from which limbs have been removed and in planaria from which parts have been removed.

regeneration blastema   *See:* **blastema**.

regenerative medicine   the process of creating living, functional tissues to repair or replace tissue or organ function lost as a result of age, disease, damage, or congenital defects.

region-specific selector genes   a class of selector genes that regulates the identity of contiguous body regions (e.g., *Hox* genes).

regional specificity   how a pattern appears in a previously similar population of cells; as applied to embryonic induction, has to do with the regional specificity of the neural structures that are produced; e.g., forebrain (archencephalic), hindbrain (deuterencephalic), and spinocaudal regions of the neural tube must all be properly organized in an anterior-to-posterior direction. It was once theorized that this regional specificity was caused by the interaction of two substances secreted by the cells of the chordamesoderm; evidence for this came from studies involving artificial tissue-specific inducers.

regulation   the ability of embryonic cells to change their fates to compensate for missing parts of the embryo or to compensate for additional parts from an extra embryo(s) as in the case of chimeras; regulation is also exhibited by slime molds, such as *Dictyostelium discoideum*.

regulative development   that type of development where the early embryo can compensate for the loss of part of the embryo; typically exhibited by echinoderm embryos, mammalian blastomeres also exhibit regulative development. The consequences of this type of development include the existence of identical twins and the ability to do preimplantation embryonic diagnosis. In this type of development, blastomeres are determined through their interactions with other cells; cytoplasmic differences between blastomeres are not great. Prospective potency is greater than prospective fate. The two major aspects of regulation are (1) the potency of an isolated blastomere is greater than its normal embryonic fate and (2) rearranged blastomeres develop according to their new locations; these hold true for the early stages of sea urchin cleavage. Eventually the blastomeres become committed to certain fates. Note that regulative development and mosaic development are the poles of a continuum. *See:* **mosaic development**.

regulatory circuit   a signaling pathway and one or more of the target genes regulated by the pathway in a given cell, tissue, or field. *See:* **hierarchy**.

regulatory evolution   evolutionary changes in gene regulation.

regulatory genes   genes that regulate the transcription of structural genes; code for regulatory proteins that bind to the regulatory sites of structural genes; mutations in regulatory genes can create large changes in morphology. It is likely that changes in development resulting in major morphological changes may involve only a relatively small number of changes in the crucial regulatory genes. *See:* **transcription factor**.

Reichert's membrane   cells from the early hypoblast (primitive endoderm) of the mouse spread out beneath the trophoblast (trophectoderm); once associated with the trophoblast, these cells secrete a thick basement membrane known as Reichert's membrane; because of its thickness and accessibility, Reichert's membrane has been used in studies of basal laminae.

relational pleiotropy   a type of pleiotropy in which several developing tissues are affected by the mutation even though they do not express the mutated gene; e.g., failure of the MITF expression causes aberrant differentiation of the retina, which causes malformation of the choroid fissure of the eye, which causes drainage of vitreous humor fluid, which causes microphthalmia. *See:* **mosaic pleiotropy.**

Remak, Robert (1815–1865)   Polish–German embryologist and anatomist; a student of Johannes Peter Müller in Berlin. In the mid-1840s, in collaboration with Müller, Remak made a major revision to the orthodox embryology of Karl Ernst von Baer. They reduced the four germ layers of von Baer to three, by taking the two middle layers as only one, and thus arrived in 1845 at the fundamental concept of three layers, ectoderm, endoderm, and mesoderm. He established the cellular nature of the egg. According to Frederick Churchill (1994), Remak, in the early 1850s, brought cell theory and germ layer doctrine together. He thought in terms of cell lineages in the context of germ layers providing the means for organizing histological destinies. A pioneer embryologist, he was one of the first to depict cell division in full, and to hold that all animal cells came from preexisting cells.

renal agenesis   the lack of formation of one or both kidneys; the latter causing death soon after birth; caused by the failure of metanephric diverticulum development.

renal tubules   the glandular tubules that elaborate the urine in the kidney.

replication   the making of copies of something, e.g., DNA.

reporter gene   a gene whose expression is used to visualize the activity of a heterologous, linked, *cis*-regulatory element *in vivo*; a gene that encodes an easily assayed product that is coupled to the upstream sequence of another gene and transfected into cells. The reporter gene can then be used to see which factors activate response elements in the upstream region of the gene of interest.

reporter protein   the easily assayed protein product of a reporter gene, e.g., green fluorescent protein. *See:* **green fluorescent protein, reporter gene.**

repressor   a transcription factor that negatively regulates the expression of a gene, often by binding directly to DNA sequences in a *cis*-regulatory element.

reproduction   the creation of new individuals; the complex process by which organisms give rise to offspring; may occur in the absense of sex, as when an amoeba divides. *See:* **asexual reproduction.**

reproductive cloning   cloning for reproductive purposes. *See:* **therapeutic cloning.**

reproductive duct: (1) in males, consists of two main portions, the contorted epididymal duct and the less contorted vas deferens; (2) in females consists of a pair of muscular tubes extending from anterior ostia to the cloaca.

Reptilia   the class of vertebrates composed of reptiles; includes four extant orders, Chelonia (turtles and tortoises), Rhynchocephalia (the New Zealand tuatara), Squamata (snakes and lizards), and Crocodilia (alligators and crocodiles).

resact   *See:* **sperm-activating peptides.**

research cloning   *See:* **therapeutic cloning.**

resegmentation   during formation of the vertebrae, the rostral segment of each sclerotome recombines with the caudal segment of the next anterior sclerotome to form the vertebral rudiment.

resorbed   *See:* **resorption.**

resorption   to break down and assimilate something previously differentiated, e.g., a minute proportion of spermatozoa that are not expelled to the exterior by ejaculation are thought to be resorbed by the male genital tract.

respecification   a phenomenon, observed during epimorphosis, by which there is dedifferentiation of adult structures to form an undifferentiated mass of cells that then becomes respecified; e.g., respecification occurs during salamander limb regeneration.

respiratory distress of the newborn   *See:* **hyaline membrane disease, respiratory distress syndrome.**

respiratory distress syndrome   symptoms, possibly including death, resulting from incomplete lung development at the time of delivery.

respiratory system   in the chick embryo, the first indication of the respiratory system is a midventral groove in the pharynx, just posterior to the fourth pharyngeal pouches, the laryngotracheal groove, which deepens rapidly and becomes separated from the pharynx except at its cephalic end; the tube thus formed is the trachea, and its opening, which persists into the pharynx, is the glottis. This describes the origin of only the epithelial lining of the trachea; supporting structures come from surrounding mesenchyme. The tracheal evagination grows caudad and bifurcates to form the pair of lung buds; the endodermal buds give rise to the epithelial lining of the bronci and air passages and chambers of the lungs, while the connective tissue stroma of the lungs comes from mesenchyme and the pleural coverings of the lungs from splanchnic mesoderm. During human development, the laryngotracheal groove develops from the floor of the pharynx during the fourth week, and the resulting tracheal tube divides during the sixth week to form the lung buds.

responder   during induction, the tissue being induced.

restitution nucleus   a single nucleus with an unreduced chromosome number as a result of a failure of meiosis or a single nucleus with the tetraploid number of chromosomes as a result of a failure of mitosis.

restriction of cell potency   in 1936, Hans Spemann suggested a "somewhat fantastical" experiment, to transfer a nucleus from a differentiated cell into an enucleated egg. In 1952, Robert Briggs and Thomas J. King developed the techniques for the frog (*Rana*); they transferred blastula stage cell nuclei into enucleated eggs and were able to obtain complete tadpoles. Spemann had already shown that blastula cell nuclei were pluripotent; what about nuclei from more advanced stages? In 1956, Briggs and King used somatic cell nuclei of tailbud-stage tadpoles (vs. germ cell nuclei of tailbud-stage tadpoles); they demonstrated (1) an apparent decrease in nuclear potency, and (2) apparently, the decrease in nuclear potency was stable and tissue specific, as demonstrated by nuclear cloning; e.g., endoderm nuclei seem to be good at making endoderm, but they are restricted in making ectoderm and mesoderm. Progressive restriction of nuclear potency during development seemed to be the general rule. However, another possible explanation for limited potency would be chromosome abnormalities. Sir John Bertrand Gurdon, using the frog *Xenopus*, used intestinal endoderm of feeding tadpoles and seemed to demonstrate that these nuclei were totipotent. However, King criticized Gurdon's work; subsequently, Gurdon cultured epithelial cells from adult frog foot webbing and by serial transplantation was able to obtain tadpoles, but these died before feeding. These amphibian cloning experiments suggested (1) a general restriction of potency occurred with progressive development and (2) the differentiated cell genome was remarkably potent. Debate persisted about the totipotency of such nuclei, but there was little doubt that they were extremely pluripotent.

rete cord   one of the deep, anastomosing strands of cells of the medullary cords of the vertebrate embryo, which form the rete testis or the rete ovarii.

rete ovarii   vestigial tubules or cords of cells near the hilus of the ovary, corresponding with the rete testis, but not connected with the mesonephric duct.

rete testis   the network of anastomosing tubules in the mediastinum testis (see Fig. 45).

reticular lamina   a loose extracellular layer formed by mesenchymal cells; together with the basal lamina, it constitutes the basement membrane.

reticulocyte   a member of the erythrocyte lineage, between the earlier erythroblast and the later erythrocyte; a reticulocyte forms when an erythroblast loses its nucleus. Reticulocytes do not synthesize globin mRNA, but they do translate existing messages into globin.

retina   the portion of the eye derived from the inner layer of the optic cup containing the photoreceptor cells and from which the optic nerve originates.

retinoblastoma   a malignant tumor of the retina that generally affects children less than 6 years of age.

retinoblastoma protein (Rb)   one factor, maintaining the G_0 state of the cell cycle; it becomes phosphorylated, and hence deactivated, in the presence of growth factors. In the absence of Rb, a transcription factor called E2F becomes active and initiates a cascade of gene activation culminating in the resynthesis of cyclins, Cdks and other components necessary to initiate the S phase of the cell cycle.

retinoic acid   a derivative of vitamin A; a small-molecule inducing factor, which can enter cells freely by diffusion; its receptor is itself a transcription factor, which upon binding retinoic acid translocates to the nucleus to activate its target genes. Retinoic acid is a morphogen controlling pattern formation in chick and amphibian limbs; it is also likely to play a role in posteriorizing the neural tube; it has been shown to activate the expression of more posterior *Hox* genes. Homeotic changes are observed when mouse embryos are exposed to teratogenic doses of retinoic acid. Human infant anomalies caused by inadvertent exposure to retinoic acid include absent or defective ears, absent or small jaws, cleft palate, aortic arch abnormalities, thymic deficiences, and abnormalities of the central nervous system. Inside the developing embryo, vitamin A and 13-*cis*-retinoic acid become isomerized to the developmentally active forms of retinoic acid, all-*trans*-retinoic acid, and 9-*cis*-retinoic acid. *See:* **retinoic acid receptors (RARs)**.

retinoic acid receptors (RARs)   a group of transcription factors, active only when they have bound retinoic acid; the RARs bind to specific enhancer elements in the DNA called retinoic acid response elements. Some *Hox* genes have retinoic acid response elements in their promoters.

retroviral vector   a retrovirus genetically engineered and used to carry genes into cells (and chromosomes) intentionally infected with the virus.

retrovirus   a virus with RNA (rather than with DNA) as its genetic material, e.g., HIV virus. After infection of a cell, the viral genome is reverse transcribed to DNA that becomes integrated into the host genome; retroviruses have been used to introduce genes into embryos.

reverse transcriptase   an enzyme used to convert mRNA into complementary DNA (cDNA), which may be used as a probe for Northern blots or for *in situ* hybridization.

Rh+, Rh−   symbols for the dominant and recessive alleles, respectively, of the Rh factor gene.

Rh incompatibility   if a pregnant woman, who is Rh−, is carrying a fetus, who is Rh+, an Rh incompatibility exists and the mother's immune system may mount an immune attack against the fetus' red blood cells. *See:* **erythroblastosis fetalis**.

rhizoid   a root-like structure lacking conductive tissues (xylem and phloem).

rhizoidal cell   one of the initial cells produced when a fern spore undergoes germination, which gives rise to a rhizoid of the fern prothallus; the other cell is the prothallial cell.

rhizome   a horizontal underground stem.

RhoB protein   may be involved in establishing the cytoskeletal conditions that promote neural crest cell migration.

rhombencephalon   the embryonic hindbrain; gives rise to the metencephalon and the myelencephalon (medulla oblongata). In the frog, the rhombencephalon includes the metencephalon and the medulla oblongata; in higher vertebrates, but not the frog, the rhombencephalon is also subdivided (see Plate 5 in the color insert).

rhombocoel   cavity of the rhombencephalon.

rhombomeres   periodic swellings that divide the rhombencephalon into smaller compartments; represent separate developmental "territories"; each rhombomere forms ganglia.

ribonucleic acid (RNA)   a type of nucleic acid; three important kinds are found in cells: messenger RNA, ribosomal RNA, and transfer RNA. RNA molecules are responsible for the translation of the genetic message into specific protein molecules. *See:* **microRNAs (miRNAs), siRNA**.

ribosomal RNA (rRNA)   a specific family of RNA molecules that, together with protein molecules, make up ribosomes, the protein-making organelles of cells.

ribosomes   are organelles found in the cytoplasm of cells and concerned with the synthesis of protein; i.e., messenger RNA molecules are translated by ribosomes. Amphibian oocytes produce as many as 10^{12} ribosomes.

ring canals   connections between the clone of cells derived from the oogonium by meroistic oogenesis in insects.

RNA interference (RNAi)   is a system within living cells that helps to ensure that genes are active as well as how active they are. Two types of small RNA molecules—microRNA (miRNA) and small-interfering RNA (siRNA)—are central to RNA interference. RNAs are the direct products of genes, and these small RNAs can bind to specific other RNAs and either increase or decrease their activity, for example, *by preventing a messenger RNA from producing a protein.* RNA interference has an important role in *directing development* as well as in gene expression in general; a naturally occurring phenomenon in which double-stranded RNAs (dsRNAs) lead to degradation of mRNAs having an identical sequence; RNAi is believed to function primarily in blocking the replication of viruses and in restricting the movement of mobile elements. *See:* **siRNA**.

RNA interference pathway   a cellular pathway of gene silencing in a sequence-specific manner at the messenger RNA level.

RNA polymerase   an enzyme that produces RNA.

RNA processing   modifications of primary RNA transcripts, including splicing, cleavage, base modification, capping, and the addition of poly-A tails; i.e., messenger RNA is not the primary transcript of a gene, but it is created from the primary transcript by processing.

RNA splicing   that part of RNA processing that splices out the introns from nuclear RNA. *See:* **alternative RNA splicing**.

robustness   *See:* **canalization**.

roof plate   the middorsal wall of the embryonic neural tube; the dorsal part of the embryonic neural tube. *See:* **floor plate**.

Rosetta stone   a block of black basalt bearing inscriptions, Egyptian hieroglyphics, Egyptian demotic (cursive development of hieroglyphic script), and Greek that eventually supplied the key to the decipherment of the Egyptian hieroglyphic script. Found in 1799, by Napoleon Bonaparte's solders near Rosetta, Egypt. The final breakthrough was by Jean-François Champollion in 1822. *See:* **homeobox**.

rostrad   in a rostral direction.

rostral   means "beak" and is an anatomic term often used to refer to positions of structures in the head for which the term "cephalic" would not be precise.

rotation   the turning of an object on its axis; plays a role in various aspects of development: e.g., (1) during frog gastrulation, the embryo undergoes rotation as a consequence of a shift in the relative position of yolk, which is relatively heavy, and, therefore, in the embryo's center of gravity; (2) when a frog egg is fertilized, the formation of a fluid-filled perivitelline space around the egg allows the egg to rotate so the heavier, yolk-filled, vegetal hemisphere will be beneath the lighter animal hemisphere; and (3) the male pronucleus and its associated centrosome often rotate as they move in the egg cytoplasm to meet the female pronucleus.

rotational cleavage   characteristic of mammalian cleavage, the two mitotic spindles of the second cleavage (from the two-cell to the four-cell stage) are at right angles to each other; one blastomere divides meridionally, and the other divides equatorially, so that at the four-cell stage, the two pairs of blastomeres are oriented at right angles to each other. Traditionally, mammalian cleavage has been difficult to study for (1) mammalian eggs are small (approximately 100 μm in diameter), (2) usually fewer than 10 eggs are ovulated at a time, and (3) development occurs within another organism; recently these difficulties have been minimized by improved microscopic techniques, superovulation, and embryo culture. The distinguishing features of mammalian cleavage are (1) relative slowness, (2) unique orientation of mammalian blastomeres, (3) marked asynchrony of early division, (4) compaction, and (5) the formation of a blastocyst.

Roux, Wilhelm (1850–1924)   German embryologist; a student of Ernst Haeckel, he is considered to be a pioneer in, and, by some, the founder of experimental embryology. According to Jane Maienschein (1994), because Wilhelm His had called for a "physiology of development," and Eduard Pflüger, Gustav Born, and others had borrowed from physiology to pursue an experimental approach to embryological questions, Roux with his program of *Entwicklungsmechanik* was not, therefore, doing anything completely new and different; and, further, Roux was considered the leader of the pack primarily because of his polemics in favor of a new program and his institutional successes.

Roux visualized a complex machine already functional in the fertilized egg that suggested preformation. To be transmitted from cell to cell, Roux reasoned, the machine must in some way break down into its component parts during the process of cell division, so the parts can be transmitted to another cell and there reassemble in some way and again become functional.

Based on his experimental evidence, of rotating eggs in their gravitational field and finding that the eggs develop normally, Roux, unlike Pflüger, concluded that eggs are self-differentiating rather than being driven by external conditions. By 1885, Roux was generating a general theory about the causes of embryonic development based on his idea of self-differentiation. He suggested that the nucleus holds all of the qualities for individual formation; offering a theory of qualitative cell division; i.e., each division actually separates off differential nuclear materials into the different daughter cells; he said the process is like producing a mosaic.

His famous experiment, the so-called half-embryo experiments in 1888, which seemed to demonstrate preformation, was carried out by killing, with a hot needle, one of the two blastomeres, of a frog embryo; only one half of an embryo developed, and Roux concluded that determination had already occurred in the two-cell stage. He recognized that his view was the leading alternative to Pflüger's hypothesis that external conditions cause differentiation.

Roux inaugurated in 1894 the physiological approach to embryology (*Entwicklungsmechanik*, developmental mechanics). Roux insisted that embryology would no longer be the servant of evolutionary studies, but embryology would assume its role as an independent experimental science. He believed that many conclusions drawn from the investigation of ontogeny (development) would throw light on the phylogenetic (evolutionary) process. Roux implied that he considered experimentation the only legitimate method for biological science. He founded his own new journal, *Wilhelm Roux's Archiv für Entwicklungsmechanik der Organismen* (*Roux's Archives of Developmental Biology*), in which he was noted for exercising a heavy editorial hand.

Roux's dictionary of embryological terms, *Terminologie der Entwicklungmechanik der Tiere und Pflanzen*, appeared in 1912.

Roux himself, in 1881, had in *Der Kampf der Theile* expressed the idea of mutual interaction between embryonic parts. *See:* **Chabry, Laurent**; **Driesch, Hans**; **Weismann, August**.

royal jelly a protein-rich food, which results in queen formation if fed to a honeybee larva for most of its larval life.

RTK (receptor tyrosine kinase) signal transduction pathway an intracellular signal transduction pathway important in development; was one of the first pathways to unite various areas of developmental biology. In migrating neural crest cells of humans and mice, the RTK pathway is important in activating the microphthalmia transcription factor (Mitf) to produce pigment cells.

RU486 (mifepristone) a synthetic steroid that blocks the progesterone receptor from binding progesterone, which stops the uterine wall from thickening, and prevents the implantation of a blastocyst; used as an alternative to surgical abortion.

rubella virus or German measles virus, is a teratogenic virus.

ruffled membranes *See:* **lamellipodia**.

Runnström, John (1888–1971) was Horstadius' advisor and proposed that the most logical explanation of Horstadius' results is that of two opposed gradients; the dual gradient model is a model for regulative development based on relative concentration gradients within the oocyte.

S

S   a stage of the cell cycle, during which DNA synthesis occurs; constitutes part of interphase.

sacculus   the lower, saclike, chamber of the membranous labyrinth of the ear.

sacral neural crest   that portion of the neural crest posterior to the trunk neural crest; together with the vagal (neck) neural crest, generates the parasympathetic (enteric) ganglia of the gut.

sagittal plane   that plane of the body that separates the left side of the body from the right side.

salamanders   animals belonging to the order Urodela of the vertebrate class Amphibia. Have been used for teaching and research in embryology and developmental biology for a variety of reasons, including their considerable powers of limb regeneration. Salamanders were used by Hans Spemann and Hilde Mangold in their experiments on the organizer, reported in their classic 1924 paper.

salivary glands   in insects, develop as paired ectodermal ingrowths at the sides of the labial segment; during development, they become drawn into the mouth cavity, where they fuse and open by a common duct on the hypopharynx. The salivary glands of *Drosophila* are often used as sources of polytene chromosomes.

sarcoma   a general term describing a cancer, that develops from transformed connective tissue cells, such as bone, cartilage, and fat cells, which originate from embryonic mesoderm.

sarcomeres   contractile units of skeletal muscle fibers made up of highly ordered arrays (reflected in their striated appearance) of contractile proteins.

satellite cells   small cells, associated with skeletal muscle fibers, which retain the capacity for cell division and may contribute to the substance of muscle fibers. Satellite cells were identified more than 40 years ago through electron microscopy. Currently, it is widely believed that satellite cells are the commited stem cells of adult skeletal muscle; their major function is to repair, revitalize, and mediate skeletal muscle tissue and growth by differentiating into myocytes; satellite cells are normally nonproliferative; they do become active, however, when skeletal muscle tissue is injured or heavily used during activities such as weight lifting or running. Satellite cells are located at the surface of the basal lamina of the myofiber.

scaphocephaly   a long, narrow skull, resulting from early closure of one of the sutures of the early skull.

scarification   the requirement for scratching or etching of the hard protective seed coats to break dormancy in some seeds.

scatter factor   a protein secreted by cells that enters the chick primitive streak, causes the breakdown of the basal lamina; converts epithelial cells into mesenchymal cells.

-schisis   this word part refers to a cleft or split; several important birth defects include such an abnormality, e.g., craniorachischisis.

schizocoely   formation of the coelom by splitting of the mesodermal layer.

Schwann, Matthias (1810–1882)   German naturalist; according to Frederick Churchill (1994), Schwann's cell theory was above all else an embryological theory of the fine structure of the body; normal differentiation of tissues and organs followed a normal *Entwicklungsgeschichte* (developmental history) of the constituent cells; abnormal growths and neoplasias followed an abnormal cellular *Entwicklunsgeschichte*. It was no accident that both Schwann and Johannes Peter Müller quickly saw the implications of cell theory for human pathology.

Dictionary of Developmental Biology and Embryology, Second Edition. Frank J. Dye.
© 2012 Wiley-Blackwell. Published 2012 by John Wiley & Sons, Inc.

Schwann cells   glial cells outside the central nervous system, which originate from the neural crest and myelinate nerves outside the central nervous system.

sclera   the tough outer coat of the eye.

sclerotium   the multinucleate, dormant, stage in the life cycle of the slime mold *Physarum plasmodium.*

sclerotome   the portion of the somite that gives rise to the vertebrae of the vertebral column. In the frog, sclerotomal cells from the inner and ventral edges of myotomes form a layer about the notochord and nerve cord, ultimately giving rise to cartilage and then bone of centra, transverse processes, and neural arches of vertebrae; in the frog, nine vertebrae are formed and the last two somites give rise to the urostyle (see Fig. 41).

scoliosis   abnormal, lateral curvature of the spine.

scrotum   the pouch containing the testes, which, with the penis, makes up the male external genitalia of most mammals. In monotremes, some insectivores, elephants, whales, and certain other mammal scrotal sacs do not develop and the testes remain permanently in the abdomen. The scrotal testis is between 4 and 7 °C cooler than the general body temperature (see Fig. 4).

Scute   an example of a transcription factor encoded by proneural genes. *See:* **achaete, proneural genes**.

scutellum   a single cotyledon found in, for example, corn grains; it absorbs nutrients from the surrounding endosperm.

seasonal polyphenism   polyphenism controlled by the seasons, e.g., the spring morph and the summer morph of the European map butterfly, *Araschnia levana*.

sebaceous glands   glands of the skin that make and secrete sebum.

sebum   the secretion of sebaceous glands, composed of fat, keratohyalin granules, keratin, and cellular debris.

second messengers   molecules that participate in relaying signals from receptors on the cell surface to target molecules inside the cell, in the cytoplasm, or in the nucleus; they relay the signals of hormones like epinephrine (adrenalin), growth factors, and others, and cause some kind of change in the activity of the cell. They greatly amplify the strength of the signal; secondary messengers are a component of signal transduction cascades. Examples include cyclic adenosine monophosphate (cAMP) and calcium ions.

second polar body   a tiny cell produced by the secondary oocyte when it undergoes meiosis II.

secondary axis   in frogs, an imaginary line joining the center of the gray crescent to the opposite side of the egg; also called the dorsoventral axis.

secondary embryonic induction   secondary embryonic induction results in the fate of cells being influenced by adjacent cells; interactions by which one tissue influences another to direct its fate specifically; important for at least two different reasons: (1) allows the embryo to position differentiated cells in precise locations in the embryos: e.g., optic cup directs a lens to form from ectoderm that is already predisposed to form a lens; and (2) sequential inductive events contribute to producing diverse cell types from relatively few distinct precursor cells (optic cup induces lens; lens induces cornea).

secondary fields   discrete units of development that are specified in the developing embryo after the primary axes are established and that give rise to appendages and organs; pattern formation within secondary fields occurs independently of other fields.

secondary follicle   an ovarian follicle with more than one layer of follicle cells but not yet containing antral vacuoles.

secondary hypoblast   in chick development, when a sheet of cells from the posterior margin of the blastoderm migrates anteriorly to join the polyinvagination islands, the secondary hypoblast is formed. The hypoblast does not contribute any cells to the embryo, but it forms portions of the extraembryonic membranes; in chick development, the secondary hypoblast directs the formation and directionality of the primitive streak. *See:* **primary hypoblast**.

secondary larvae   larvae that posses the same basic body plan as the adult, e.g., caterpillars and frog tadpoles, both of which have the same axes and are organized on the same pattern as the adult. *See:* **larva, primary larvae**.

secondary mesenchyme cells   cells that emerge from the innermost tip of the wall of the archenteron in sea urchin gastrulae and that, by contraction of their filopodia, are attached to the inside of the wall of the blastocoel, affect the third stage of invagination. These cells give rise to mesodermal organs. *See:* **gastrulation**.

secondary neurulation   a solid cord of cells sinks into the embryo and, then, hollows out (cavitates) to form the neural tube; neurulation in fishes is exclusively secondary. *See:* **primary neurulation**.

secondary sex determination   that which affects the bodily phenotype outside the gonads.

secondary tissues   plant tissues produced by the activity of lateral meristems, e.g., cork.

secondary vesicles   the five early subdivisions of the embryonic brain. In human embryos, form during the fifth week. *See:* **primary vesicles**.

seed   a sexual reproductive structure formed by seed plants, consisting of the plant embryo (derived from fertilization of the egg), endosperm (nutritive tissue, derived from fertilization of the binucleate central cell of the embryo sac), and seed coat (derived from the integuments that surround the nucellus) (see Plate 11 in the color insert).

seed coat   the protective covering of the seed that develops from the integuments enclosing the megasporangium.

seed dormancy   when a mature seed is released from the plant, it becomes dehydrated and enters a period of dormancy. The mature embryo is developmentally arrested—metabolism is minimal, and growth of the embryo does not resume until suitable conditions of water, oxygen, temperature, and light are present. The plant hormone, abscisic acid, has been implicated in this developmental arrest.

seed germination   the resumed development of a seed, including the resumption of development of the plant embryo, which occurs when seed dormancy is broken; the postembryonic phase of plant development begins with germination. The environmental conditions that break seed dormancy and allow the seed to germinate vary from one species of plant to another.

Seessel's pocket   *See:* **foregut**.

segment-polarity genes   patterning genes in *Drosophila* concerned with patterning segments and parasegments; they fix the positions of the parasegment boundaries and establish the final segment boundaries of the larval epidermis. The effect of their mutant forms, generally, is to upset the anteoposterior polarity of the segments. They are a diverse group of genes that bear no obvious relation to each other in their protein products or mechanism of operation. Segment polarity genes are activated in response to pair-rule gene expression; each is expressed in 14 transverse stripes; these genes act in a cellular rather than in a syncytial environment. Segment polarity genes are expressed in restricted domains within each parasegment. Maintenance of a parasegment boundary depends on an intercellular signaling circuit being set up between adjacent cells on either side of the boundary and involves interactions between segment polarity genes. Some segment polarity genes (*hedgehog* and *wingless*) encode secreted signaling proteins, whereas others (*patched*) encode receptors for these signaling molecules and components of the signaling pathways involved in maintaining parasegment and segment boundaries.

segmental plate   the bands of, as yet, unsegmented paraxial mesoderm in bird embryos.

segmental zones of mesoderm   in the early chick embryo, the thicker (than it is further laterad) mesoderm adjacent to the midline of the embryo; at approximately 22–23 hours, the first somites will begin to form out of this mesoderm.

segmentation   (cleavage) In insects, early in the development of the embryo (two-layered germ band), it becomes divided (segmentation) by transverse furrows into a series of segments that ultimately number 20 in all. Segmentation is a gradual process beginning anteriorly and extending backward. The embryo is at first divisible into a protocephalic or primary head region and into a protocormic or primary trunk region. As development progresses, the first 3 protocormic segments become added to the protocephalic region. The next 3 body segments are grouped to form the thorax, and the remaining 11 segments, together with the telson, constitute the abdomen. Each of these, except the first segment and the tail piece or telson, develops a pair of outgrowths or embryonic appendages: (1) the first (ocular) segment is formed of the large procephalic lobes; (2) the first pair of appendages or antennae belongs to the second segment, while the very small second pair of appendages is transitory and soon disappears; (3) the third, fourth, and fifth pairs of appendages grow, respectively, into the mandibles, maxillae, and labium; (4) pairs six, seven, and eight of embryonic appendages are usually larger and more conspicuous; they are the forerunners of the thoracic legs; and, finally, (5) there follow 11 more pairs of abdominal appendages, of which the last pair becomes the cerci and the remaining pairs are usually resorbed before hatching. The presence of these evanescent limb rudiments can only be interpreted as being an indication of a many-legged ancestral stage. In *Drosophila*, the first visible signs of segmentation in the embryo are transient grooves that appear on the surface of the embryo after gastrulation; these grooves define the parasegments, of which there are 14 and they are the fundamental units in segmentation of the embryo. *See:* **cleavage**.

segmentation cavity   the blastocoel. *See:* **blastocoel**.

segmentation genes   in *Drosophila*, these genes mediate the transition from specification to determination; these genes divide the early embryo into a repeating series of segmental primordia along the anterior-posterior axis.

segmentation nucleus   *See:* **cleavage nucleus**.

segments   a series of morphologically similar units comprising the body of an organism. In *Drosophila*, the cell lineage restriction within the segments is between anterior and posterior regions; the segment is thus divided into anterior and posterior compartments, with *engrailed* expression defining the posterior compasrtment. *See:* **parasegments**.

segregate   means "to separate."

selective affinity   the sorting out of cells of different germ layers (or derivatives of a single germ layer, e.g., neural plate and epidermis) from each other after they are artificially aggregated; the cells sort themselves out into their proper embryonic positions; selective affinities change during development.

selector gene   in *Drosophila*, a gene that confers a particular identity on a region or regions (i.e., a gene whose activity is sufficient to cause cells to adopt a particular fate), by controlling the activity of other genes, and which continues to act for an extended period. Selector genes can control the development of a region such as a compartment and, by controlling the activity of other genes, give the region a particular identity. Selector genes must be switched on continuously throughout development to maintain the required phenotype. Selector genes of the Antennapedia and bithorax complexes are expressed in the ectoderm and in the underlying somatic and visceral mesoderm, but not in the endoderm from which the gut develops. A selector gene is a gene that controls cell fate. *See:* ***engrailed*, visceral mesoderm**.

self-determination   determination not dependent on interactions between groups of cells, but, rather, as a result of morphogens contained within the cells themselves, *See:* **Spemann, Hans**.

self-incompatibility   the failure of pollen from an individual plant to germinate and/or grow on the stigma of the same plant. *See:* **intraspecific incompatibility**.

self-pollination   transfer of pollen from the anthers to the stigma of the same flower or to the stigma of another flower on the same plant.

semaphorin proteins   also, in addition to the ephrin proteins, guide axon growth cones by selective repulsion; observed throughout the animal kingdom, they are responsible for steering many axons to their targets. Semaphorins are a large family of chemorepellents.

semelparous   describes a species or population in which individuals experience only one bout of reproduction over the course of a typical lifetime, e.g., salmon.

semen   a seed; the complex mixture of cells (spermatozoa) and liquid (seminal plasma) that makes up the product of the male reproductive system. The average pH of human semen is basic, between 7.2 and 7.8, reflecting the relative contibutions of the acidic prostatic fluid and alkaline secretion from the seminal vesicles.

semicircular canals   those three portions of each inner ear that allow us to orient ourselves in space and thereby maintain equilibrium (see Plate 13 in the color insert).

semiconservative replication   a method of copying something in which each new copy consists of a half of a previous copy, which has acted as a template; e.g., when DNA replicates, it undergoes semiconservative replication.

semi-dominant mutation   a mutation that affects the phenotype when just one allele carries the mutation but where the effect on the phenotype is much geater when both alleles carry the mutation.

seminal plasma   the fluid portion of semen. In humans, the bulk of the seminal plasma comes from the seminal vesicles (60%) and the prostate gland (30%).

seminal receptacles   in female insects, seminal receptacles receive sperm from the male and store it for fertilization.

seminal vesicles   one of four kinds of male auxiliary sex glands in humans; these paired glands make a substantial contribution to semen. In male frogs, seminal vesicles store sperm. Seminal vesicle secretion is rich in fructose and, in many species, the long-chain fatty acids known as prostaglandins. In male insects, seminal vesicles store sperm (see Fig. 4).

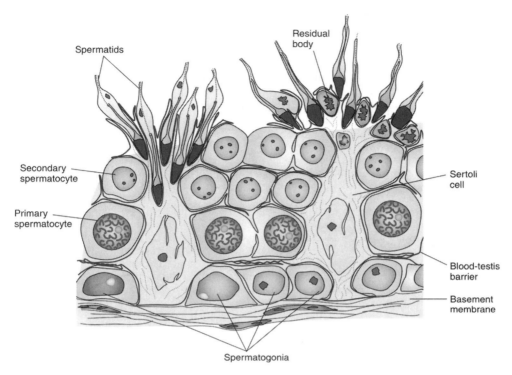

Figure 39. Cell types in the seminiferous tubules. A diagrammatic section through a portion of the wall of a seminiferous tubule. Reprinted from Frank J. Dye, *Human Life Before Birth*, Harwood Academic Publishers, 2000, fig. 5-3, p. 36.

seminiferous   seed-bearing.

seminiferous tubules   found in the testes, the convoluted portions of these tubules are the sites of spermatogenesis. In many species, including the human, the tubules are arranged in lobules separated by bands of fibrous tissue, but in others, for example, most rodents, there are no subdivisions; the 30-odd tubules in the rat, being tightly coiled, lie apparently at random throughout the organ. The seminiferous tubules are not penetrated by any blood vessels or nerves for these are confined to the spaces between the tubules. Both ends of every seminiferous tubule open into the rete testis. The intercellular spaces between pairs of Sertoli cells are divided into a basal compartment, between the basement membrane and cell junctions between the Sertoli cells, and an adluminal compartment between these cell junctions and the lumen of the tubule. Spermatogonia, as they rest on the basement membrane, exist in the basal compartment, while primary spermatocytes and later spermatogenic cells are in the adluminal compartment (see Figs. 15, 39, and 45). *See:* **testicular lobe, tubuli recti**.

seminism   this theory was based on the idea that the generative principle resides in the male and female "semen." Aristotle believed that the male semen represented the impulse and the female "semen" was the substance on which it operated in producing an embryo.

semperflorous   flowering throughout the year.

senescence   a developmental program leading to death; a decline with age in reproductive performance, physiological function, or probability of survival; the impairment of function associated with aging.

sense plate   a thickened region on the surface of the frog neurula, at the anterior end of the embryo, just ventral to the transverse neural fold. The following surface features appear on the sense plate: dorsally, a pair of optic bulges; medially, the stomodaeal invagination; bilaterally, a pair of mandibular arches; and ventrally, a pair of oral suckers.

sensory cranial nerves   those cranial nerves, such as optic nerves, which carry only sensory nerve impulses into the brain.

sepal   the next to the outermost floral part of a flower. *See:* **calyx**.

septum   a partition that divides a chamber.

septum primum   the first partition of the atrial region of the early developing heart.

septum secundum   the second partition of the atrial region of the early developing heart.

septum transversum   the primordium of the diaphragm (see Figs. 6 and 20).

serial homologs   repeated structures of a single organism that share a similar developmental origin.

serial transplantation   as applied to the question of the totipotency of somatic nuclei, enucleated frog oocytes are injected with nuclei of older somatic cells; before these embryos died, the nuclei are transplanted back into new enucleated oocytes, and so on.

serine/threonine kinases   enzyme-linked receptors of importance in development; these kinases phosphorylate proteins at serine or threonine amino acid residues. *See:* **kinases, tyrosine kinases**.

sero-amniotic cavity   the cavity between the serosa (chorion) and the amnion is the sero-amniotic cavity, a part of the extraembryonic coelom. *See:* **amnion**.

sero-amniotic raphe   growth of the head fold of the amnion, lateral amniotic folds, and tail fold of the amnion results in their meeting and fusion above the embryo, resulting in the formation of a scar-like thickening, the sero-amniotic raphe. *See:* **amnion**.

serosa   also called the chorion. The serosa is derived from the outer limbs of the amniotic folds (somatopleure). In insect eggs, a layer of epithelium formed beneath the vitelline membrane; excepting the germ band, the remainder of the blastoderm forms the serosa. *See:* **amnion**.

serotinous   late; late in flowering or leafing; with flowers developing after the leaves are fully developed. *See:* **precocious**.

Sertoli cells   somatic cells found in seminiferous tubules, which assist spermatids undergoing cytodifferentiation (spermiogenesis) into spermatozoa (gametes). The Sertoli cells apparently do not divide after puberty, although just before then, there seems to be a burst of mitotic activity (see Fig. 39). *See:* **seminiferous tubules**.

seven cardinal movements of labor   those movements undergone by the fetus during the process of childbirth, i.e., engagement, descent, flexion, internal rotation, extension, external rotation, and expulsion (see Fig. 40).

sex   the combining of genes from two different individuals into new arrangements; may occur in the absense of reproduction, as in conjugation exhibited by paramecia.

sex cells   gametes; those cells that actually undergo fertilization.

sex chromosomes   those chromosomes, other than the autosomes, that play a direct role in sex determination: the X and Y chromosomes in mammals.

sex cords   cordlike masses of epithelial tissue that invaginate from the germinal epithelium of the gonad and give rise to seminiferous tubules and rete testes in the male and primary ovarian follicles and rete ovarii in the female.

sex determination   the process by which genetic sex is specified; human females have an XX chromosome makeup in their cells and males an XY chromosome makeup in their cells. Normally, human eggs carry one X chromosome, but a given sperm may carry either an X or a Y chromosome. Therefore, at the time of fertilization, sex is determined. If an X-bearing sperm fertilizes the egg, an XX chromosome (female) zygote will be produced. If a Y-bearing sperm fertilizes the egg, an XY chromosome (male) zygote will be produced.

sex glands   glands that make up parts of the reproductive systems and whose secretions play a direct role in reproduction. In the female, Bartholin's glands and Skene's glands. In the male, seminal vesicles, prostate gland, bulbourethral glands (Cowper's glands), and the glands of Littré.

sex hormones   or sex steroids, those hormones that play dramatic and essential roles in reproduction, e.g., estrogens, progesterone, and testosterone.

sex-linked genes   genes (alleles) found on the X chromosome.

sex-linked inheritance   inheritance of characteristics whose genes are located on the X chromosome.

sex-linked mutations   mutations present on a sex chromosome; generally on the X chromosome in mammals.

sexual dimorphism   a difference between the phenotypes of females versus those of males within a species.

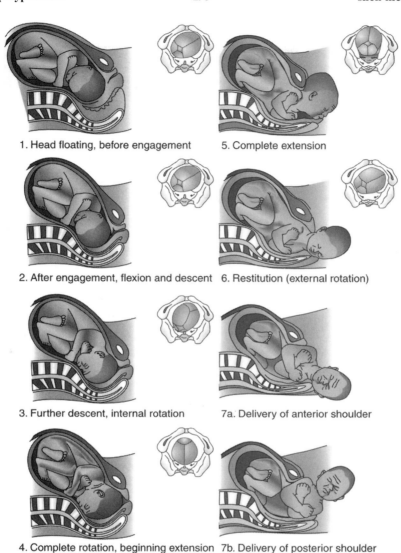

1. Head floating, before engagement

5. Complete extension

2. After engagement, flexion and descent

6. Restitution (external rotation)

3. Further descent, internal rotation

7a. Delivery of anterior shoulder

4. Complete rotation, beginning extension

7b. Delivery of posterior shoulder

Figure 40. The seven cardinal movements of labor. Reprinted from Frank J. Dye, *Human Life Before Birth*, Harwood Academic Publishers, 2000, fig. 11-2, p. 92.

sexual polyphenism phenotypes elicited by environmental social interactions; for example, species of fish, e.g., marine gobys and the blue-headed wrasse, can change from female to male if the male of the group dies, i.e., change their sex depending on the gender makeup of their social group.

sexual reproduction reproduction that involves two individuals and meiosis (i.e., recombination of genes), i.e., reproduction involving gametes; involves the union of two distinct processes, sex and reproduction. *See:* **reproduction, sex**.

shell in the posterior part of the oviduct, or uterus, the chicken egg becomes surrounded by a solid, porous, calcareous shell (of carbonates and phosphates of calcium and magnesium).

shell membrane actually, two closely apposed layers, which come to surround the albumin of the chicken egg in the posterior part of the oviduct; the shell membrane is permeable to liquid albumin so the "egg" still grows in the posterior part of the oviduct. Within the blunt end of the "egg" is a small space, air space, between the closely apposed layers of the shell membrane.

shoot   a young stem or branch.

short-day plants   plants that flower when the night is longer than a critical length. They cannot flower under long days or if a pulse of artificial light is shone on the plant for several minutes during the middle of the night; they require a consolidated period of darkness before floral development can begin. Natural nighttime light, such as moonlight or lightning, is not of sufficient brightness or duration to interrupt flowering. Examples include chrysanthemums, poinsettias, and common duckweed.

short-germ development   a type of insect development in which the blastoderm is short and forms only anterior segments; the posterior segments are formed after completion of the blastoderm stage and gastrulation; most segments are formed from a cellular blastoderm, and posterior segments are added by growth in the posterior region. *Tribolium* is a short-germ insect. Although *Drosophila* is a long-germ insect, there is good evidence that the same genes and developmental processes are involved in the patterning of *Tribolium* and *Drosophila*. However, in certain wasps, development apparently does not depend on maternal information to specify the body axes and in this respect resembles the development of the early mammalian embryo. *See:* **long-germ development**.

Siamese twins   *See:* **conjoined twins**.

signal transduction   converting a signal from one form to another; e.g., converting a cyclic adenosine monophosphate (cAMP) signal to a kinase signal as an intracellular signal transduction pathway progresses into the interior of a cell. *See:* **intracellular signal transduction**.

signaling center   a localized region of the embryo that exerts a special influence on surrounding cells and thus determines how they develop, e.g., Spemann organizer and Nieuwkoop center.

silencers   "negative enhancers;" enhancers that inhibit transcription.

simplex uterus   in humans, fusion of embryonic mullerian ducts gives rise to a single uterus; consequently, humans are said to have a simplex uterus.

single nucleotide polymorphism   a polymorphism caused by the change of a single nucleotide. *See:* **polymorphism**.

sinus rhomboidalis   in the early chick embryo, a space, the lateral boundaries of which are formed by the neural folds converging toward the midline; Hensen's node and the primitive pit lie in the floor of this region.

sinus terminalis   or terminal sinus or marginal sinus; in the chick embryo, the peripheral boundary of the area opaca vasculosa.

sinus venosus   the caudal-most portion of the early cardiac tube, into which open the vitelline, allantoic, and common cardinal veins (see Fig. 20).

siRNAs   these small RNAs induce the degradation of mRNAs that contain the same sequence; mammalian cells can be made to engage in RNAi by treatment of cells with small (21 nt) RNAs; siRNA is involved in the RNA interference (RNAi) pathway, where it interferes with the expression of a specific gene. *See:* **RNA interference (RNAi)**.

sister chromatids   those chromatids of the same chromosome in a pair of homologous chromosomes. *See:* **nonsister chromatids**.

site-specific mutagenesis   an *in vitro* technique in which an alteration is made at a specific site in a DNA molecule, which is then reintroduced into a cell.

situs inversus   when the normal arrangement of bilateral symmetry, situs solitus, is inverted (by mutation or experimentation).

situs solitus   the normal arrangement of bilateral symmetry: in animals with bilateral symmetry, this symmetry is not exact; e.g., in mammals, the cardiac apex, stomach, and spleen are on the left and the liver, vena cava, and greater lung lobation are on the right.

skeletal muscle   one of three general types of muscles found in the body and generally is that muscle under conscious control. Also called voluntary or striated muscle.

skeletogenic mesenchyme   an alternative name for the primary mesenchyme of sea urchin gastrulation because it will form the larval skeleton. *See:* **primary mesenchyme**.

Skene's glands   a type of female auxiliary sex gland that releases its secretions into the vestibule of the vulva.

skin   the largest organ of the body, which provides a protective covering for the organism. It is made up of two layers: the outer epidermis derived from ectoderm and the inner dermis derived from mesoderm.

skull   the skeleton of the head.

slime molds   a group of organisms, classified as protists, and generally regarded as being of two types: (1) true slime molds, with syncytial plasmodia and exemplified by *Physarum polycephalum*, and (2) cellular slime molds, with cellular plasmodia and exemplified by *Dictyostelium discoideum*.

Slit protein   a chemorepulsive protein, which provides guidance by repulsion; e.g., in *Drosophila*, the Slit protein, secreted by neural cells in the midline, acts to prevent most neurons from crossing the midline from either side.

slug   the pseudoplasmodium of *Dictyostelium discoideum*. *See:* **grex, slime molds**.

slug protein   protein expressed by cells destined to become neural crest; one of its functions is to activate factors that dissociate the tight junctions between cells.

Smad signal transduction pathway   an intracellular signal transduction pathway important in development. In vertebrates, this pathway seems to be activated by the TGF-ß superfamily ligand Nodal in those cells responsible for the formation of the mesoderm and for specifying the left-right axis in vertebrates.

small-interfering RNA (siRNA)   sometimes known as short interfering RNA or silencing RNA, is a class of double-stranded RNA molecules, 20–25 nucleotides in length, that play a variety of roles in biology. Most notably, siRNA is involved in the RNA interference (RNAi) pathway, where it interferes with the expression of a specific gene; siRNAs also act in shaping the chromatin structure of a genome. *See:* **microRNAs (miRNAs)**.

small nuclear RNAs (snRNA)   small RNA molecules, found in the nucleus of the cell, which combines with proteins to form the spliceosome.

small toolkit   *See:* **molecular parsimony**.

smooth chorion   *See:* **chorion laeve**.

smooth muscle   one of three general types of muscles found in the body; involuntary muscle, that is, not under conscious control.

soft palate   *See:* **palate**.

somatic cell nuclear transfer (SCNT)   the nucleus from a donor somatic cell is transferred into an enucleated egg cell, generally, as part of a cloning technique.

somatic cells   all the cells of the body except the germ cells; cells not directly involved in reproduction.

somatic mesoderm   also called parietal mesoderm; the dorsal layer of mesoderm derived from the lateral mesoderm. In *Drosophila*, somatic mesoderm gives rise to the main body (see Figs. 28 and 41).

somatic mutation   a mutation that may occur in a cell at any stage of development and may be important in the life of the individual animal, but it cannot affect the next generation. *See:* **germ-line mutation**.

somatic mutation hypothesis   holds that cancers are caused by mutations in the DNA. *See:* **epigenetic progenitor model**.

somatoplasm   a term that refers to somatic cells collectively.

somatopleure   a composite layer consisting of somatic mesoderm and ectoderm. *See:* **splanchnopleure**.

somite bulges   visible surface bulges on the flanks of the trunk and tail of the embryo, during the middle of the embryonic period, indicative of the underlying, developing somites.

somites   paired aggregates of mesodermal cells derived from the somitic mesoderm; they appear in a cephalocaudal direction with such regularity that their appearance is used to time early development in many species of animals; segmented subdivisions of vertebrate mesoderm that give rise to vertebrae and associated processes, selected muscles, and the dermis. In frogs, a 5.5-mm tadpole may contain 45 somites; 13 from just back of the auditory capsules to the base of the tail. After hatching, the first 2 pairs disappear and those in the tail are lost during metamorphosis, leaving 11 well-defined somites in the body region. In the chick embryo, 52 pairs of somites are present after 6 days of incubation, but the last 10 pairs are purely embryonic (see Plates 2a and 2b in the color insert as well as Figs. 6, 9, and 41).

somitic mesoderm   the portion of the mesoderm that becomes segmented into pairs of somites. Also called dorsal, segmental, or paraxial mesoderm.

somitogenesis   somite formation.

somitomeres   whorls of paraxial mesoderm cells that become compacted and bound together by an epithelium and separate from the presomitic paraxial mesoderm to form individual somites.

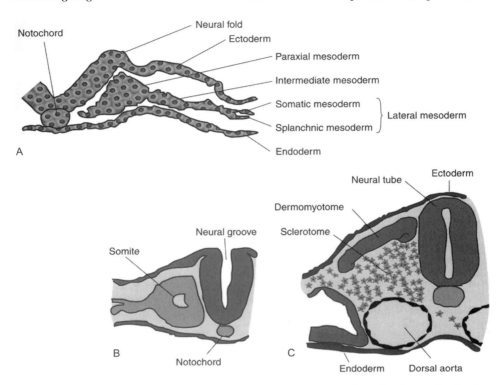

Figure 41. Earlier (A) and later (B and C) diagrammatic transverse sections showing somite development. In each case, only one somite of each bilaterally symmetric pair is shown. Reprinted from Frank J. Dye, *Human Life Before Birth*, Harwood Academic Publishers, 2000, fig. 14-3, p. 113.

sonic hedgehog   one of three homologues of the *Drosophila hedgehog* gene, it is the most widely used of the three vertebrate homologues; a gene essential for formation of the mammalian neural tube. Made by the notochord, the *sonic hedgehog* protein is responsible for inducing the floor plate cells and the motor neurons in the neural tube. It is also responsible for inducing the sclerotome in the somites, mediating the formation of the left-right axis in chicks, initiating the anterior-posterior (thumb-pinky) axis in limbs, and inducing the polarized axis of the gut.

sonogram   the picture resulting from the prenatal diagnostic procedure of ultrasonography (use of ultrasound, i.e., high-frequency sound waves).

sorocarp   the "fruiting body," of cellular slime molds, including *Dictyostelium discoideum*. The sorocarp consists of three parts: the sorus, which contains the spores; the sorophore, which functions as a stalk; and the basal disc, which anchors the sorocarp to the substratum.

sorophore   the stalk of a sorocarp. *See:* **sorocarp**.

sorus   (1) in ferns, a clustered group of sponangia; (2) in the life cycle of cellular slime molds, such as *Dictyostelium discoideum*, a ball of spores found at the top of the sorocarp.

***SOX9* gene**   a gene found throughout the vertebrates; may be the older and more central sex determination gene (compared with *SRY*), although in mammals it became activated by its relative *SRY*. *See:* ***DAX1* gene, *SRY* gene**.

Spallanzani, Lazaro (1729–1799)   Italian biologist and naturalist, described and sketched (in 1780) the furrows formed by the first two cleavage divisions in the egg of a toad; he was also the first to demonstrate artificial fecundation; however, he maintained that semen deprived of animalculae still possessed fecundating properties.

speciation   the formation of a new species.

specific transcription factors   *upstream transcription factors* are proteins that bind somewhere upstream of the initiation site to stimulate or repress transcription. These are roughly synonymous with *specific transcription factors* because they vary considerably depending on what recognition sequences are present in the proximity of the gene, e.g., c-myc and Oct-1. *See:* **general transcription factors, transcription factors**.

specification   the early part of commitment to cell fate; where the fate of a cell depends on environmental cues, but it is flexible and can still be altered in response to environmental signals. *See:* **determination, differentiation**.

specification map   a diagram of an embryo showing what the cells have been programmed to do by that particular developmental stage; the specification of a region need not be the same as its fate in normal development. *See:* **competence map, fate map**.

Spemann, Hans (1869–1941)   German zoologist; recipient of the Nobel Prize in Physiology or Medicine, in 1935, for his discovery of the organizer effect in embryonic development; wrote the classic book, *Embryonic Development and Induction* (1938, Yale University Press, New Haven, CT).

According to Margaret Saha (1994), during his student days at Heidelberg, Karl Gegenbaur exercised the most profound influence over Spemann; another potent influence on Spemann was Gustav Wolff, who, in 1895, published his results concerning regeneration of the lens from the upper margin of the iris after extirpation (Wolffian lens regeneration). Spemann maintained that it was Wolff's results that initially led him to undertake his first lens induction experiments and that his knowledge of lens regeneration from the iris contributed significantly to his formulation of the "double assurance" concept and his "embryonic field" ideas. Spemann did his doctoral work under Theodor Boveri's direction, a descriptive embryological study of a nematode.

Interestingly, according to Saha, Spemann held a lifelong conviction that at least certain aspects of development could never be reduced to cellular analysis; furthermore, although Spemann was not the first scientist to introduce the concept of induction into embryological thought, he, in the course of his lens induction studies, popularized and refined the concept, rendering it one that might serve as a unifying principle for vertebrate development. According to Saha, by the turn of the century, the concepts of "dependent differentiation," "correlative interactions," and "induction" had entered the realm of general scientific consciousness; these concepts would be molded by Spemann into a more refined and precise analytical tool—one that could be applied to yet other aspects of determination in order to create a more unified picture of embryonic development.

According to Saha, in a series of experiments published in 1901, Spemann provides the earliest example of an inductive event to be *demonstrated experimentally*; not only is the induction of the lens of the eye the first instance of induction to be experimentally demonstrated, but it also came to serve as the classic textbook paradigm for general embryonic induction: Saha states that this 1901 paper remains one of the most significant and seminal in the history of embryology; i.e., prior to this publication, it was still possible that all the examples of correlative development cited by Curt Herbst, Wilhelm Roux, and others were nothing more than vague reciprocal relationships, incapable of being separated and analyzed, necessary to the development of any complex organism, but following Spemann's lens induction experiments, it seemed that in this instance and perhaps others, it might prove feasible to dissect these interactions with the hope of actually constructing the type of hierarchy envisioned by Herbst. Spemann, according to Saha, within 5 years after the publication of this paper endowed the concept of induction with more precision and drew critical distinctions that created the context for subsequent studies of induction.

In a 1912 monograph, Spemann confirmed his earlier results demonstrating that in certain species the optic vesicle was essential for lens formation, in other species its role was less than negligible, while the majority of species investigated lay somewhere in between. He believed that two mechanisms of lens determination were operative, with their relative contributions varying among species: (1) induction by the optic vesicle and (2) the presence of self-determination tendencies in the presumptive lens ectoderm. He abandoned the simplistic view of the American embryologist, Warren Lewis, that lens determination was the result of a one-step process.

Using ligature experiments, he demonstrated nuclear equivalence in newt cleavage. He took a baby's hair and lassoed the zygote in the plane where the first cleavage was expected; he concluded that early amphibian nuclei were genetically identical and that each was capable of giving rise to an entire organism.

Spemann demonstrated asymmetry in the amphibian egg; if one of the first two blastomeres receives the gray crescent, only that blastomere will form a normal embryo. The first cleavage normally bisects the gray crescent equally into the two blastomeres. It seemed that gray crescent cytoplasm is essential for proper embryonic development; cells of the dorsal blastopore lip are somehow "programmed" to invaginate into the blastula.

In 1918, Spemann demonstrated that the presumptive neural ectoderm of the newt becomes determined between the early gastrula and the late gastrula stages; namely, if presumptive neural ectoderm is transplanted, during the early gastrula stage, to a region that normally forms epidermis, the transplant will form epidermis (in accordance with its new surroundings), but if the same transplantation is carried out during late gastrula stage, the transplant will form a secondary neural plate (not in accordance with its new surroundings). Thus, within the time separating early and late gastrulae the potencies of these groups of cells had become restricted to their eventual paths of differentiation. According to Saha, one should regard Spemann's mutually symbiotic interests, lens induction and determination during gastrula stages, as outgrowths of the same fundamental concern for the mechanisms underlying the problem of restriction of developmental potential in a given region of the embryo.

As part of this series of experiments, Spemann also transplanted a portion of the upper blastopore lip to the trunk region of another embryo and obtained a number of neural tubes; to explain this phenomenon, he invoked the same two mechanisms that he did for lens determination, self-differentiation and induction. He suggested that the dorsal lip might serve as a (self-)differentiation center, speculating that determination spread in wave-like fashion, posterior to anterior, entirely within the sheet of ectoderm. He also hypothesized that a signal from the underlying mesoderm might induce neural structures. It was as a footnote to a 1921 paper that Spemann first reported preliminary data on the organizer experiments.

In 1924, Spemann and Hilde Mangold (his graduate student), published the results of their experiments using the salamanders *Triturus taeniatus* (darkly pigmented) and *Triturus cristatus* (lightly pigmented), demonstrating the self-differentiation of the dorsal lip of the blastopore; namely, if this tissue were transplanted, during the early gastrula stage, to a presumptive epidermal region of a second early gastrula, the recipient embryo would undergo two invaginations: the expected (primary) invagination and an extra (secondary) invagination caused by the transplant; furthermore, two embryonic axes would develop and two fused salamander embryos would result. The dorsal lip of the blastopore is the only self-differentiating region in the early gastrula. This self-differentiating tissue of the dorsal lip of the blastopore was referred to as the organizer, and this experiment is considered by many to be the most important experiment carried out in embryology (developmental biology). Spemann and Mangold's paper heightened interest in embryonic induction and led to a school of research into the nature of the organizer. According to Johannes Holtfreter, this famous paper was written on the basis of five barely differentiated specimens.

For Spemann, these results once again confirmed that determination was not a simple one-step process; as with lens determination; two mechanisms appeared to be operative. He concluded his analysis by remarking on reciprocal interactions between the induced and inducing tissues, suggesting that such reciprocal interactions may play a large role, in general, in the development of harmonious equipotential systems. His work on lens induction and neural induction exercised a reciprocal effect on each other, with the conceptual framework derived from his lens induction experiments applied to neural induction and the terminology of the organizer experiment applied to lens induction.

Spemann concluded that the principle of progressive determination through organizers of ever rising degrees (e.g., the optic cup was an "organizer of the second degree") possessed a general validity, at least among amphibian species. Following the publication of Holtfreter's first paper documenting the ability of a dead organizer to induce neural tissue, Spemann became disillusioned and effectively ceased active research. In his *Embryonic Development and Induction*, Spemann reveals the potent effect that the conceptual framework provided by his lens induction studies exercised over his thought on induction and determination in general.

According to Saha, Spemann postulated a predisposition of the responding ectoderm prior to the action of the inducer and, as a corollary to this, the idea of induction as a multistep process. He viewed induction not only as a subjacent tissue inducing an overlying one, but also in terms of signals traveling laterally through a single tissue layer. Saha believes that Spemann's most perceptive insight was his insistence that induction, or, more appropriately, determination, was a complex, multistep process involving a delicate interplay between the inducing and induced tissue—a conviction that first originated from his study of lens induction. *See:* **Herbst, Curt.**

Spemann organizer (Spemann–Mangold organizer)   the source of dorsalizing signals that both induce the neural plate and pattern the mesoderm. *See:* **organizer**; **Spemann, Hans.**

sperm   a shorthand term for spermatozoa.

sperm-activating peptides   peptides that cause dramatic and immediate increases in mitochondrial respiration and sperm motility, e.g., resact, isolated from sea urchin egg jelly.

sperm bank   a collection of sperm kept in cryogenic storage.

sperm fertility   estimates of the fertile life of spermatozoa, in the female reproductive tract, for a number of species are mouse, 6 hours; cow, 28–50 hours; human, 28–48 hours; and the bat, 135 days!

sperm survival   in the uterine cavity is short because phagocytosis by leucocytes begins a few minutes after mating. In certain bats, uterine spermatozoa are known to maintain full fertility for several months. Evidence indicates that certain mice can experience two successive pregnancies from one series of copulations. In many species, spermatozoa reaching the ampullary portion of the fallopian tube are propelled into the peritoneal cavity.

sperm transport   the time required for sperm transport from the caput to the cauda is fairly constant at 3–5 days in a variety of species. The total sperm transport time from the seminferous tubule to the exterior in sexually active men is approximately 10–14 days. Spermatozoa that are not expelled to the exterior by ejaculation are probably voided in the urine, and only a minute proportion are thought to be resorbed by the male genital tract.

spermateliosis   *See:* **spermiogenesis.**

spermatheca   in female insects, stores spermatozoa received during mating; later the spermatozoa are liberated over the eggs after the latter have passeed down the oviduct. Also called the receptaculum seminis (seminal receptacle). The spermatheca opens on to the dorsal wall of the median oviduct or of the vagina.

spermatic cord   a composite structure including blood vessels, lymphatic vessels, nerves, and the vas deferens or sperm duct proper.

spermatids   each spermatid develops, by meiosis II, from a secondary spermatocyte; the spermatids undergo cytodifferentiation, called spermiogenesis, to produce spermatozoa (see Figs. 39 and 42).

spermatocytes   each primary spermatocyte develops, by growth, from a spermatogonium; this cell, in turn, gives rise to a total of two secondary spermatocytes, resulting from meiosis I. Each secondary spermatocyte undergoes meiosis II to produce two spermatids. Primary spermatocytes undergo prophase I of meiosis, during which crossing over occurs, and anaphase I of meiosis, during which random segregation of chromosomes, and reduction of chromosome number, occurs (see Figs. 39 and 42).

spermatogenesis   the type of gametogenesis undergone by males and resulting in the production of sperm (see Fig. 42).

spermatogonia   (sing., spermatogonium) male germ cells that act as stem cells and that may undergo meiosis and cell differentiation to become spermatozoa (gametes). Spermatogonia are spermatogenic (sperm-producing) cells. Various types of spermatogonia are recognized in humans, i.e., A_1, A_2, A_3, A_4, intermediate, and B. It is thought that each of the A types are stem cells, capable of renewal; intermediate spermatogonia are committed to becoming spermatozoa and divide to form type B spermatogonia; type B spermatogonia are the precursors of spermatocytes (see Fig. 39).

spermatophore   a structure, deposited in water by the male of some species of salamanders, which bears spermatozoa and is picked up by the cloacal lips of the female of the species, e.g., as with the spotted salamander, *Ambystoma maculatum.* In insects, a kind of capsule, produced by male accessory glands, which encloses the spermatozoa; it is deposited in the bursa copulatrix or in the vagina of the female, where the spermatozoa ultimately become freed.

Spermatophyta   seed plants.

spermatophyte   plants reproducing by seeds.

spermatozoon   (pl. spermatozoa) the male gamete. In many animals, the spermatozoon is flagellated, but in the nematode worm, *Ascaris megalocephala,* the spermatozoon is amoeboid. In some plants, such as ferns, spermatozoa have multiple flagella, but in others, such

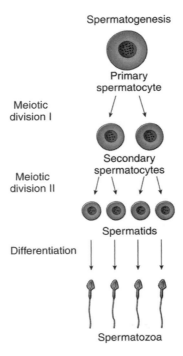

Spermatogenesis

Figure 42. Spermatogenesis. Reprinted from Frank J. Dye, *Human Life Before Birth*, Harwood Academic Publishers, 2000, fig. 6-2, p. 46.

as flowering plants, spermatozoa are delivered to the egg by the elongation of the pollen tube (see Plate 16 in the color insert and Fig. 42).

spermicide   a chemical designed to intentionally kill sperm; a birth control substance.

spermiogenesis   a type of cell differentiation by which spermatids are transformed into spermatozoa. Also called spermateliosis (see Fig. 42).

spermist   one who believed that the new individual is already preformed in the spermatozoon and that development involves essentially the growth of the preformed organism.

Sperry, Roger W. (1913–1994)   Nobel laureate (for physiology or medicine, 1981, shared with David Hubel and Torsten Wiesel); developmental neurobiologist, helped to establish the means by which nerve cells come to be connected in particular ways in the central nervous system.

spicule   a small needle-shaped body, e.g., the calcium carbonate spicules of the skeleton of the pluteus larvae of sea urchins.

Spiegelman, Sol   (1914–1983) microbiologist; saw (in 1947) an essential similarity between the induction of new cell types in the embryo and the induction of new enzymes in the microorganism. By the late 1950s, a number of researchers believed that microbes were an excellent and easily studied model for embryonic differentiation. During the 1960s, Spiegelman epitomized, in a good aphorism, the new developmental biology, formerly embryology, which was being transformed by molecular genetics, "Synthesize the right proteins in the right place, at the right time, and everything else follows."

spina bifida   a type of birth defect resulting from incomplete fusion of the neural folds in the formation of a defective neural tube. *See:* **posterior neuropore**.

spinal canal   the fluid-filled lumen of the spinal cord; during the early formation of the spinal cord, the laterally compressed neural canal becomes lined by ependymal cells and the bulk of the lateral walls consists of neuroblasts and glial cells (see Fig. 43).

spinal cord   the portion of the nervous system derived from a part of the neural tube, which, with the brain, makes up the central nervous system (see Plate 5 in the color insert as well as Figs 7 and 43).

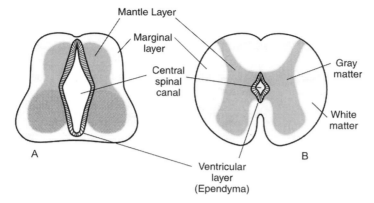

Figure 43. Spinal cord development. Two successive stages, earlier (A) and later (B) in spinal cord development. Reprinted from Frank J. Dye, *Human Life Before Birth*, Harwood Academic Publishers, 2000, fig. 13-2, p. 106.

spinal ganglia ganglia, derived from the neural crest, that flank the spinal cord in pairs and are connected to the spinal cord by means of the dorsal roots of the spinal nerves.

spinal nerves mixed (sensory and motor components) nerves of the peripheral nervous system. Their sensory components originate in spinal ganglia, and their motor components originate in the ventral portion of the spinal cord. In the frog, the tadpole has 40 or more pairs of spinal nerves; in the adult, only the anterior 10 pairs remain.

spindle a transient organelle found in dividing eukaryotic cells, the function of which is the equitable distribution of chromosomes to daughter cells (see Fig. 29).

spindle fibers fibers, made up of microtubules, that radiate from each pole of the spindle of a dividing cell and attach to the chromosomes or interdigitate with spindle fibers from the opposite pole; they provide for the elongation of the spindle and the movement of the chromosomes toward the poles of the spindle (see Fig. 29).

spinocaudal in the context of regional specificity of neural structures, that are produced during development of the central nervous system, the spinocaudal region is that portion of the neural tube caudal to the forebrain (archencephalic) and hindbrain (deuterencephalic) regions of the neural tube.

spinous layer *See:* **Malpighian layer**.

spiracle an opening for the passage of air or water in the respiration of various animals; in frogs, the single opening, on the left side, from the gill chamber to the outside environment, formed between hatching and metamorphosis. *See:* **tracheal system**.

spiral cleavage a cleavage pattern in which, at the eight-cell stage, the animal tier of blastomeres is not directly above the vegetal tier of blastomeres, but the animal tier of blastomeres lies above the cleavage furrows between the vegetal blastomeres. This cleavage pattern is a consequence of the longitudinal axes of the mitotic spindles being at oblique angles, rather than at parallel and perpendicular angles, to the animal-vegetal axis of the egg; characteristic of annelids, turbellarian flatworms, nemertean worms, and all molluscs except cephalopods. The direction of spiral cleavage has been studied in the snail, *Limnaea peregra*; the orientation of right-handed (dextral) and left handed (sinestral) snail coiling can be traced to the orientation of the mitotic spindles at the second cleavage; the direction of cleavage is controlled by a single pair of genes; however, the direction of cleavage is not determined by the genotype of the developing snail but by the genotype of the snail's mother. *See:* **maternal effect genes**.

splanchnic mesoderm also called visceral mesoderm; the ventral layer of mesoderm derived from the lateral mesoderm (see Figs. 20 and 41).

splanchnopleure a composite layer consisting of splanchnic mesoderm and endoderm. *See:* **somatopleure**.

spliceosome a complex made up of small nuclear RNAs and protein, through which splicing of nRNA is mediated.

splicing isoforms   different proteins encoded by the same gene, resulting from differential RNA processing; e.g., alternative RNA splicing enables the alpha-tropomyosin gene to encode brain, liver, skeletal muscle, smooth muscle, and fibroblast forms of this protein.

spontaneous abortion   the spontaneous expulsion of the conceptus from the womb. Also called a miscarriage.

spontaneous generation (abiogenesis)   the belief that life can originate without preexisting life, e.g., frogs originating from mud, mice from putrefied matter, and insects from dew. *See:* **biogenesis**.

sporangiophore   a stalk bearing sporangia.

sporangium   literally, "a spore container"; produced by a variety of organisms, including ferns, mosses, and slime molds.

spore   a reproductive cell resulting from meiotic cell division in a sporangium, representing the first cell of the gametophyte generation: (1) in plants, a cell that germinates to produce, by mitosis, the gametophyte generation; (2) in *Dictyostelium discoideum*, one of the two cell types produced during the life cycle of this organism; spores function in reproduction (see Plate 15 in the color insert). *See:* **stalk cells**.

spore germination   the beginning of the development of a spore (see Plate 15 in the color insert).

spore mother cells   cells found in the sporangia of certain plants (e.g., ferns) that undergo meiosis to produce spores. Also called sporocytes.

sporic meiosis   occurs during an intermediate stage in the life cycle, as in most plants where it occurs between the sporophyte and gametophyte generations.

sporocarp   a specialized structure containing sporangia, as in the water fern *Marsilea*.

sporocytes   *See:* **spore mother cells**.

sporophyll   a modified leaf that develops sporangia.

sporophyte   literally, "spore plant"; a multicellular, diploid plant that produces, by meiosis, spores. Ranges in conspicuousness from the diminutive brown plant supported atop the green moss plant, to the visible fern plant, to the macroscopic angiosperms, such as the visible oak tree (see Plates 16 and 17 in the color insert).

sporophyte generation   in the typical plant life cycle, exhibiting alternation of generations, the sporophyte is the multicellular, diploid, spore-producing generation.

Spratt, Nelson T.   carried out marking experiments (in the 1940s) with chick embryos and found that (1) certain cells in the surface layer of the area pellucida move toward the primitive streak, (2) many of the cells sink to a deep-lying position in the thickened zone and then move inward and take part in the formation of endoderm (other cells enter the endoderm by delamination), (3) cells push in from Hensen's node to form the notochord, and (4) caudal to Hensen's node, mesoderm extends out from the primitive streak between the ectoderm and the endoderm.

***SRY* gene**   sex-determining region of the Y chromosome; this gene, found specifically in mammals, seems to be the gene that encodes the human testis-determining factor. *See:* ***DAX1* gene, *SOX9* gene**.

stabilizing selection   selection that acts to keep a character constant in a population. *See:* **genetic assimilation**.

stadia   the intervals between the ecdyses (molts).

stadium   an interval between two ecdyses (molts).

staging of development   in the chick embryo, because of the regular addition of somites, as the embryo increases in age, the number of somites (pairs) is the most reliable criterion of stage of development; better than hours of incubation. As development progresses, new somites are added caudal to those formed first. *See:* **incubation**.

stalk cells   in *Dictyostelium discoideum*, one of the two cell types produced during the life cycle of this organism; stalk cells have a vegetative, support function; they hold the sorus above the substratum. *See:* **spore**.

stamen   male part of a flower, the next to innermost of the floral organs of a flower.

staminate flowers   imperfect flowers lacking carpels (pistils).

stasis   lack of movement; as in the cessation of blood circulation.

statoblasts   bodies produced by bryozoans, during asexual reproduction, that can survive after the maternal organism dies off during an unfavorable season (e.g., winter and drought). When conditions become favorable, the statoblasts develop into new individuals. *See:* **gemmules**.

STD   sexually transmitted disease.

stem   the portion of the plant axis bearing nodes, leaves, and buds and usually found above ground.

stem cell   a cell that undergoes mitotic cell division to give rise to the same type of cell. At some point, stem cells leave the pool of mitotically dividing cells to begin a process of cell differentiation. Stem cells are, in effect, an embryonic population of cells, continually producing cells that can undergo further development within an adult organism. The path of differentiation that a stem cell descendant enters depends on the molecular milieu in which it resides; e.g., erythrocytes, granulocytes, neutrophils, platelets, and lymphocytes shared a common precursor cell, the pluripotential hematopoietic stem cell.

stem cell factor (SCF)   a paracrine protein that promotes cell division in numerous stem cell populations; mice lacking SCF or its receptor (c-Kit protein) are sterile (no germ cells), white (no pigment cells), anemic (no red blood cels), and immunodeficient (no lymphocytes).

stem cell mediated regeneration   stem cells allow an organism to regrow certain organs or tissues, e.g., hair and blood cells) that have been lost.

stem cell niche   an environment (regulatory microenvironment) that provides a milieu of extracellular matrices and paracrine factors that allows certain cells residing within it to remain relatively undifferentiated; regulates stem cell proliferation and differentiation.

stemness   essential characteristics of a stem cell that distinguishes it from ordinary cells.

stenosis   refers to a narrowing, e.g., pulmonary stenosis; a stenosis can have disastrous developmental effects depending on what is being narrowed.

stereoblastula   a blastula without a blastocoel.

stereotaxis   cells sense the physical topography of their substrate; may play a role in guidance of migratory cells. *See:* **chemotaxis, contact inhibition, haptotaxis.**

stereotropism   *See:* **contact guidance.**

sterile   infertile, not capable of reproducing, as a stamen that does not bear pollen or a flower that does not bear seed.

Stevens, Nettie M. (1861–1912)   demonstrated a critical correlation (as did Edmund Beecher Wilson) between nuclear chromosomes and organismal development: in insects, XO or XY embryos became male; XX embryos became female.

Steward, F. C. (1904–1993)   conducted classic experiment (in the 1950s) demonstrating that whole plants could be regenerated from single somatic cells dissected from a carrot; his experiments demonstrated the totipotency of carrot phloem cells.

stigma   the uppermost portion of the pistil of a flower; on which pollen grains land and germinate their pollen tubes.

stomach   the portion of the digestive tube that receives food from the esophagus and briefly stores and partially digests the food before passing it on to the small intestine. In the frog, shortly after hatching, a portion of the foregut, between the future glottis and the opening of the bile duct, elongates; the posterior part of this elongation gives rise to the stomach.

-stomia   a word part referring to the mouth, e.g., macrostomia.

stomodaeal invagination   the invagination that divides the sense plate of the frog embryo into a bilateral pair of mandibular arches.

stomodaeum   also called stomodeum; anterior ectoderm-lined portion of the alimentary canal; in the frog, forms at the dorsal end of the stomodaeal invagination. In insects, an ingrowth of ectoderm, just behind the antennae, that becomes the embryonic fore-intestine.

stomodeum   *See:* **stomodaeum.**

stratification   the requirement for chilling (5 °C) to break dormancy in some seeds.

stratum corneum   the outermost layer of the epidermis, composed of keratinized cells; the cells are flattened sacs of keratin protein.

stratum germinativum   the innermost layer of the epidermis, which contains mitotic stem cells required for the continual replenishment of the epidermis. *See:* **basal layer.**

stratum granulosum   an intermediate layer of the epidermis consisting of cells containing many granules; these are granules of the protein keratin. Cells of this layer do not divide but differentiate into keratinocytes.

stratum lucidum   a translucent layer of the epidermis, consisting of transparent cells.

stratum spinosum   a layer of the epidermis, consisting of cells with conspicuous intercellular bridges.

strengths of adhesion   a quantitative aspect of cell-sorting behavior; e.g., if cell type A is more adhesive than cell type B, then B will eventually surround A.

striated striped.

strobilus a cone-like cluster of sporophylls on an axis, found on some cryptograms such as *Lycopodium* and *Equisetum*.

stroma the framework, usually of connective tissue, of an organ, gland, or other structure, as distinguished from the parenchyma.

stromal cells cells found in stroma.

structural genes genes that code for proteins.

structural genomics the study of the sequences of genomes. *See:* **functional genomics**.

strypsin a trypsin-like protease, located on trophoblast cell membranes of mouse embryos, which lyses a hole in the fibrillar matrix of the zona pellucida, allowing the blastocyst to hatch from the zona.

***Styela* (*Cynthia*)** a urochordate, the eggs of which possess a natural pattern of pigmentation and distribution of yolk that makes it possible to follow the developmental fates of the naturally "marked" regions of the egg cytoplasm.

style the part of the pistil between the upper stigma and the lower ovary, through which the pollen tube grows to reach the embryo sac in the ovary.

stylopod the proximal bones (humerus/femur) of the vertebrate limb, adjacent to the body wall. *See:* **autopod, zeugopod**.

subcaudal pocket *See:* **subcaudal space**.

subcaudal space the space between the tail and the blastoderm of the early chick embryo.

subcephalic pocket *See:* **subcephalic space**.

subcephalic space the space between the head and the blastoderm of the early chick embryo.

subgerminal space the space formed between the blastoderm and the underlying yolk mass in the early chick embryo.

suction method a method of abortion used early in pregnancy, where the conceptus is literally aspirated (sucked) out of the uterus.

sulcus limitans a longitudinal groove (one of two) that divides the neural tube (bilaterally) into dorsal and ventral halves.

superficial cleavage exemplified by insect eggs, wherein a large mass of centrally located yolk confines cleavage to the cytoplasmic rim of the egg.

superior vena cava a major venous (vein) blood vessel returning blood from the head and neck (parts of the body superior to the heart) to the right atrium of the heart.

supernumerary breast extra breast; breast in excess of the normal two for a human; may develop anywhere along the mammary ridges.

supernumerary chromosome a chromosome present in addition to the normal chromosome complement; e.g., the chromosomal basis of Down's Syndrome is trisomy 21; i.e., there are three copies of chromosome 21, rather than the normal two copies (disomy).

supernumerary nipples the presence of extra nipples, e.g., more than two nipples in humans.

supernumerary sperm extra sperm; sperm in excess of the one involved in fertilization of the egg; in some species (e.g., birds), supernumerary sperm may penetrate the yolky egg but disintegrate before they can complete the fertilization process (fusion of pronuclei).

superovulate to ovulate more than the normal number of eggs, under the influence of hormonal stimulation of the ovary.

superovulation the artificial stimulation, by administration of hormones, of the ovulation of a greater than normal number of ova (actually oocytes) from the ovary. In the frog, one protocol involves the intraperitoneal injection of pituitary glands or pituitary gland extract. In the mouse, one protocol involves the sequential intraperitoneal injection of a pregnant mare's serum and human chorionic gonadotropin, with an intervening interval of 48 hours.

supine lying on one's back.

supporting cells cells that aid and/or protect the cells carrying out a specific function, e.g., the supporting cells in taste buds, the Sertoli cells of the seminiferous tubules, and the follicle cells of the ovarian follicles.

suppressor gene a mutant gene that masks the effect of another mutant, thus causing partial or total restoration of the normal phenotype.

surfactant lipid molecules (lecithin and sphingomyelin) found in the mature lungs that allow terminal alveoli to expand with each inhalation.

surrogate mother a woman carrying a pregnancy when the conceptus is not hers.

suspensor the larger of the two cells produced by the first division of the angiosperm zygote, the basal cell, gives rise to the suspensor, which is attached to the embryo proper; plays a role in providing nutrition to the embryo. The suspensor may be a source of hormones. In the absence of the embryo proper, the suspensor may develop as an embryo.

sutures   the narrow spaces between the bony plates of the skull, which are most conspicuous at approximately the time of birth.

Swammerdam, Jan (1637–1680)   Dutch naturalist and microscopist, he was a preformationist. His pioneering work on insects laid the foundations of modern entomology. At the theoretical level, he developed a new argument in support of the preformationist position, based on the nature of insect metamorphosis; with the aid of a microscope, he suceeded in identifying structures belonging to butterflies in pupae and caterpillars. The caterpillar, he insisted, was not changed into a butterfly, but, it grew by the expansion of parts already formed. Cleavage of the egg was first observed by Swammerdam in the 17th century (but published in 1738); he saw the first furrow of the frog's egg in cleavage.

sweat glands   skin glands derived from the epidermis of the skin, with pores that open onto the surface of the skin. The evaporation of their secretions, sweat, is an important part of dissipation of heat by the body.

Sylvian aqueduct   or aqueduct of Sylvius, is the portion of the ventricular system of the brain, which connects the third and fourth ventricles and through which cerebrospinal fluid circulates; the cavity of the midbrain that connects the cavities of the forebrain and hindbrain.

symbiont   a member of a symbiosis.

symbiosis   living together, in close association, of dissimilar organisms; the larger organism is referred to as the host, and the smaller organism is referred to as the symbiont.

symbiotic regulation of development   symbionts' alteration of development of their hosts, e.g., *Wolbachia* effects on many different organisms, ranging from transforming genetically male embryos into females (as in the common pillbug *Armadillidium vulgare*) to their being necessary for its host's completion of normal oogenesis (as in the parasitic wasp *Asobara tabida*).

symplastic   continuity allowing electrical continuity and the flow of small molecules between cells; provided by plasmodesmata. *See:* **apoplastic continuity**.

synapomorphy   a shared, derived character; in a phylogenetic analysis, synapomorphies are used to define clades and distinguish them from outgroups.

synapse   the specialized junction formed when an axon contacts its target; the junction between two nerve cells or between a nerve cell and a muscle fiber.

synapsis   the pairing of homologous chromosomes during prophase I of meiosis. Generally limited to chromosomes undergoing pairing during prophase I, certain cells possess chromosomes that undergo somatic (not in the germ line) synapsis, e.g., the polytene chromosomes of the larval salivary glands of *Drosophila melanogaster*.

synaptonemal complex   (also synaptinemal complex) a protein complex, visible with the electron microscope, that is the physical basis of the pairing of homologous chromosomes (synapsis) during meiosis; it is assembled during zygotene as homologous chromosomes pair up, and it is disassembled during diplotene as homologous chromosomes separate.

syncytial   the adjective for syncytium.

syncytial blastoderm   the blastoderm of the *Drosophila* embryo is a syncytium before it becomes a cellular blastoderm at approximately 320 minutes after the deposition of the egg. Transplantation experiments demonstrate that syncytial nuclei are each equivalent and totipotent, but cells of the cellular blastoderm are precisely determined; determination of the syncytial-stage nuclei depends on the region of the egg into which they migrate. Pole plasm seems to contain the morphogenetic determinants for germ cell production.

syncytial cable   an elongated syncytial structure, formed by fusion of primary mesenchyme cells during gastrulation, which forms a skeletal rod of the skeleton of the pluteus larvae of the sea urchin.

syncytial specification   in addition to specification of cell fate by neighboring cells and by soluble molecules secreted at a distance from the target cells, insects use a third means, syncytial specification, to commit cells to their fates. Here, interactions occur not between cells, but between parts of one cell, as early insect embryos are syncytia.

syncytiotrophoblast   the outer, syncytial, highly invasive component of the trophoblast (see Fig. 44).

syntrophoblast   the outer synctial layer of the trophoblast that forms the outermost fetal element of the placenta. *See:* **syncytiotrophoblast**.

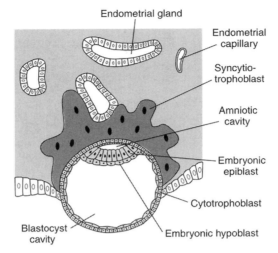

Endometrial gland

Endometrial capillary

Syncytio-trophoblast

Amniotic cavity

Embryonic epiblast

Cytotrophoblast

Embryonic hypoblast

Blastocyst cavity

Figure 44. The trophoblast. Early in implantation. Reprinted from Frank J. Dye, *Human Life Before Birth*, Harwood Academic Publishers, 2000, fig. 8-5, p. 65.

syncytium   a multinucleated mass of cytoplasm; loosely, a multinucleated cell; an embryo or tissue containing nuclei that are not separated by cell (plasma) membranes.

syndactyly   the adhesion of fingers or toes; webbed fingers or toes; a type of birth defect involving fusion of the digits. Note that during development, the digits are "sculpted" out of a hand plate or foot plate by specific, genetically programmed cell death.

syndetome   derived from the dorsal-most sclerotome cells, which express the *scleraxis* gene and generate the tendons.

syndrome   several malformations or pathologies that occur concurrently, e.g., Down's syndrome.

synergids   the two cells that flank the egg in the embryo sac, which may be evolutionary remnants of the archegonium (the female sex organ of mosses and ferns).

syngamy   the union of gametes in fertilization.

synteny   the occurrence of two genes on the same chromosome.

syphilis   a bacterial, sexually transmitted disease.

systemic circulation   the general blood circulation of the body, exclusive of the pulmonary and placental circulations.

systems biology   views an organism as an integrated and interacting network of genes, gene products, cellular components, physiological coordinating systems, and ecological agents; this approach has a long-standing history in developmental biology.

T

T-box factors   transcription factors; have a DNA-binding domain similar to the prototype gene product known as "T" in the mouse and as Brachyury in other animals, e.g., the endodermal determinant VegT and the limb identity factors Tbx4 and Tbx5.

T cells or T lymphocytes   a subset of lymphocytes (which are a subset of white blood cells) that carry out cell-mediated immunity.

tadpole   a name given to an immature (larval) stage in the development of some animal species, e.g., amphibians, and tunicates. *See:* **larva**.

tail bud   also called end bud; in chick embryos, at the end of the second day, made up of the remains of the primitive streak, essentially Hensen's node and a short portion of the primitive streak (see Fig. 9).

tail fold   one of four body folds; the tail fold sculpts the tail end of the embryo out of the caudal region of the embryonic disc (see Fig. 6).

tapetum   nutritive tissue in a sporangium, especially an anther.

taste cells   those cells found in a taste bud that are receptors for the sense of taste.

TATA box   a nucleotide sequence, which is part of a gene promoter, to which RNA polymerase attaches.

Tatum, Edward Lawrie (1909–1975)   American biochemist; developed, with George Wells Beadle, the idea that specific genes control the production of specific enzymes (one gene-one enzyme hypothesis), virtually creating the science of biochemical genetics. Beadle and Tatum shared the Nobel Prize for Physiology or Medicine in 1958 with Joshua Lederberg. *See:* **one gene-one polypeptide**.

taxon   a taxonomic entity of any rank, such as order, family, genus, or species. *See:* **phylum**.

Tbx4   a transcription factor that seems to be critical in instructing the chick limbs to become hindlimbs.

Tbx5   a transcription factor that seems to be critical in instructing the chick limbs to become forelimbs.

tectum   a roof or covering; the tectum of the mesencephalon is its dorsal part, including the superior and inferior colliculi and adjacent areas. *See:* **corpora quadrigemina**.

telencephalon   the cephalic-most vesicle of the early five-vesicle brain; gives rise to the cerebral hemispheres; gives rise to the olfactory lobes (smell), hippocampus (memory storage), and the cerebrum ("intelligence") (see Figs. 7 and 8).

teleosts   the largest group of bony fishes; this is the dominant group of modern fish. The teleosts belong to the subclass Actinopterygii (ray-finned fish); the other extant subclass of bony fish are the Sarcopterygii (lobe-finned fish, including lungfish and the coelacanth). The superorder Teleostei shares the subclass Actinopterygii with two other superorders, the superorder Chondrostei (including sturgeons and paddlefish) and the superorder Holostei (including bowfin and gars).

teloblasts   in leeches and other annelids, give rise to the segmental structures.

telocoel   cavity of the telencephalon; actually the cavity is divided into three parts: two lateral telocoels, the cavities of the lateral telencephalic vesicles (early cerebral hemispheres), and one median telocoel, the cavity of the rest of the telencepahalon.

telolecithal eggs   a type of egg that has a concentration gradient of yolk distributed along its animal-vegetal axis (highest concentration at the vegetal pole); frog eggs and chicken eggs are telolecithal (see Figs. 12 and 14). *See:* **homolecithal eggs**.

telomerase   an enzyme that repairs the ends of chromosomes, telomeres, so that chromosomes do not shorten each time a cell divides; somatic cells of the body do not exhibit telomerase activity, and their chromosomes do shorten with cell divisions; this has been

Dictionary of Developmental Biology and Embryology, Second Edition. Frank J. Dye.
© 2012 Wiley-Blackwell. Published 2012 by John Wiley & Sons, Inc.

proposed as a possible mechanism of aging. However, there is no correlation between the telomere length and the life span of an animal (humans have much shorter telomeres than mice). Germ cells and cancer cells do exhibit telomerase activity, and it has been proposed that this contributes to their immortality. *See:* **immortality of the germplasm**.

telomeres repeated DNA sequences at the ends of chromosomes.

telophase the stage of mitosis or meiosis that begins when the chromosomes reach the poles of the spindle (see Fig. 29).

teloplasm the cytoplasm in the eggs of leeches and other annelids that is involved in the specification of the blastomere that will give rise to the teloblasts.

telson the hindmost segment of the arthropod abdomen; in insects, it is present only in the embryo; a marker for the posterior terminus of the *Drosophila* embryonic axis.

temperature affects development in several ways; e.g., the rate of early frog development is influenced by the temperature of the environment, human sperm development requires a temperature lower than that of body temperature (37°C) so human testes are found in the cooler scrotum, and there are temperature-sensitive mutants that will only show specific phenotypic characteristics within a specific range of temperature.

temperature sensitive mutations those that display the phenotype at a nonpermissive (usually high) temperature and do not show the phenotype at the permissive (usually low) temperature; develop from changes in the conformation of the protein product that are sensitive to changes in temperature in the range compatible with embryonic survival.

template a guide to the form of a piece being made. *See:* **semiconservative replication**.

tepal a sepal or petal.

teratocarcinoma a type of tumor in which germ cells become embryonic stem (ES) cells; these ES cells not only divide, but they can also differentiate into a wide variety of tissues, e.g., gut and respiratory epithelia, muscle, nerve, cartilage, and bone; such tumors can give rise to most of the tissue types in the body.

teratogen a substance (but also including radiation) that causes abnormal development of an organism; that which causes a birth defect, e.g., some viruses, some kinds of radiation, and some chemicals; those exogenous agents responsible for disruptions (see Fig. 13).

teratogenesis the development of congenital malformations.

teratology the study of birth defects; the study of how environmental agents disrupt normal development.

teratoma a neoplasm composed of bizarre and chaotically arranged tissues foreign embryologically as well as histologically to the area in which the tumor is found.

terminal cell one of the two cells resulting from the first division of the angiosperm zygote, which gives rise to the embryo proper. *See:* **basal cell**.

terminal gene group in *Drosophila* embryos, the set of maternal genes whose proteins generate the extremes of the anterior-posterior axis; mutations in these terminal genes result in the loss of the unsegmented extremities of the organism: the acron and the most anterior head segments and the telson and the most posterior abdominal segments. *See:* **anterior organizing center, posterior organizing center**.

terminal sac period a period during lung development when the alveolar ducts give rise to terminal air sacs.

tertiary follicle an ovarian follicle with antral vacuoles or a forming antrum.

test cells extraembryonic cells that surround tunicate eggs and embryos; contain large amounts of mycosporine pigments that attenuates ultraviolet (UV) radiation. *See:* **mycosporines**.

test tube baby a loosely used expression for babies resulting from *in vitro* fertilization (IVF).

testes (singular testis) the male gonads (see Figs. 15 and 45).

testicular dysgenesis syndrome a syndrome characterized by disorganized testis development, testicular germ cell tumors, and low sperm count; may be attributed, at least partially, to endocrine disruptors, e.g., phthalates.

testicular feminization as a result of a sex-linked mutation, tissues that normally respond to androgens (male hormones, primarily testosterone) do not; as a consequence, affected individuals, although carrying the XY sex chromosome makeup, are phenotypically females.

testicular lobe the insect testis, here using the grasshopper as a model, consists of many lobes, the pointed ends of which open into vasa deferentia, or sperm ducts, Each lobe consists of numerous compartments (cysts) that are separated from each other by connective tissue partitions (septa). At the blunt (apical) end of each lobe are numerous primordial germ cells (spermatogonia) all undergoing mitosis. As a result of this continual increase

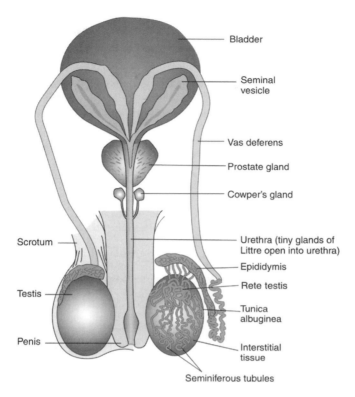

Figure 45. The male gonads. Reprinted from Frank J. Dye, *Human Life Before Birth*, Harwood Academic Publishers, 2000, fig. 5-2, p. 35.

in cell number, there is a pinching off of groups of these presumptive germ cells into newly forming cysts; each group is composed of cells in exactly the same stage of development. Cysts from the blunt to the pointed end of each testicular lobe show progressive stages of maturation; many mature spermatozoa are observed at the region where the testicular lobe joins the vas deferens; also, along the same vector, one may distinguish a germarium or zone of spermatogonia, a zone of maturation, where the spermatocytes undergo reduction division, and a zone of transformation, where spermatids are converted into spermatozoa.

testis-determining factor   a protein encoded by the *SRY* gene (normally found on the mammalian Y chromosome) that organizes the gonad into a testis rather than an ovary.

testosterone   a male sex hormone (androgen) responsible for the development of male secondary sex characteristics and necessary for spermatogenesis. Testosterone secreted by the fetal testis masculinizes the fetus, stimulating the formation of the penis, scrotum, and other portions of the male anatomy, as well as inhibiting development of the breast primordia.

tetrad   collectively, the four chromatids found in a synapsed pair of homologous chromosomes during prophase I of meiosis.

tetraploid number   four times the haploid number of chromosomes: 92 for humans.

tetraploidy   exhibiting the tetraploid number of chromosomes; not consistent with postnatal human life, although sometimes found in a mosaic individual.

tetrapod   the four-limbed terrestrial classes of vertebrates, including amphibians, reptiles, birds, and mammals.

TGF-α   See **transforming growth factor α (TGF-α)**.

TGF-β   See **transforming growth factor β (TGF-β)**.

TGF-β receptor   there are a number of receptors for the transforming growth factor β (TGF-β) superfamily; their specificity for different factors is complex and overlapping.

thalidomide   a chemical teratogen that is especially detrimental to limb formation.

thallophytes   literally, "thallus plants"; plants with a simple plant body or thallus.

thallus   a simple plant body, lacking much cell specialization; a plant body that is not obviously differentiated into stems, roots, and leaves; algae traditionally have been classified as thallophytes.

theca externa   one of two connective tissue layers surrounding the mature graafian follicle in humans; this is the outer, fibrous layer.

theca folliculi   in mammals, the theca folliculi is made up of the theca externa and the theca interna; in chickens, consists of loosely arranged cells surrounding the theca granulosa of the chicken egg ("yolk") and continuous with the vascular stroma of the ovary.

theca granulosa   a layer of follicle cells surrounding the vitelline membrane of the chicken egg ("yolk").

theca interna   one of two connective tissue layers surrounding the mature graafian follicle in humans; this is the inner, cellular layer, with blood capillaries.

thecal cells   steroid hormone-secreting ovarian cells surrounding the follicle that differentiate from the mesenchyme cells of the ovary.

thecodonts   reptiles, abundant during the Triassic Period, considered to be ancestors of the birds.

Theory of Epigenesis   the concept that development is not a process of unfolding and growth in size of preformed structures, but, rather, it is one in which certain elements increase in number and gradually become molded into the form of layers that later give rise to the organic structure of the organism.

therapeutic cloning   nuclear material from a donor somatic cell is transferred into an enucleated egg cell and ESCs (embryonic stem cells) are harvested from the resulting blastocyst; if the ESCs can be coaxed to differentiate into functioning tissues, the new cells might be introduced into the donor to regenerate diseased tissue and even organs. An objective of therapeutic cloning is to derive immunologically compatible embryonic stem cells.

therapsids   reptiles considered to be the ancestors of the mammals.

Theria   a subclass of mammals, the live-bearing mammals; includes two infraclasses, Metatheria and Eutheria.

third ventricle   the main cavity of the forebrain, exclusive of the cerebral hemispheres. *See:* **lateral ventricles**.

Thompson, D'Arcy W. (1860–1948)   Scottish zoologist; professor of biology at Dundee and St. Andrews Universities; he interpreted the forms of organs and biological structures based on the physical forces acting on them during development, and he was able to demonstrate mathematically that the superficial differences between related animals could be accounted for by differential growth rates; authored *On Growth and Form* (1917).

threshold concentration   that concentration of a chemical signal or morphogen that can elicit a particular response from a cell.

threshold concept   the means of coordinating metamorphic events seems to be a difference among tissues and organs in their responsiveness to different amounts of hormone.

threshold effect   a specific response to a chemical signal that only occurs above or below a particular threshold concentration of the signal.

thrifty phenotype hypothesis   altered maternal nutrition leads to a cascade of consequences mediated by the phenotypic plasticity of the fetal organs.

thymopharyngeal duct   the third pharyngobranchial duct; it may elongate between the pharynx and the thymus.

thymus gland   an endocrine gland that plays a role in the development of the immune system, especially in the development of T-lymphocytes. In the chick embryo originates from diverticula of the posterior faces of the third and fourth pharyngeal pouches.

thyroglossal duct   *See:* **thyroid gland**.

thyroid gland   an endocrine gland that produces the hormones thyroxine (that regulates metabolism and growth) and calcitonin (that helps to regulate calcium levels in the blood). In the frog, develops before hatching as a median longitudinal evagination from the floor of the pharynx; later it separates entirely from the pharynx and divides into two lateral parts that eventually become vascular. In the chick embryo, the thyroid gland develops from the floor of the pharynx as a median diverticulum between the first and second pairs of pharyngeal pouches. By the end of the fourth day, the thyroid evagination is saccular and retains connection with the pharynx by a narrow opening at the root of the tongue, the thyro-glossal duct.

thyroid stimulating hormone (TSH, thyrotropin)   synthesized and released from the anterior pituitary gland, under stimulation by the hypothalamus of the brain; controls the thyroid gland and stimulates synthesis of triiodothyronine (T_3) and thyroxin (T_4). In the absence of T_4 and/or T_3, metamorphosis in the frog will not occur.

thyropharyngeal duct   the fourth pharyngobranchial duct.

thyroxine   the metamorphic changes of frog development are all brought about by the secretion of the hormones thyroxine (T_4) and triiodothyronine (T_3) from the thyroid gland during metamorphosis.

tight junctions   specialized contacts formed between cells that establish partitions between isolated compartments of the body. During compaction of early mammalian embryos, tight junctions form between the cells of the trophoblast; these cells with their tight junctions seal the blastocoel off from the embryo's environment. *See:* **gap junctions, slug protein**.

Tinman   the transcription factor that is active in specifying the heart tube of *Drosophila*; homologous to Nkx2-5 in vertebrates. *See:* **Nkx2-5**.

tip cells   certain endothelial cells that can respond to the vascular endothelial growth factor (VEGF) signal and begin "sprouting" to form a new vessel during angiogenesis.

tissue engineering   a regenerative medicine approach whereby a scaffold is generated, seeded with stem cells, and is used to replace an organ or part of an organ.

tissue organization field (TOC) hypothesis   an epigenetic hypothesis. Cancer develops from disruptions in tissue organization; mutations are not required for a cell to become cancerous; a cell that remains normal in one tissue context may become cancerous in another; and such cancer cells can revert back to normal growth patterns if placed back in their original environment or if instructed to differentiate. This hypothesis holds that normal interactions between cells inhibit the normal propensity of a cell to proliferate, and that cancers develop from the disruption of these repressive interactions; e.g., the growth of mammary epithelial cells seems to be kept in place not only by the *extracellular matrix*, but also by the paracrine factor transforming growth factor-ß (TGF-ß) secreted by the *stomal cells*. *See:* **epigenetic progenitor model, random epigenetic drift, somatic mutation hypothesis**.

toad   any of several species of the amphibian order Anura, especially in the family Bufonidae.

Toll   a maternal gene in *Drosophila* that encodes a receptor protein, which is only activated in the future ventral region of the embryo, by a ligand, called the spätzle fragment, which is only produced in the ventral perivitelline space. Activation of the *Toll* receptor generates a signal that causes a maternal gene product in the adjoining cytoplasm, the dorsal protein encoded by the *dorsal* gene, to enter nearby nuclei; this dorsal protein is a transcription factor with a vital role in organizing the dorso-ventral axis of the embryo.

tonsils   two masses of lymphoid tissue; each between folds of tissue, at the back of the throat (tonsillar region).

tool kit (of developmental biology)   the collection of signals, signal transduction pathways, and gene regulatory mechanisms known to operate in most animals.

tooth germ   the embryonic rudiment, or anlagen, of a tooth.

tornaria   hemichordate larva of the class Enteropneusta (acorn or tongue worms).

torpedo stage embryo   a late stage in the embryogenesis of angiosperms; upright cotyledons can give the embryo a torpedo shape; all three primary meristems are clearly visible. By this stage, the suspensor is degenerating. *See:* **globular stage, heart stage embryos**.

torsion   a twisting of the body. In the chick embryo, beginning at approximately 38 hours, flexion and torsion begin; because the yolk constitutes a barrier to flexion, the chick embryo must twist around on its side (i.e., undergo torsion); torsion begins in the cephalic region and progresses caudad; as torsion progresses, the yolk no longer impedes flexion. By 4 days, the entire body of the chick embryo has turned, by torsion, through 90° and the embryo lies with its left side on the yolk; the body of the embryo has become undercut and is attached to the yolk by a slender stalk. Torsion is characteristic only of embryos developing on the surface of a very large yolk.

torso   a maternal gene in *Drosophila* that plays a key role in specifying the structures at the extreme ends of the anteroposterior axis of the embryo (e.g., the acron at the anterior end and the telson at the posterior end). *Torso* encodes a receptor protein, which is uniformly distributed throughout the egg plasma membrane, but it is only activated at the ends of the fertilized egg because the protein ligand for the receptor is only present there. The ligand for the torso protein, which distinguishes the termini, is synthseized and secreted by both posterior and anterior follicle cells, but not by other follicle cells.

torus transversus   in the frog, a bulge in the floor of the brain anterior to the optic recess; gives rise to the ventral anterior commisure.

totipotent   describes a cell that is capable of giving rise to all of the cell types of an organism; the zygote is a totipotent cell, as is each of the blastomeres of a four-cell sea urchin embryo and the phloem cells of an adult carrot.

toxicogenomics   attempts to explain how the entire genome is involved in biological responses to toxicants and stressors; measures specific changes in gene expression, species to species, as a result of exposure to drugs and other environmental agents, and provides a genetic snapshot of individuals' unique susceptibility to cancers and other diseases.

toxoplasmosis   a protozoan, parasitic disease that can be teratogenic to a fetus being carried by an infected mother.

trabeculae carneae   muscular extensions of the interior of the ventricles of the heart that coalesce to form part of the interventricular septum and the papillary muscles and tendinous cords that operate the valves of the heart.

trace fossils   the impressions left in sediments by the meanderings and burrowing activities of ancient animals.

trachea   the windpipe portion of the respiratory system.

tracheal system   in insects, the tracheae develop from paired lateral ingrowths, near the bases of the appendages, on the meso- and meta-thorax and on the first eight abdominal segments. The mouths of these invaginations become the spiracles, and at their inner ends, anterior and posterior longitudinal extensions meet and fuse to form the main tracheal trunks.

tracheophytes   those plants exhibiting true vascular tissue, xylem and phloem.

trans-**acting**   a term that describes a genetic element, such as a repressor gene or transcription factor gene, that can be on a separate chromosome and still influence another gene; these trans-acting genes function by producing a diffusible substance that can act at a distance.

trans-**activating domain**   that domain of a transcription factor concerned with activating or suppressing the gene's transcription; activation domains of transcription factors often contain many acidic amino acids forming an "acid blob," which accelerates the formation of the general transcription complex.

transcription   the synthesis of RNA using DNA as a template.

transcription factor   a molecule that attaches to DNA at a gene regulatory site and by so doing influences the rate of transcription of a specific gene; a protein that regulates the transcription of genes, often, but not exclusively, by binding to *cis*-regulatory elements (promoters and enhancers). Transcription factors come in two general types: basal transcription factors (e.g., TFIID, TFIIA, etc.), which attach to gene promoters, and cell-specific transcription factors (e.g., Pax6, Mitf, etc.), which bind to gene enhancers. *Trans*-regulatory factors are usually transcription factors. There are four major families of transcription factors: homeodomain, basic helix-loop-helix (bHLH), basic leucine zipper (bZip), and zinc finger (Table 3). *See:* **general transcription factors, MITF, Pax6, specific transcription factors**.

transcription factor domains   the three major domains are a DNA-binding domain, a *trans*-activating domain and a protein-protein interaction domain.

Table 3. Examples of Major Transcription Factor Families/Subfamilies*

Families	Examples	Functions
Homeodomain:		
Hox*	Hoxa1	Axis formation
Pax*	Pax6	Eye development
Basic helix-loop-helix	MyoD	Muscle specification
Basic leucine zipper	C/EBP	Fat cell differentiation
Zinc finger:		
Standard*	Engrailed	Drosophila segmentation
Nuclear hormone receptors*	Estrogen receptor	Secondary sex determination

transcription initiation complex collectively, the RNA polymerase and associated basal transcription factors that attach to a gene promoter.

transcriptional regulation regulation of transcription of RNA from DNA, under the control of RNA polymerase, transcription factors, promoters, and enhancers.

transcriptome the complete set of RNAs transcribed from a genome; the sum of all the different transcripts an organism can make in its lifetime. *See:* **genome, proteome**.

transcriptomics the global study of an organism's transcripts.

transdifferentiation also the use of transcription factors to directly transform one differentiated cell type into another. *See:* **metaplasia**.

transfection a method for incorporating substances (e.g., genes) directly into cells by incubating the cells in a solution that makes them "drink in" the substance. *See:* **electroporation, microinjection**.

transfer RNA (tRNA) a class of RNA molecules that carries amino acids to their correct positions (as specified by the genetic code) along mRNA on the ribosome.

transferrin an iron-transport protein that is necessary for mitosis in all dividing cells; proliferation of the salamander limb regeneration blastema is dependent on the presence of nerves; transferrin is a candidate for a neural-derived mitotic factor involved in such regeneration.

transforming growth factor α (TGF-α) an autocrine growth factor that stimulates development of the epidermis; made by the basal cells and stimulates their own division. *See:* **keratinocyte growth factor**.

transforming growth factor β (TGF-β) a family of protein growth factors, including the activins, bone morphogenetic proteins (BMPs), decapentaplegic protein (in *Drosophila*), and Vg1 protein (in *Xenopus*); their actions include potentiating or inhibiting responses of most cells to other growth factors, regulating the differentiation of some cell types, and acting as inductive signals in embryonic development. TGF-β was originally discovered as a mitogen secreted by "transformed" (cancer-like) cells.

transgene a foreign gene transplanted into an organism, making the recipient a transgenic organism.

transgenerational continuity the inherited epigenetic state, i.e., DNA methylation patterns, of certain genes, e.g., *Agouti* gene and *Axin* gene in mice. This suggests that the DNA methylation pattern is not erased between generations. *See:* **methylation erasure**.

transgenic organism organism having an exogenous gene inserted into its genome. *See:* **transgene**.

translation the synthesis of proteins using mRNA as a template.

translational regulation regulation of translation of mRNA already existing in the cytoplasm; e.g., mRNA in the egg may not be translated until after fertilization.

translocated chemical messengers chemicals made in one part of a plant that move to another part. *See:* **auxin**.

transmission genetics the study of the transmission of genes from one generation to the next. *See:* **developmental genetics**.

transplant tissue removed from any portion of the body and placed in a different site

transplantation the artificial removal of part of an organism and its replacement in the body of the same or of a different individual.

transplantation experiment involves replacing one part of an embryo with a portion from a different embryo.

transposable element a naturally occurring mobile region of DNA that can integrate itself into the genome of an organism; can be used to introduce genes into an organism, e.g., P elements of *Drosophila*.

trans-**regulation** *trans*-acting; a factor, usually a soluble protein, produced at a gene distant from the chromosome region that it affects.

trans-**regulatory factor** a factor, usually a soluble protein, produced by a gene distant from the chromosome region that it affects; are usually transcription factors that bind to *cis*-regulatory elements (promoters and enhancers). *See:* *cis*-**regulatory element**.

transverse plane that plane of the body that separates the anterior end of the body from the posterior end.

trefoil stage the two-cell embryo stage in some molluscs, e.g., *Illyanassa* or *Dentalium*, where the polar lobe attached to one of the blastomeres (the CD blastomere) is so large and is attached by such a narrow stalk that the embryo appears to consist of three blastomeres.

trehalose   a protective sugar used to preserve proteins during the dehydration accompanying dormancy in some species.

Trembley, Abraham (1710–1784)   Swiss zoologist, discovered (in 1740) regeneration of the freshwater polyp, *Hydra*. Embryologists, of his time, saw in the polyp conclusive proof against the preformationists who claimed that embryos were minute but preformed individuals.

trichoblasts   those cells of the plant epidermis that are hair cell precursors. *See:* **atrichoblasts**.

trichomes   the hairs on the leaves and stems of plants, or a hair-bearing cell in plant epidermis.

tricuspid valve   the valve between the right atrium and the right ventricle of the heart, formed from endocardial cushion tissue.

triiodothyronine   *See:* **thyroxine**.

trimester   one third of the duration of pregnancy.

triplets   the offspring of a single pregnancy simultaneously carrying three conceptuses; three siblings that happen to occupy the same uterus (womb) at the same time.

triploblasts   animals composed of three germ layers, including all bilaterians; only three phyla are not triploblastic: Porifera (sponges), Cnidaria (e.g., *Hydra*, jellyfish, and sea anemones) and Ctenophora (comb jellies).

triploid number   three times the haploid number of chromosomes: 69 for humans.

triploidy   exhibiting the triploid number of chromosomes; not consistent with postnatal human life, although sometimes found in a mosaic individual.

trisomy   three bodies; three copies of a given chromosome rather than the normal two copies; a number of human developmental abnormalities involve trisomy of different chromosomes; e.g., see trisomy 21. Except for trisomy 1, trisomies of all human autosomes have been found in spontaneous abortuses.

trisomy 21   the presence of three (rather than of the normal two) copies of chromosome 21; results in offspring with Down's syndrome. *See:* **karyotype**.

tritocerebrum   *See:* **neuromeres**.

trizygotic   of three zygotes; as with fraternal triplets.

Trochaea Theory   Berthold Hatschek (in 1878) first used the term "trochophore" for the larva of a polychaete, *Polygordius* sp. He described this larva as having an apical organ, pre-oral and post-oral ciliated bands flanking a groove, a complete gut, and a pair of protonephridia. Based on this study, Hatschek (in 1878) developed the theory that the trochophore was the larval form of a *hypothetical* animal, Trochozoon. Trochozoon, which in its adult form resembled a trochophore-like rotifer, was the proposed ancestor of most bilaterian metazoan taxa, and the trochophore larva was hence plesiomorphic for these taxa. *See:* **plesiomorphic**.

trochophore   a pear- or top-shaped organism, the larval form of many aquatic invertebrates; this larva in modified form is found in many invertebrates, such as polychaetes, rotifers, mollusks, and others; it is supposed to signify a common ancestry of coelomate Protostomia.

trophectoderm   *See:* **blastocyst**.

trophoblast   the outer of the two parts of the blastocyst, which does not contribute to the embryo but only to extraembryonic structures (see Plate 7 in the color insert).

trophocytes   nutritive cells of the ovary or testis of an insect; the nurse cells of the insect ovariole.

trophoderm   *See:* **blastocyst**.

truncoventricular region   the region of the early developing heart that gives rise to the ventricles and the truncus arteriosus.

truncus arteriosus   the embryonic arterial trunk between the bulbous arteriosus and the ventral aorta in anamniotes and the early stages of amniotes; the arterial trunk that develops from the heart of the embryo; gives rise to the aorta and the pulmonary trunk (see Fig. 20).

trunk neural crest   located between the cardiac neural crest and the sacral neural crest; give rise to melanocytes, dorsal root ganglia, sympathetic ganglia, the adrenal medulla, and the nerve clusters surrounding the aorta.

tubal ligation   a birth control method involving the cutting and tying of the fallopian tubes.

tubal pregnancy   *See:* **ectopic pregnancy**.

tube cell   one of two cells in a pollen grain, the tube cell directs the growth of the pollen tube of the germinated pollen grain. *See:* **generative cell.**

tubercles   small nodules.

tuberculum posterius   the most anterior part of the floor of the brain in the frog embryo; it is a landmark used to determine the posterior boundary of the forebrain.

tubuli recti   the straight portions of the seminiferous tubules. *See:* **seminiferous tubules.**

tubulin   the protein that forms the microtubules of the cytoskeleton and spindle.

tumor angiogenesis factors   factors secreted by microtumors, including vascular endothelial growth factor (VEGF), FGF2, placenta-like growth factor, and others, which stimulate mitosis in endothelial cells and direct cell differentiation into blood vessels in the direction of the tumor.

tumor-suppressor genes   genes that cause a cell to become malignant when both alleles of the gene have been inactivated. *See:* **oncogenes.**

tunica   meristems in flowering plants are organized into layers that encompass cells in both the central and peripheral zones; the outer layer or layers are referred to as the tunica. *See:* **corpus.**

tunica albuginea   the tough connective tissue covering of the testis or the connective tissue layer surrounding the cortex of the ovary (see Figs. 15 and 45).

tunicates   also known as urochordates, are members of the subphylum Tunicata or Urochordata, a group that is classified within the phylum Chordata. The Tunicata (Urochordata) encompasses four classes, including the Ascidiacea. *See:* **ascidians.**

Turing, Alan Mathison (1912–1954)   co-founder of computer science; the mathematician who cracked the German "Enigma" code during World War II; proposed the reaction-diffusion system of pattern generation; wrote the influential paper, *The Chemical Basis of Morphogenesis* (1952). *See:* **morphogenesis, reaction-diffusion model.**

Turing model   *See:* **reaction-diffusion model.**

Turner's syndrome   a syndrome, including a female phenotype, short stature, and, possibly, mental retardation; resulting from an XO sex chromosome constitution (see Fig. 46).

twins   the offspring of a single pregnancy simultaneously carrying two conceptuses; two siblings that happen to occupy the same uterus (womb) at the same time. *See:* **conjoined twins, fraternal twins, identical twins.**

tympanic cavities   the cavities of the middle ears.

tyrosine kinases   enzyme-linked receptors of importance in development; phosphorylate proteins at tyrosine amino acid residues. *See:* **kinases, serine/threonine kinases.**

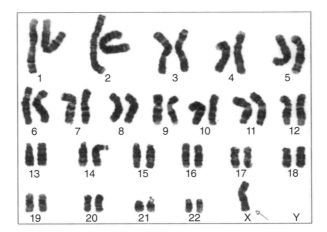

Figure 46. Karyotype of a patient with Turner's syndrome; note the presence of only one sex chromosome (X). (Courtesy of Jacqueline Burns, PhD, the Danbury Hospital.) Reprinted from Frank J. Dye, *Human Life Before Birth*, Harwood Academic Publishers, 2000, fig. 19-5, p. 165.

U

ultimobranchial bodies   the small, endocrine, structures that originate as terminal outpock-etings from each side of the embryonic vertebrate pharynx; can produce the hormone calcitonin.

ultrasonography   a noninvasive, prenatal diagnostic procedure that uses high-frequency sound waves (ultrasound) to create images of the embryo or fetus.

ultrasound   (see Fig. 1). *See:* **ultrasonography**.

umbilical arteries   the pair of blood vessels that carries blood from the fetus through the umbilical cord out to the placenta; the blood carried is deficient in food and oxygen (see Fig. 35).

umbilical cord   the cord-like structure that attaches the belly of the fetus to the placenta and through which materials exchanged between fetus and mother flow (see Figs. 6, 21, 36, and 47).

umbilical duct   *See:* **vitelline duct**.

umbilical hernia   a hernia occurring through the umbilical ring, which may be congenital as a result of imperfect closure of the umbilical ring; the umbilical ring is a dense fibrous ring surrounding the umbilicus at birth. An umbilical hernia may also be acquired later in life.

umbilical ring   a dense fibrous ring surrounding the umbilicus at birth.

umbilical vein   the blood vessel that carries blood to the fetus through the umbilical cord from the placenta. The blood carried is rich in food and oxygen (see Fig. 35).

umbilicus   the naval; a naval-like structure, as the hilum of a seed.

unicellular   composed of a single cell; the only stage in the human life cycle that is unicel-lular is the fertilized egg (zygote).

unipotent stem cells   adult stem cells that are involved in regenerating a specific type of cell; e.g., spermatogonia are unipotent stem cells that give rise to only spermatozoa.

unity of type   the similarities among organisms; homologies. *See:* **descent with modification, homologous**.

unsegmented mesoderm   the bands of, as yet, unsegmented paraxial mesoderm in mam-malian embryos.

up-mutation   a mutation, usually in a promoter, that results in more expression of a gene. *See:* **down-mutation**.

upregulation   process that increases ligand/receptor interactions as a result of an increase in the number of available receptors; epidermal growth factor (EGF)-induced upregulation of cyclin D1 protein in NIH 3T3(M17) fibroblasts. *See:* **downregulation**.

upstream   (1) in terms of a signal transduction pathway (cascade), upstream would be closer to the cell surface/receptor; e.g., in the MAK-kinase pathway Ras is upstream from Raf; (2) in terms of a gene, other genes that regulate the gene in question; e.g., the HOM cluster genes are subject to generalized regulation by upstream genes, such as activators like *trithorax* and receptors like *polycomb*. *See:* **downstream**.

urachus   a cord or tube of epithelium connecting the apex of the urinary bladder with the allantois; its connective tissue forms the median umbilical ligament.

Urbilateria   the hypothetical last common ancestor of all bilaterians.

ureotelic organisms   organisms, such as most terrestrial vertebrates, including many kinds of adult frogs, which excrete urea; during the metamorphosis of many kinds of frogs (e.g., *Rana*), the liver develops the enzymes of the urea cycle. *See:* **ammonotelic organisms**.

ureter   the pair of tubes that carries urine from the kidneys to the urinary bladder (see Fig. 4).

Dictionary of Developmental Biology and Embryology, Second Edition. Frank J. Dye.
© 2012 Wiley-Blackwell. Published 2012 by John Wiley & Sons, Inc.

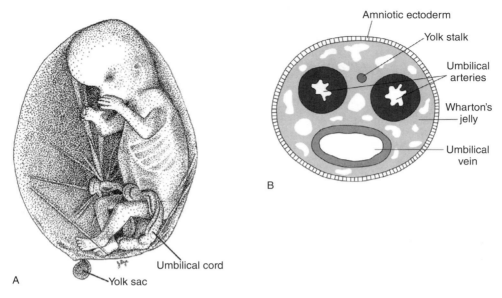

Figure 47. The umbilical cord. (A) The human fetus within its amniotic sac. (B) Cross section of the umbilical cord at term. Reprinted from Frank J. Dye, *Human Life Before Birth*, Harwood Academic Publishers, 2000, fig. 10-6, p. 86. Original artwork (A) by John C. Dye.

ureteric buds   paired epithelial branches induced by the metanephrogenous mesenchyme to branch from each of the paired nephric ducts; will form the collecting ducts, renal pelvis, and ureters.

urethra   the duct running from the urinary bladder to the outside of the body. In females, it is a purely urinary duct leading to the urethral opening into the vestibule; in males, it is a urogenital duct carrying both urine and semen to the urogenital opening on the surface of the glans penis (see Fig. 4).

urinary bladder   the sac-like organ that receives and temporarily stores urine until the time of urination.

urine   liquid nitrogenous waste formed from the circulatory system and excreted by the kidneys.

uriniferous tubules   tubules found in the kidneys, which participate in the formation of urine.

urochordates   marine animals placed in the subphylum, Urochordata, of the phylum Chordata; they generally possess a notochord in their tails (and only in their tails) during part of their life cycle; also called tunicates (because they are surrounded by a tough tunic); they include sea squirts, larvaceans, and thaliaceans.

Urodela   tailed amphibians; the salamanders.

urogenital ducts   ducts that carry the products of both the urinary system (urine) and the reproductive system (spermatozoa), e.g., uriniferous tubules in male frogs and the penile urethra in male humans.

urogenital membrane   the membrane found between the labiscrotal swellings of the early external genitalia; its rupture opens the urethral groove.

urogenital ridge   either of the paired ridges from which the urinary and genital systems develop.

urogenital sinus   the part (anterior) of the cloaca into which the embryonic urogenital ducts open. The early urogenital sinus (upper) in males gives rise to the bladder and the prostatic urethra; in females, it gives rise to the bladder and the urethra. The definitive urogenital sinus (lower) gives rise to the lower urethra in males; in females, it gives rise to the vestibule.

urorectal septum   the embryonic, horizontal, connective-tissue septum that divides the cloaca into the rectum and primary urogenital sinus.

urostyle   an unsegmented bone representing several fused vertebrae and forming the posterior part of the vertebral column in Anura.

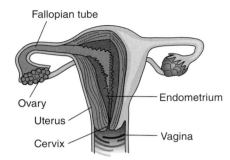

Fallopian tube

Ovary

Uterus

Cervix

Endometrium

Vagina

Figure 48. The uterus. A partially dissected uterus showing the surface and interior of the organ. Reprinted from Frank J. Dye, *Human Life Before Birth*, Harwood Academic Publishers, 2000, fig. 5-9, p. 40.

uteric bud   *See:* **metanephric diverticulum**.

uterine cavity   the cavity or lumen of the body (corpus) of the uterus. The young human embryo arrives in the uterus in the morula stage, still encased in the zona pellucida. For a period of approximately 3 to 4 days, the embryo remains unattached in the uterine cavity; before implantation, there is a hollowing out of the morula to form the blastodermic vesicle (blastocoel). Escape of the embryo from the zona pellucida allows rapid expansion of the trophoblast layer and permits these cells for the first time to make contact with the uterine lining (see Fig. 22).

uterine cycle   a component of the menstrual cycle; the function of the uterine cycle is to provide the appropriate environment for the developing blastocyst.

uterine horn   one of the two principle parts of the mouse bicornuate uterus, into each of which blastocysts will implant.

uterine tube   the oviduct in mammals, also called the fallopian tube.

uterus   the female reproductive organ concerned with the reception, retention, nurturing, and expulsion of the conceptus, in mammals. Generally, in vertebrates, protection and care of the egg or of the embryo during a part or all of its development is the main function of the uterus. In frogs, the uteri store eggs just prior to their extrusion. In chickens, the uterus is where the "egg" has a shell added to its accessory membranes (see Figs 16, 21, 22, 34, and 48).

utriculus   the part of the membranous labyrinth into which the semicircular canals open.

uvula   the small piece of flesh that hangs down from the soft palate at the back of the throat.

V

vagal neural crest the neck neural crest; together with the sacral neural crest, generates the parasympathetic (enteric) ganglia of the gut.

vagina the female organ of intercourse, which also serves as part of the birth canal during childbirth. By contrast with the seminal plasma pH of 7.2–7.8, the vagina is very acid, being approximately pH 5.7 in women (see Figs. 16, 22, and 34).

valproic acid (VA) a teratogen that blocks folate (vitamin B9) from being absorbed by the embryo, leading to neural tube defects; decreases the level of *Pax1* transcription in chick somites, causing malformations of the vertebrae and ribs; is implicated in certain cases of autism; and apparently inhibits the activity of histone deacetylases, crucial regulators of gene expression.

varicose veins veins that have become abnormally dilated and tortuous.

vas deferens (pl. vasa deferentia) the sperm duct proper in males; the distal portion of the reproductive duct in males. In male insects, there are two lateral ducts or vasa deferentia; these join a median ejaculatory duct; there is frequently a pair of seminal vesicles, formed by the enlargement of a part of each vas deferens (see Figs. 15). *See:* **ductus deferens**.

vasa efferentia approximately a dozen fine ducts that carry sperm from each frog testis to uriniferous tubules, which, during breeding season, are functional urogenital ducts.

vascular cambium a plant lateral meristem, the activity of which results in the formation of vascular tissues, i.e., xylem and phloem. The vascular cambium produces xylem toward the center of a plant part and phloem toward the periphery of a plant part; vascular cambium is less active in herbaceous plants than in woody plants.

vascular endothelial growth factor (VEGF) a family of proteins involved in vasculogenesis; apparently enables the differentiation of angioblasts and their multiplication to form endothelial tubes. *See:* **angiopoietin-1, FGF2**.

vascular spiders minute reddened elevations of the skin from which there is branching of radicles (little roots); commonly exhibited by pregnant women.

vasculogenesis a type of blood vessel development that occurs during embryogenesis, whereby endothelial cells are born from progenitor cells. *See:* **angiogenesis**.

vasectomy a surgical contraceptive procedure that entails cutting the vasa deferentia and tying (ligating) the cut ends.

vasovasostomy the surgical reversal of a vasectomy.

vault an arched structure; as in the vault of the vagina.

vector a substance used to carry genes into cells and chromosomes. *See:* **retroviral vector, transposable element**.

vegetal determinant one of two determinants found in the fertilized *Xenopus* egg; becomes established during oogenesis as a result of mRNA localization to the vegetal cortex and causes formation of the endoderm; this is the source of a mesoderm-inducing signal. *See:* **dorsal determinant**.

vegetal gradient one of two gradients of principles that are mutually antagonistic, but that must interact with each other if normal sea urchin development is to occur. The vegetal gradient is centered at the vegetal pole. *See:* **vegetalized**.

vegetal hemisphere that half of the egg that has the vegetal pole at its center.

vegetal plate the flattened region surrounding the vegetal pole of the early sea uchin gastrula, which subsequently undergoes invagination.

vegetal pole that point on the surface of an oocyte, egg, zygote, or embryo directly opposite (180°) the animal pole (see Plate 3 in the color insert).

Dictionary of Developmental Biology and Embryology, Second Edition. Frank J. Dye.
© 2012 Wiley-Blackwell. Published 2012 by John Wiley & Sons, Inc.

vegetalization   the process of becoming vegetalized; factors that cause the sea urchin embryo to become vegetalized include lithium ions, sodium azide, and dinitrophenol. *See:* **dorsalization**.

vegetalized   a term used in the context of experimental embryology of, primarily, sea urchins; when the sea urchin develops to excess those parts pertaining to the vegetal gradient, e.g., the gut. If the animal gradient is weakened or suppressed, the vegetal gradient becomes preponderant and the embryo is vegetalized. *See:* **animalized, vegetal gradient**.

vegetalizing factors   those molecules that cause vegetalization.

vegetative   not directly concerned with sexual reproduction. The vegetative parts of flowering plants are roots, stems, and leaves, as opposed to flowers that are fertile parts of flowering plants. *See:* **fertile**.

vegetative propagation   a type of asexual reproduction in plants; production of a new plant from a part of another plant, such as a stem. Vegetative propagation may be artificial, as in taking cuttings of geraniums or the production of new carrot plants from individual totipotent carrot phloem cells, or natural, as in the bulblet fern, *Cystopteris bulbifera*. *See:* **asexual reproduction**.

veins   blood vessels that carry blood to the heart. Although this blood is generally deficient in food and oxygen, the umbilical vein carries replenished blood from the placenta into the fetus.

vellus   short and silky hair that replaces the lanugo, which is usually shed before birth. *See:* **lanugo**.

venae cavae   (sing. vena cava) the major venous blood vessels returning blood to the heart (right atrium). *See:* **cardinal veins**.

venter   the belly.

ventrad   in a ventral direction.

ventral   referring to the venter or belly of an animal.

ventral furrow   the inward folding of the prospective mesoderm, making up the first movement of *Drosophila* gastrulation.

ventral mesocardium   a membrane, derived from a double layer of splanchnic mesoderm, which forms beneath the developing heart in the pericardial region of the coelom. This membrane disappears almost as soon as it forms.

ventricles   small cavities. (1) heart: two of four chambers of the heart, they pump blood out of the heart; (2) brain: the fluid-filled spaces within the brain through which cerebrospinal fluid circulates (see Figs. 7 and 20). *See:* **fourth ventricle, lateral ventricles, third ventricle**.

ventricular layer   the inner layer of the spinal cord, made up of ependymal cells and adjoining the cerebrospinal fluid-filled spinal canal; a layer of dividing cells lining the lumen of the neural tube, from which neurons and glia are formed (see Fig. 43).

ventricular septal defect   also called VSD, a congenital anomaly characterized by an abnormal opening in the ventricular septum, between the right and left ventricles.

Veratrum californicum   California false hellebore. *See:* **cyclopia**.

vernal   flowering in the spring.

vernalization   temperature-dependent chromatin changes. A period of chilling that can enhance the competence of shoots and leaves to perceive or produce a flowering signal. Also, the phenomenon whereby the seeds of certain plants need to be exposed to cold temperatures in order to flower.

vernation   the arrangement of leaves within the bud.

vernix caseosa   a whitish, cheesy, coat on the skin of the fetus made up of secretions of sebaceous glands (sebum), periderm cells, and lanugo.

vertebrae   (sing. vertebra) the elements of the vertebral column or "backbone."

vertebral column   the "backbone" of the vertebrates; that portion of the axial skeleton derived from the sclerotomes of somites responsible for the bony protection of the spinal cord.

vertebrate body plan   the fundamental vertebrate body plan consists of a (1) head (cephalic) end, (2) trunk (middle) region, (3) dorsum (uppermost or dorsal region), (4) venter (lower or ventral region), and (5) tail (caudal) end. A transverse section through the trunk of the vertebrate body reveals basically five hollow tubes (all oriented around the notochord): an epidermal tube, a neural tube, an enteric (gut) tube, and two mesodermal tubes.

vertebrate characteristics   there are four vertebrate characteristics: (1) a notochord, at least in the embryo; (2) presence of a pharynx with pouches or slits in its walls, at least in the embryo; (3) a dorsal, tubular nervous system; and (4) a vertebral column.

vertebrates animals placed into the subphylum Vertebrata, of the phylum Chordata, because they possess, in addition to a notochord, a vertebral column. The development of neural crest cells (and the epidermal placodes that give rise to the sensory nerves of the face) distinguish the vertebrates from the protochordates. *See:* **vertebrate characteristics**.

vertical division *See:* **meridional division**.

vertical transmission the transfer of symbionts from one generation to the next through the germ cells, usually the eggs; e.g., certain clams produce oocytes that contain microbial symbionts. *See:* **horizontal transmission**.

vesicle a small sac; usually containing fluid.

vestibular glands (1) greater, such as Bartholin's glands, and (2) lesser, such as Skene's glands. *See:* **Bartholin's glands, Skene's glands**.

vestibule the almond-shaped opening that makes up part of the vulva or external genitalia of the female, into which the urethra and vagina open.

vestigial an organ or structure that is much reduced and likely nonfunctional.

Vg1 a gene whose expression in the Nieuwkoop center of the frog blastula and its homo-logue's expression in the chick posterior marginal zone strengthens the identification of these two organizing regions with each other.

Vg1 **mRNA** is tethered to the vegetal yolk mass during oogenesis and remains in the vegetal hemisphere during cleavage. This message is initially present throughout the oocyte and is translocated into the vegetal cortex in a two-step process: First, microtubules bring the *Vg1* mRNA into the vegetal hemisphere, and second, microfilaments are responsible for anchoring the *Vg1* message to the cortex.

Vg1 protein in the frog embryo, can induce dorsal mesoderm formation in the cells above it. After fertilization, the Vg1 protein is made throughout the vegetal hemisphere of the blastula, in an inactive form that must be cleaved to become active; activated Vg1 protein: (1) induces dorsal mesoderm in animal cap cells, (2) induces an embryonic axis when injected into ventral vegetal cells, and (3) rescues the dorsal axis of ultraviolet (UV)-irradiated eggs when injected into dorsal vegetal cells. It is possible that the combination of Vg1 and some product specified by the *Siamois* gene can induce the specification of the dorsal mesoderm and its differentiation into notochord. The mature Vg1 protein seems to be critical in the functioning, and perhaps the establishment, of the Nieuwkoop center of amphibians; Vg1 is also observed in the homologous region of the chick embryo. In animal cap cultures, the Vg1 protein can induce the transcription of the *goosecoid* gene.

viability ability to live.

vis essentialis an unknown force (essential force) postulated to exist by Wolff, Friedrich Kaspar (1733–1794), which, acting like gravity or magnetism, would organize embryonic development.

viscera the internal organs, especially of the abdominal cavity.

visceral relating to viscera (internal organs), as in visceral pericardium or visceral peritoneum.

visceral arches branchial arches.

visceral furrows branchial furrows.

visceral grooves branchial grooves.

visceral mesoderm in *Drosophila*, visceral mesoderm generates the smooth muscle sur-rounding the gut, while the somatic mesoderm gives rise to the main body muscles. HOM gene expression in the visceral mesoderm seems to induce a pattern in the gut endoderm; the visceral mesoderm may pattern the endoderm by transfer of positional information from one germ layer to another by means of extracellular signals. *See:* **splanchnic mesoderm**.

viscerocranium that portion of the cranium that surrounds the oral cavity, pharynx, and upper respiratory tract. *See:* **neurocranium**.

vital stains/dyes in general, to stain cells and their parts, it is necessary to fix (kill) the cells; a few dyes can stain living cells and their parts; such dyes are called vital dyes or vital stains. Historically, vital stains have been used by embryologists and developmental biologists to produce fate maps for organisms as diverse as amphibians and slime molds. *See:* **fate map**.

vitalism the theory that certain animal functions are dependent on a special form of energy or force, the vital force, distinct from any other of the physical forces.

vitamin A (retinol) teratogenic in megadose amounts; causes babies to be born with disrup-tions similar to those produced by retinoic acid. Vitamin A deficiency leads to retarded development and growth.

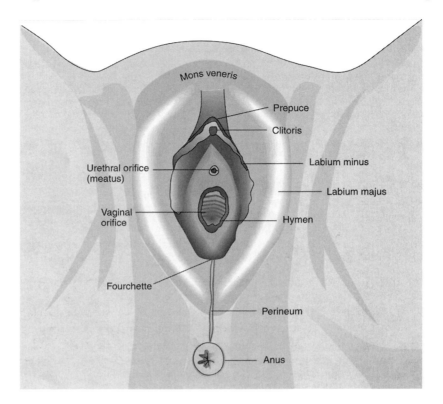

Figure 49. The female external organs of reproduction. Reprinted from Frank J. Dye, *Human Life Before Birth*, Harwood Academic Publishers, 2000, fig. 5-6, p. 38.

vitamin B₁₂   *See:* **folic acid.**

vitellarium   the part of the insect ovariole formed by the backward extension of the wall of the germarium; oocytes are liberated in order of their age into the vitellarium.

vitelline duct   the passageway between the yolk sac and the midgut of the embryo; also known as yolk duct, omphalomesenteric duct, and umbilical duct.

vitelline envelope   *See:* **vitelline membrane.**

vitelline membrane   a noncellular, glycoprotein coat found around the eggs of certain animals, e.g., amphibians, birds, and sea urchins; the homologous egg coat in mammals is the zona pellucida. Frog and chick eggs are enveloped by a vitelline membrane. In chick development, the vitelline membrane is essential for epiboly of the ectoderm during gastrulation. Insect eggs have a vitelline membrane, a product of the egg itself, just beneath the chorion.

vitelline vascular plexus   the maze of blood vessels on the yolk surface; of chick embryos, for example.

vitelline veins   also known as omphalomesenteric veins; any of the embryonic veins in vertebrates uniting the yolk sac and the sinus venosus; their proximal fused ends form the portal vein.

vitellogenesis   the formation of yolk in the ooplasm of an oocyte.

vitellogenin   the major yolk component in frog eggs, a 470-kDa protein synthesized by the liver, carried to the ovary by the bloodstream, and incorporated into the oocyte by micropinocytosis. In the mature oocyte, vitellogenin is split into two smaller proteins: phosvitin and lipovitellin, which are packaged together into membrane-bounded yolk platelets.

viviparity   birth of young as opposed to the laying of eggs.

viviparous   in plants, sprouting on the parent plant, as the bulblets forming in some inflorescences. *See:* **viviparity.**

viviparous mutants   single-gene mutants, in corn (maize, *Zea mays*), that lack the ability to make abscisic acid or exhibit a reduced sensitivity to the hormone; the embryos of which do not become dormant and germinate directly on the cob.

Vogt, Walter (1888–1941)   brought to prominence the significance of Edwin Grant Conklin's work on the chordate phylum as a whole, i.e., the concept of organ-forming, germinal areas. He stained different parts of the amphibian blastula with vital dyes, e.g., Nile-blue sulfate, Bismark brown, and neutral red. He then developed fate maps for amphibian embryos.

Volvox   a genus of colonial, freshwater, green algae studied by developmental biologists interested in cell differentiation and the origin of multicellularity.

vulva   the female external genitalia (see Fig. 49).

W

Waddington, Conrad Hal (1905–1975) English embryologist and geneticist, professor of animal gentics at Edinburgh University; proposed (in 1932) that the hypoblast influenced the orientation of the chick embryonic axis. Attempted to find developmental mechanisms for producing new species; he, as did Richard Goldschmidt, looked at homeotic mutations in flies as models for drastically new phenotypes and formulated the notion of competence transfer to explain certain aspects of morphological evolution. Waddington was largely ignored because, among other things, his work was misinterpreted as supporting the inheritance of acquired traits. He studied the effects of chemical messengers in inducing embryonic cells to form particular tissues during development, and he was especially concerned with the ways in which both genes and environmental influences control the development of embryos. He believed that an explanation of developmental canalization required supplementing conventional gene theory with an "epigenetic theory," i.e., one in which discrete and separate entities of classic genetics would be displaced by collections of genes that could "lock in" development through their interactions. His *Organisers and Genes* (1940) covers Mendelian genetics and experimental embryology, and he also wrote a standard textbook of embryology, *Principles of Embryology* (1956). *See:* **canalization, competence transfer, genetic assimilation**.

Warynski, Stanislas (1858–1888) Swiss physiologist; published, with Hermann Fol, (in 1884) a new experimental methodology of embryo manipulation, the direct method, which involved placing an instrument directly on a precise point of the embryo. By this method, Fol and Warynski, could produce specific malformations in chick embryos by precisely controlling the point of injury caused by their instruments; thus, they could better understand normal chick development by observing the relationship between experimentally produced embryonic defects and the resulting malformations. *See:* **Dareste, Camille**.

Weismann, August (1834–1914) German biologist; was one of the great synthesizers of biology; in 1887, he recognized that the doubling of the chromosome number that occurs in fertilization would lead to unmanageable numbers of chromosomes unless a means was at hand to counteract this buildup (meiosis, of course, is that process). Proposed the first testable alternative to Charles Darwin's theory of pangenesis. In 1883, he proposed the germplasm theory, which maintained that different nuclear determinants entered different cells and only in the nuclei of germ cells were all types of determinants retained; his hypothesis proposed the continuity of the germplasm and the diversity of the somatoplasm, that the germ line was totally independent of the somatic cells. Edmund Beecher Wilson claimed that his textbook grew out of Weismann's hypothesis.

Weismann came to conclusions similar to those of Wilhelm Roux, but he went on to offer a larger theory of development.; his theory was based on the assumption that physical hereditary units exist within the nucleus, and he postulated a mechanism for the separation of those units. Weismann's theory offered three levels of units: biophores, arranged in packets called ids; ids that determine each particular characteristic of the developing embryo; and idants, corresponding to the chromosomes that are the smallest units actually visible. At each cell division, idants divide into parts that differ from each other and then move into the daughter cells; the ids, discrete units located along the lengths of the idants, contain for each characteristic, a set of biophores that undergo competition and those that prevail will determine what the resulting cell will become. *See:* **immortality of the germplasm**.

Weismann barrier the concept, developed by August Weismann, that the germ plasm is insulated from environmental influences.

Dictionary of Developmental Biology and Embryology, Second Edition. Frank J. Dye.
© 2012 Wiley-Blackwell. Published 2012 by John Wiley & Sons, Inc.

Went, Fritz (1903–1990)   carried out the crucial experiment (in 1923), as a young Dutch botany student, which proved that plants also possess growth hormones; he named the substance that he had discovered "auxin."

Wharton's jelly   the connective tissue matrix of the umbilical cord; surrounds the umbilical cord blood vessels (see Fig. 47).

white matter   those portions of the central nervous system that have a large concentration of nerve fibers and therefore appear white in the fresh state (see Fig. 43).

Whitman, C. O. (1842–1910)   determined (in 1878) the first cell lineage, for the leech *Clepsine*; his pioneering studies were soon followed by extremely detailed determinations of cell lineages in other embryos.

wholist organicism   the concept that the properties of the whole cannot be predicted solely from the properties of the component parts and that the properties of the parts are informed by the whole. According to S. F. Gilbert (2006), embryology has traditionally espoused wholist organicism as its ontology (model of reality) while maintaining a reductionist methodology (experimental procedures). *See:* **reductionism**.

Wieschaus, Eric F. (1947–)   corecipient, with Edward B. Lewis (United States) and Christiane Nüsslein-Volhard (Germany), of the 1995 Nobel Prize for Physiology or Medicine for discovery of genes that control the early stages of the body's development.

Wiesel, Torsten (1924–)   Nobel laureate (for physiology or medicine, 1981, shared with David Hubel and Roger W. Sperry); demonstrated that there is competition between the retinal neurons of each eye for targets in the cortex and that their connections must be strengthened by experience.

Wilson, Edmund Beecher (1856–1939)   American zoologist; specialized in cytology, embryology, and experimental morphology. He wrote the classic book, *The Cell in Development and Heredity*, 3rd edition (1925, Macmillan, New York). In the early 1890s, Wilson worked with Hans Driesch at the Zoological Station in Naples. Wilson, Driesch, and others did not gravitate toward Wilhelm Roux's emphasis on nuclear division to explain development; rather, they continued to look at the internal structure and the patterns of cell division for clues to the causes of embryonic differentiation.

In 1898, at Woods Hole, Wilson presented a lecture that was a landmark in the use of embryonic homologies to establish phylogenetic relationships. Wilson (1904) carried out isolation experiments with *Patella coerulea*, a mollusc, and found that the blastomeres possess within themselves all the factors that determine (1) the form and rhythm of cleavage and (2) the differentiation that they undergo, wholly independent of their relationship to the remainder of the embryo. He also carried out experiments (in 1904) where he removed polar lobes from *Dentalium*, a mollusc, and determined that its mosaic development was based on the segregation of specific morphogenetic determinants into specific blastomeres; e.g., Wilson concluded that the polar lobe cytoplasm contained the mesodermal determinants. *See:* **Lillie, F. R.**; **Weismann, August**.

Wilson, H. V. (1863–1939)   according to Gerald B. Grunwald (1994), in a 1907 paper, Wilson describes the original observations that led to the studies often cited as the first experiments in the field of cell adhesion; and this paper is also cited as an experimental foundation for the analysis of adhesive selectivity (Wilson's dissociated sponge cells went on to reconstitute sponges but only within a species and not between species); the paper was concerned with some phenomena of coalescence and regeneration in sponges. Wilson obtained his sponge cells by straining fresh normal sponges through bolting cloth, and so, as Grunwald relates, by expression through a fine silk cloth was the field of cell adhesion born. Grunwald states that Wilson believed that aggregates of sponge cells formed a syncytium and redifferentiation occurred. Wilson's work was later cited by many embryologists as one of the foundations of the field of embryonic cell adhesion studies.

Wilson's principles of teratology   James Wilson (in 1956) put forth six principles regarding teratogenesis.

Winged helix proteins   transcription factors; have a 100-amino-acid winged helix domain that forms a DNA-binding region; known as "Fox" proteins, e.g., Forkhead in *Drosophila* embryonic termini and HNF3ß in the vertebrate main axis.

wingless   a segment polarity gene in *Drosophila*; a mutation of it results in the whole of the ventral abdomen being covered in denticles, with the denticle pattern reversed in the posterior half of each segment. *Wingless* encodes a secreted signaling protein and is related to the *Wnt* gene in vertebrates. Wingless and hedgehog proteins are highly conserved developmental signaling molecules.

winter bud   a hibernating vegetative shoot.

Wnt   a family of inducing (paracrine) factors, the founding member of which was discovered as an oncogene in mice and as the *wingless* mutation in *Drosophila*.

Wnt receptors   are called frizzleds.

Wnt signal transduction pathway   an intracellular signal transduction pathway used in different cells in different ways; additionally, its components can have more than one function in the cell; involved in segment polarity in *Drosophila*, axis formation in *Caenorhabditis elegans*, and axis formation in *Xenopus*.

Wolbachia   a genus of bacteria.

Wolff, Friedrich Kaspar   (1733–1794)   German anatomist; founder of modern embryology; several embryological structures are named after him. A century after William Harvey, he became the most celebrated proponent of epigenesis; Wolff was able to observe microscopically the building up of the chick embryo and he saw no evidence of an encapsulated chick in the egg.

Ranked by some as the foremost student of embryology prior to Karl Ernst von Baer. His *Theoria Generationis* was published in 1759; consisting of three parts devoted to (1) development of plants, (2) development of animals, and (3) theoretical considerations. He contended that the organs of animals make their appearance gradually and that he could actually follow their successive stages of formation. Wolff was the discoverer of the primitive kidneys or "Wolffian bodies." According to William A. Locy (1908), because he assumed a total lack of organization in the beginning, he was obliged to make development "miraculous" through the action on the egg of a hyperphysical agent; from a total lack of organization, he conceived of its being lifted to the highly organized product through the action of a *vis essentialis corporis*.

In 1768–1769, he published his best work in embryology on the development of the intestine; of which von Baer said, "It is the greatest masterpiece of scientific observation which we possess." Again, according to Locy, although Wolff's investigations for *Theoria Generationis* did not reach the level of Marcello Malpighi's, those of the paper of 1768 surpassed them and held the position of the best piece of embryological work up to that of Christian Pander and von Baer.

Wolff's *De Formatione Intestinorum* rather than his *Theoria Generationis* embodies his greatest contribution to embryology; in it he foreshadows the idea of germ layers in the embryo, which, under Pander and von Baer, became the fundamental conception in structural embryology—he laid the foundation for germ layer theory. Wolff foreshadowed germ layer theory by showing that the material out of which the embryo is constructed is, in an early stage of development, arranged in the form of leaf-like layers. Locy recognizes Wolff as the foremost investigator in embryology before von Baer.

Wolffian ducts   also called mesonephric ducts; paired ducts that develop during vertebrate development; in human males, they give rise to the vasa deferentia, and in human females, they are relatively undeveloped. In frogs they give rise to the kidney ducts.

womb   the uterus.

wound epidermis   *See:* **apical ectodermal cap**.

X

X chromosome   the chromosome responsible for the human female phenotype when it is present in the absence of a Y chromosome.

X chromosome inactivation   inactivation of one (normally) X chromosome in each female mammalian cell; a means of achieving X chromosome dosage compensation.

X-linked genes   *See:* **sex-linked genes**.

xenoestrogens   estrogen-like compounds in the environment, e.g., phytoestrogens (plant estrogens), diethylstilbesterol (DES), and chemicals leached from plastics. *See:* **developmental estrogen syndrome, estrogen disruptors**.

xenogamy   pollination between flowers of separate plants.

Xenopus   a genus of anuran (frog) amphibinas used extensively for teaching and research in embryology and developmental biology; note that the fate map of *Xenopus* is fundamentally different from that of *Rana*, especially in that the presumtive mesoderm does not map to the surface of the *Xenopus* blastula.

Xist **gene**   a gene found on the inactive X chromosome of humans; its transcript does not encode a protein, stays in the nucleus, and interacts with the inactive X chromatin; the *Xist* gene is an excellent candidate for the initiator of X inactivation.

Xwnt8   a member of the Wnt family of growth and differentiation factors; an anti-neuralizing protein, it inhibits neural induction; secreted from the nonorganizer mesoderm. *See:* **BMP4**.

xylem   the component of plant vascular tissue that is made up of water-conducting cells. *See:* **cambium, procambium, tracheophytes, vascular cambium**.

Dictionary of Developmental Biology and Embryology, Second Edition. Frank J. Dye.
© 2012 Wiley-Blackwell. Published 2012 by John Wiley & Sons, Inc.

Y

Y chromosome the chromosome responsible for the human male phenotype when present in a pair with the X chromosome, XY.

yellow crescent on fertilized *Styela* eggs, a yellow crescent occurs naturally; by following the fate of this yellow material, it has been determined that the yellow crescent of the fertilized egg consists of prospective mesoderm.

yolk nutritive material deposited in the developing egg (oocyte) as it differentiates in the ovary of the mother. It provides nutrition for the early developing embryo. Also, significantly, yolk has an inhibitory effect on cleavage; eggs are classified according to their amount and distribution of yolk. Yolk amount and distribution are major determinants of cleavage patterns, blastocoel shapes and positions, and gastrulation patterns. The "yolk" of the chicken egg is actually the egg or oocyte, a single cell, which contains an enormous amount of stored food (yolk) compared with the amount of "active" cytoplasm; bird eggs in general are macrolecithal, heavily telolecithal, and undergo meroblastic discoidal cleavage, as a result of the inhibitory effect on cleavage of the large amount of yolk. In chicken eggs, yolk comes in two varieties: (1) white yolk, with a light yellow appearance, and (2) yellow yolk, with a darker yellow appearance (see Plate 3 in the color insert).

yolk cell the uncleaved, yolk-containing portion of the telolecithal egg that underlies the blastoderm.

yolk cells cleavage nuclei of the early insect embryo that have not migrated to the peripheral periplasm but remain behind surrounded by contiguous cytoplasm.

yolk duct the small opening by which the diminished midgut opens ventrally into the yolk sac (see Fig. 6).

yolk nuclei those nuclei found in yolk, beyond the blastoderm; characteristic of macrolecithal, heavily telolecithal, eggs undergoing cleavage. These nuclei precede the expansion of the blastoderm. *See:* **periblast**.

yolk nucleus in young ovarian follicles of chickens, a dense mass of yolk granules, which grows in size to become the latebra. *See:* **latebra**.

yolk platelet literally, "a little plate"; the small organelles found in the yolk of eggs in which nutritive yolk is stored. *See:* **yolk**.

yolk plug the blastopore of late-gastrula amphibian embryos is not a patent opening, but it is, rather, occluded by a mass of yolk-laden macromeres, referred to collectively as the yolk plug (see Plate 4 in the color insert).

yolk plug stage the late gastrula stage in amphibain development when a yolk plug is present (see Plate 4 in the color insert).

yolk sac one of the four extraembryonic membranes formed during the development of higher vertebrates, including humans. It contains yolk in most cases but generally not in mammals; in humans, it is the site of the first appearance of the primordial germ cells and the first blood cells. In the chick, the yolk sac is the first extraembryonic membrane to make an appearance; the extraembryonic extension of the splanchnopleure forms a sac-like investment for the yolk; the intraembryonic splanchnopleure, through a series of changes, establishes a completely walled gut. Splanchnopleure is involved in the formation of the subcephalic fold, which also produces the foregut, and the subcaudal fold, which also produces the hindgut; the gut in between is the midgut. As the embryo is constricted off the yolk by the body folds, the foregut and hindgut increase at the expense of the midgut. Eventually, the diminished midgut opens ventrally by a small opening into the yolk sac; this opening is the yolk duct, and its wall is the yolk stalk. The yolk stalk is the boundary between the intraembryonic splanchnopleure of the gut and the extraembryonic

Dictionary of Developmental Biology and Embryology, Second Edition. Frank J. Dye.
© 2012 Wiley-Blackwell. Published 2012 by John Wiley & Sons, Inc.

splanchnopleure of the yolk sac. The omplalomesenteric arteries and veins are brought together and traverse the yolk stalk side by side. By the 19th day, the remains of the yolk sac are enclosed within the body wall of the embryo. Higher mammals have virtually no yolk but a sizable yolk sac, suggesting that morphology is more conservative than physiology. The blood vessels characteristically associated with the yolk sac in a functional condition are still the source of the first blood corpuscles to enter the circulation. In pig embryos, during its period of greatest development, the yolk sac supplies nutritive material to the embryo, by absorption from the uterus into the vitelline circulation; later the allantois takes over this function and the yolk sac rapidly decreases in size, becoming a shriveled sac buried in the belly stalk (see Figs. 5, 6, and 19).

yolk stalk   the cord-like structure that attaches the belly of the fetus to the yolk sac; the wall of the yolk duct (see Fig. 47).

yolk syncytial layer (YSL)   a single large yolk "cell," containing many nuclei, underneath the zebrafish blastodisc. *See:* **deep layer, enveloping layer**.

Z

zeugopod   the middle bones (radius-ulna/tibia-fibula) of the vertebrate limb. *See:* **autopod, stylopod**.

zinc finger   a distinct class of transcription factors in which a DNA-binding polypeptide loop or "finger" forms through a coordination complex of zinc with cysteine and histidine residues at the base of the loop.

zinc finger family   one of four major families of transcription factors; includes the nuclear hormone receptor subfamily; transcription factors in this family function in secondary sex determination, e.g., glucocorticoid receptor, estrogen receptor, testosterone receptor, and so forth.

zinc finger proteins   transcription factors: a large and diverse group of proteins, e.g., Krüppel in the early *Drosophila* embryo, WT-1 in the kidney, and Krox20 in the rhombomeres of the hindbrain.

zona-binding proteins   numerous proteins on the sperm that bind to ZP3 on the zona pellucida.

zona pellucida   a noncellular egg coat produced in the ovary during oogenesis and found around the ovulated mammalian egg. The mammalian zona pellucida, secreted by the growing oocyte, is functionally equivalent to the vitelline membrane in invertebrates. The zona pellucida plays two major roles in fertilization: (1) It binds the sperm, and (2) it initiates the acrosome reaction after the spermatozoon is bound (see Plate 7 in the color insert and Fig. 23).

zona proteins (ZP1, ZP2, ZP3)   the three major glycoproteins of the mouse zona pellucida; ZP3 binds the sperm and, once the sperm is bound, initiates the acrosome reaction; ZP2 binds to the acrosome-reacted sperm; ZP1 holds the zona pellucida together.

zona radiata   the zona pellucida of some species, e.g., chickens, has fine striations running through its thickness and is referred to as the zona radiata.

zona reaction   a process in mammals analogous to the cortical reaction in many other animals; as a result of which the glycoprotein zona pellucida is chemically altered, during fertilization, and becomes refractory to supernumerary sperm.

zone of differentiation   that part of a plant characterized by cellular specialization (differentiation), e.g., in a growing root tip, the region just behind the zone of elongation where cells are differentiating; in growing root tips, the presence of root hairs is an indicator of the zone of differentiation. *See:* **meristem, zone of elongation**.

zone of elongation   that part of a plant characterized by cellular elongation, e.g., in a growing root tip, the region just behind the meristematic region where cells produced in the meristem are elongating; a major contributor to growth in length of the root tip. *See:* **meristem, zone of differentiation**.

zone of junction   in the chick blastula, the marginal area of the blastoderm where the cells are not yet detached from the yolk.

zone of overgrowth   the narrow superficial rim of marginal periblast cells of the early chick embryo beyond the zone of junction.

zone of polarizing activity (ZPA)   the organizer at the posterior margin of the developing vertebrate limb that regulates anteroposterior polarity.

zoophilous   animal-pollinated.

zootype   the common constellation of expression domains for developmental genes found in the animal body plan (the expession domains are active around the phylotypic stage for each of the main animal groups); the totality of regional specification genes active in all animals; this cryptic anatomy of developmental gene expression patterns defines what an

Dictionary of Developmental Biology and Embryology, Second Edition. Frank J. Dye.
© 2012 Wiley-Blackwell. Published 2012 by John Wiley & Sons, Inc.

animal is; refers to a pattern of expression of the *Hox* genes and certain other genes along the antero-posterior axis of the embryo that is characteristic of all animal embryos.

ZP ZP1, ZP2, and ZP3 are the glycoproteins that make up the zona pellucida of mouse eggs. Mouse spermatozoa specifically bind to ZP3 and ZP3 also initiates the acrosome reaction. As the anterior part of the sperm plasma membrane (which has the ZP3 binding proteins) is shed, the sperm undergoes a secondary binding to the zona; this is accomplished by proteins in the inner acrosomal membrane that bind specifically to ZP2. ZP1 crosslinks strands made of dimers of ZP2 and ZP3. In pigs, sperm proacrosin mediates secondary binding and proacrosin becomes the protease acrosin, known to be involved in digesting the zona pellucida. *See:* **acrosome reaction**.

ZPA *See:* **zone of polarizing activity (ZPA)**.

zygote the fertilized egg.

zygote intrafallopian transfer (ZIFT) artificial transfer of zygotes into the normal site of fertilization in the ampulla of the fallopian tube.

zygotene the second stage of prophase I of meiosis, during which the chromosomes begin to undergo synapsis.

zygotic genes those genes expressed by the embryo itself; the embryo's genes as opposed to maternal genes expressed in the embryo. In early *Drosophila* development, most of the zygotic genes, first activated along the anteroposterior and dorsoventral axes set up by maternal genes, encode transcription factors, which, then, activate more zygotic genes. The expression of zygotic genes along the dorsoventral axis is controlled by the graded concentration of intranuclear dorsal protein. In the ventral region of the embryo, the dorsal protein has two main functions, it activates certain genes and it represses the activity of other genes; e.g., in the ventral-most region, the zygotic genes *twist* and *snail* are activated (expression of both is required for mesoderm development and gastrulation), but the genes *decapentaplegic*, *tolloid*, and *zerknüllt* are repressed by the dorsal protein (so their activity is confined to more dorsal regions; *zerknüllt* apparently specifies the amnioserosa, and *decapentaplegic* is a key gene in the specification of pattern in the dorsal part of the dorsoventral axis). *See:* **maternal genes**.

zygotic meiosis is where meiosis is at the beginning of the life cycle; as in the green alga, *Chlamydomonas*, where the zygote is the only diploid cell in the life cycle and it undergoes meiosis.

References

Churchill, F. B., The Rise of Classical Descriptive Embryology. In: Gilbert, S. F., editor, *A Conceptual History of Modern Embryology*. The Johns Hopkins University Press, Baltimore, Maryland (1994).

Gardner, E. J., *History of Biology*, 2nd edition, Burgess Publishing Co., Minneapolis (1965).

Gilbert, S. F. *Developmental Biology*, 8th edition, Sinauer, Sunderland, Massachusetts (2006)

Gilbert, S. F. and D. Epel, *Ecological Developmental Biology*, Sinauer Associates, Inc., Sunderland, Massachusetts (2009).

Goldschmidt, R., *The Material Basis of Evolution*, Yale University Press, New Haven, Connecticut (1940).

Grosser, A. E., The Culinary Alchemy of Eggs. American Scientist (March/April 1983).

Grunwald, G. B., The Conceptual and Experimental Foundations of Vertbrate Embryonic Cell Adhesion Research. In: Gilbert, S. F., editor, *A Conceptual History of Modern Embryology*. The Johns Hopkins University Press, Baltimore, Maryland (1994).

Hamburger, V., *The Heritage of Experimental Embryology: Hans Spemann and the Organizer*. Oxford University Press, New York (1988).

Harrison, R. G., Embryology and its relations. Science 85: 369–374 (1937).

Hickman, C. P, *Integrated Principles of Zoology*, 3rd ed., C. V. Mosby Co., Saint Louis, Missouri (1966).

Holtfreter, J., Reminiscences on the Life and Work of Johannes Holtfreter. In: Gilbert, S. F., editor, *A Conceptual History of Modern Embryology*. The Johns Hopkins University Press (1994).

Locy, Wm. A., *Biology and its Makers*, Henry Holt and Co., New York (1908).

Maienschein, J., The Origins of *Entwicklungsmechanik*. In: Gilbert, S. F., editor, *A Conceptual History of Modern Embryology*. The Johns Hopkins University Press, Baltimore, Maryland (1994).

Oppenheimer, J. M., Curt Herbst's Contributions to the Concept of Embryonic Induction. In: Gilbert, S. F., editor, *A Conceptual History of Modern Embryology*. The Johns Hopkins University Press, Baltimore, Maryland (1994).

Saha, M., Spemann Seen through a Lens. In: Gilbert, S. F., editor, *A Conceptual History of Modern Embryology*. The Johns Hopkins University Press, Baltimore, Maryland (1994).

Slack, J. M. W. et al. (1993) The zootype and the phylotypic stage. Nature 361, 490–492.

Weismann, A. *The Germplasm: A Theory of Heredity*, Translated by W. N. Parker and H. Ronnfeld. Walter Scott, Ltd., London (1883).

Wolpert, L., *The Triumph of the Embryo*. Oxford University Press, New York (1991).

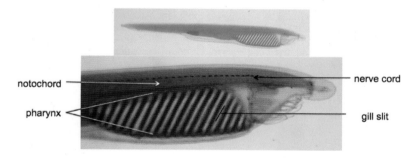

notochord

pharynx

nerve cord

gill slit

Plate 1. Amphioxus, fixed and stained. Photos by the author.

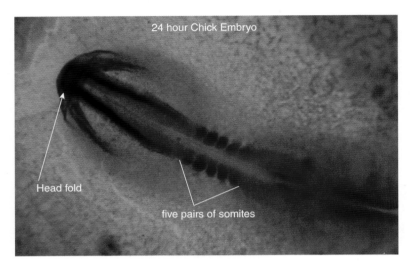

Plate 2a. Chick embryo, 24-hour, fixed and stained. Photo by the author.

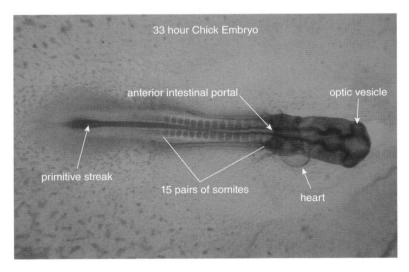

Plate 2b. Chick embryo, 33-hour, fixed and stained. Note the linear axis of the embryonic body. Photo by the author.

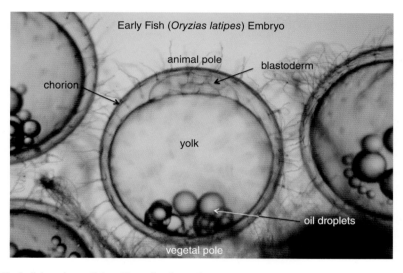

Plate 3. Early fish embryo, living. Photo by the author.

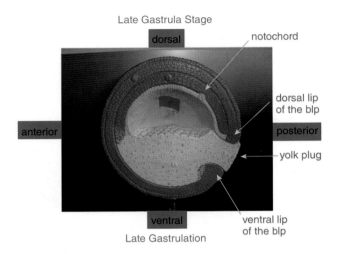

Late Gastrula Stage

dorsal

notochord

anterior

dorsal lip
of the blp

posterior

yolk plug

ventral

ventral lip
of the blp

Late Gastrulation

Plate 4. Frog gastrula model. Photo by the author.

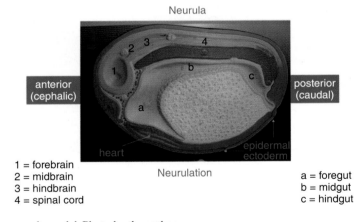

Neurula

anterior
(cephalic)

posterior
(caudal)

heart

epidermal
ectoderm

Neurulation

1 = forebrain
2 = midbrain
3 = hindbrain
4 = spinal cord

a = foregut
b = midgut
c = hindgut

Plate 5. Frog neurula model. Photo by the author.

Plate 6. Gametes. Photo by the author.

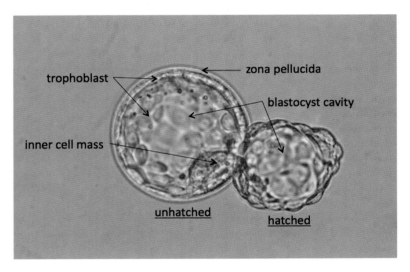

Plate 7. Mouse hatching blastocyst. Photo by the author.

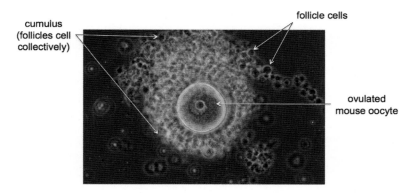

Plate 8. Mouse oocyte. Photo by the author.

Plate 9a. Placenta: maternal side. Photo courtesy of Jack Wolk, MD, the Danbury Hospital.

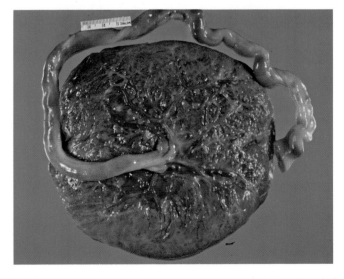

Plate 9b. Placenta: fetal side. Photo courtesy of Jack Wolk, MD, the Danbury Hospital.

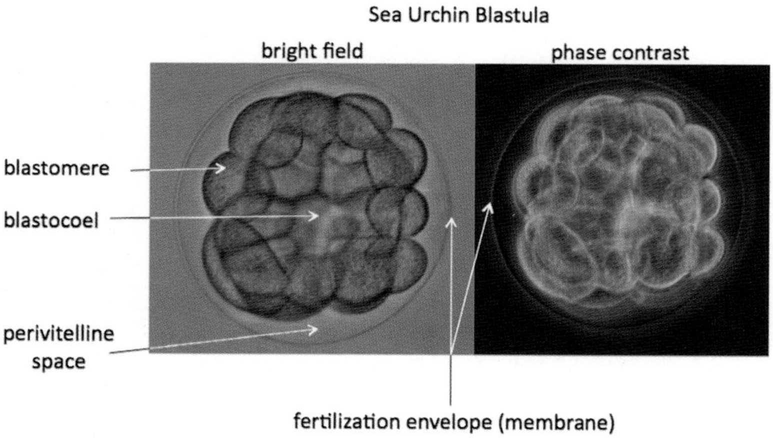

Plate 10. Sea urchin blastula. Photos by the author.

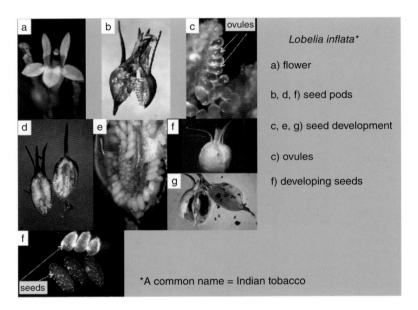

Plate 11. Seed pod development. Photos by the author.

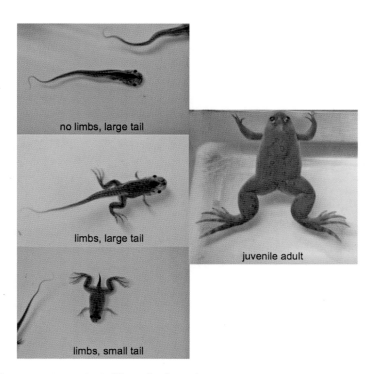

no limbs, large tail

limbs, large tail

limbs, small tail

juvenile adult

Plate 12. *Xenopus* metamorphosis. Photos by the author.

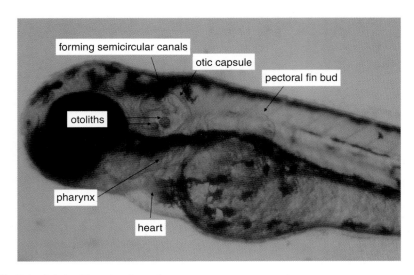

Plate 13. Zebrafish fry. Photo by the author.

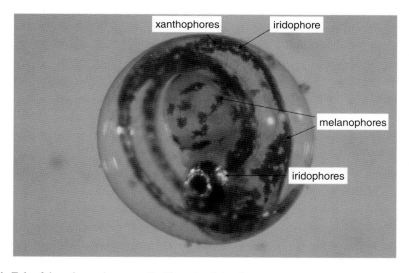

Plate 14. Zebrafish embryo pigment cells. Photo by the author.

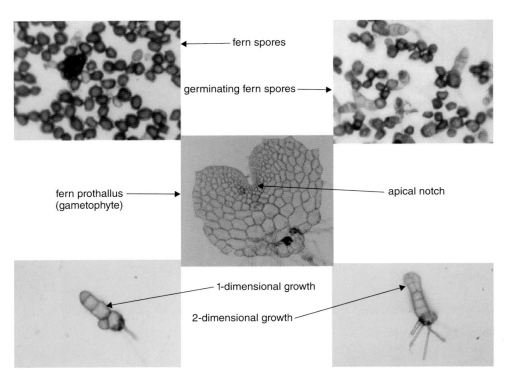

fern spores

germinating fern spores

fern prothallus
(gametophyte)

apical notch

1-dimensional growth

2-dimensional growth

Plate 15. Early fern development; spores to prothallus. Photos by the author.

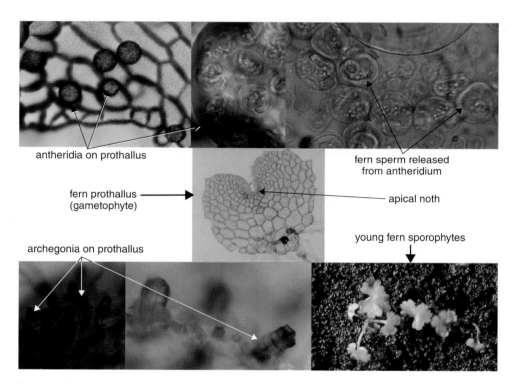

antheridia on prothallus

fern sperm released from antheridium

fern prothallus (gametophyte)

apical noth

archegonia on prothallus

young fern sporophytes

Plate 16. Early fern development: gametangia. Photos by the author.

Plate 17. Moss: alternation of generations. (a) moss sporophytes; (b) gametophytes. Photos by the author.

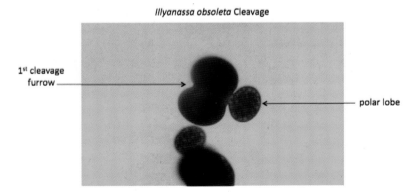

Illyanassa obsoleta Cleavage

1st cleavage furrow

polar lobe

Plate 18. Mud snail cleavage. Photo by the author.